D1357127

DR004627

CORNWALL COLLEGE

Principles of Plant Science

Environmental Factors and Technology in Growing Plants

Principles of Plant Science

Environmental Factors and Technology in Growing Plants

Dennis R. Decoteau
The Pennsylvania State University

Upper Saddle River, New Jersey 07458

Executive Editor: Debbie Yarnell
Development Editor: Kate Linsner
Production Editor: Cindy Miller and Holly Henjum, Carlisle Publishers Services
Production Liaison: Janice Stangel
Director of Manufacturing and Production: Bruce Johnson
Managing Editor: Mary Carnis
Manufacturing Manager: Ilene Sanford
Manufacturing Buyer: Cathleen Peterson
Creative Director: Cheryl Asherman
Marketing Manager: Jimmy Stephens
Cover Design Coordinator: Christopher Weigand
Cover Design: Kevin Kall
Cover Photo: Courtesy of Kathy Denchak, Penn State University
Composition: Carlisle Communications, Ltd.
Printing and Binding: Phoenix Color Corp.

Copyright © 2005 by Pearson Education, Inc., Upper Saddle River, New Jersey 07458. Pearson Prentice Hall. All rights reserved. Printed in the United States of America. This publication is protected by Copyright and permission should be obtained from the publisher prior to any prohibited reproduction, storage in a retrieval system, or transmission in any form or by any means, electronic, mechanical, photocopying, recording, or likewise. For information regarding permission(s), write to: Rights and Permissions Department.

Pearson Prentice Hall™ is a trademark of Pearson Education, Inc.
Pearson® is a registered trademark of Pearson plc
Prentice Hall® is a registered trademark of Pearson Education, Inc.

Pearson Education Ltd.
Pearson Education Singapore, Pte. Ltd.
Pearson Education Canada, Ltd.
Pearson Education—Japan

Pearson Education Australia PTY, Limited
Pearson Education North Asia Ltd.
Pearson Educación de Mexico, S.A. de C.V.
Pearson Education Malaysia, Pte. Ltd.

10 9 8 7 6 5 4 3 2 1
ISBN: 0-13-016301-5

*To my Pennsylvania family
(Chris and Megan)
for their patience, understanding, and support,*

and

*To my New Hampshire family
(Mom, Bobby, Hank, Priscilla, Donald, and Carol)
for their constant support and encouragement
in bringing up the youngest.*

ROSEWARNE
LEARNING CENTRE

Contents

PART 3   ENVIRONMENTAL FACTORS THAT INFLUENCE PLANT GROWTH AND CROP PRODUCTION TECHNOLOGIES   87

Preface

Principles of Plant Science: Environmental Factors and Technology in Growing Plants was written to provide a unique plant science text that emphasizes understanding the role of the environment in plant growth and development instead of the more traditional focus topics of analyzing the industries and surveying important crops. By emphasizing the scientific principles associated with the biological effects that the various environmental factors have on plant development, I hope that the reader will be better equipped to understand current and emerging technologies that modify the environment to improve plant production. The overriding philosophy of this text is that conceptualization and evaluation are more important than memorization, especially when trying to understand and direct biological responses and organisms.

Key Features

To set the stage and provide some background information, *Principles of Plant Science: Environmental Factors and Technology in Growing Plants* begins with an overview of the plant sciences, including the role of plants in the development of societies, industries, and science. Its emphasis in plant science is on non-forest agricultural crops. A primer on plant growth and development follows that includes information on photosynthesis and respiration, plant hormones, and ecology. The influence of the environment on agricultural plant production constitutes the remainder of the presented material and is the primary emphasis of the text.

The environmental factors that affect plant growth and development are discussed in one of two broad categories: the aerial environment or the rhizosphere (Fig. P.1). The factors covered in the aerial environment section include irradiance (light), temperature, gases, air pollutants, and mechanical disturbances. The factors covered in the rhizosphere section include water, mineral nutrients, soil organisms, and allelochemicals. Although some of the environmental factors (e.g., temperature, water, atmospheric gases) may affect both the aerial environment and the rhizosphere, for discussion they were placed in the section that their effect was considered greater.

Material on each environmental factor covered is presented in three sections. The first section provides background information on the specific environmental factor. Appropriate physical definitions and explanations are presented. The second section discusses how the environmental factor affects plant growth. Direct effects on biological processes and specific effects on plants are discussed. The third section illustrates how agricultural technologies affect the environmental factor, especially for the benefit of plant growth and production.

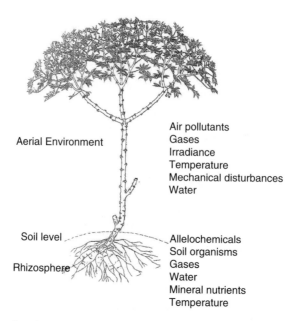

Aerial Environment

Air pollutants
Gases
Irradiance
Temperature
Mechanical disturbances
Water

Soil level

Rhizosphere

Allelochemicals
Soil organisms
Gases
Water
Mineral nutrients
Temperature

FIGURE P.1   Examples of environmental factors that can affect plant growth and development.

Target Audience and Uses of the Book

This book is designed for use in plant science (or horticulture) courses that would be taken before a student enrolls in the various advanced plant production courses such as agronomy, crop science, vegetable crops, small fruits, pomology, and floriculture. The material may also be utilized for plant growth and development or applied introductory plant physiology courses taught at universities, junior colleges, or community colleges.

The primary emphasis of this book is on the environmental factors and their role in plant growth and production. Its intent is to provide sufficient introductory material on the various environmental factors and examples of effects of these environmental factors on plant growth and development to facilitate further discussions and study.

Sources of Additional Information

Complete and exhaustive discussions on the fields of plant science and the influence of the various environmental factors on plant growth are beyond the scope of any single textbook. Further information on the plant sciences can be gleaned from other introductory textbooks on plant science, crop science, and horticulture. Information from local extension services should serve as important sources of current location-specific marketing and crop production information. More in-depth information on the environment and plant growth can be obtained from plant physiology, plant ecology, and plant ecophysiology texts and/or articles on specific topics in scientific journals.

Acknowledgments

Several colleagues provided important feedback. I am grateful to Craig Anderson, University of Arkansas; Mark Bennett, Ohio State University; Rebecca Darnell, University of Florida; Robert Gough, Montana State University; Wallace G. Pill, University of Delaware; Robert P. Rice, Jr., California Polytechnic State University—San Luis Obispo; Mark Rieger, University of Georgia; and Sudeep Vyapari, Sam Houston State University, for their invaluable assistance.

My colleagues at the Pennsylvania State University are thanked for their support. Special thanks go out to the many students that I have taught for allowing me to learn along with them. Dr. E. Jay Holcomb, Professor of Floriculture, is gratefully acknowledged for co-teaching with me the "Environmental Effects on Horticultural Crop Growth" course that is taught in the Department of Horticulture at Penn State University. Finally, I would like to thank all the editors and professionals at Prentice Hall for helping me pull this together and keeping me on track.

About the Author

Dr. Dennis R. Decoteau is a Professor and the Head of the Department of Horticulture at The Pennsylvania State University. His previous academic appointments include serving as Professor and Chair of the Department of Horticulture at Clemson University. Dr. Decoteau received his B.S. degree in Environmental Studies from the University of Maine at Fort Kent and his M.S. and Ph.D degrees in Plant Science with an emphasis in Environmental Plant Physiology from the University of Massachusetts. He has won numerous teaching and research awards, including the Outstanding Teacher Award in Horticulture from Clemson University, the Outstanding Research Paper on Teaching from the American Society for Horticultural Sciences, the L. M. Ware Distinguished Research Award from the Southern Region of the American Society for Horticultural Science, and the IV Congreso International De Neuvas Technologias Agricolas Award. Dr. Decoteau is the current Vice President, Education Division, and member of the Board of Directors for the American Society for Horticultural Science and has taught several undergraduate and graduate courses, including "Environmental Effects on Horticultural Crop Growth." He has written extensively in the scientific press and for the general public and is the author of *Vegetable Crops*, also published by Prentice Hall.

INTERNET RESOURCES

AGRICULTURE SUPERSITE

This site is a free online resource center for students and instructors in the field of Agriculture. Located at http://www.prenhall.com/agsite, this site contains numerous resources for students including additional study questions, job search links, photo galleries, PowerPoint™ slides, *The New York Times eThemes* archive, and other agriculture-related links.

On this supersite, instructors will find a complete listing of Prentice Hall's agriculture texts as well as instructor supplements that are available for immediate download. Please contact your Prentice Hall sales representative for password information.

The New York Times eThemes of the Times for AGRICULTURE and The New York Times eThemes of the Times for AGRIBUSINESS

Taken directly from the pages of *The New York Times*, these carefully edited collections of articles offer students insight into the hottest issues facing the industry today. These free supplements can be accessed by logging onto the Agriculture Supersite at: http://www.prenhall.com/agsite.

AGRIBOOKS: A CUSTOM PUBLISHING PROGRAM FOR AGRICULTURE

Just can't find the textbook that fits your class? Here is your chance to create your own ideal book by mixing and matching chapters from Prentice Hall's agriculture textbooks. Up to 20% of your custom book can be your own writing or come from outside sources. Visit us at: http://www.prenhall.com/agribooks.

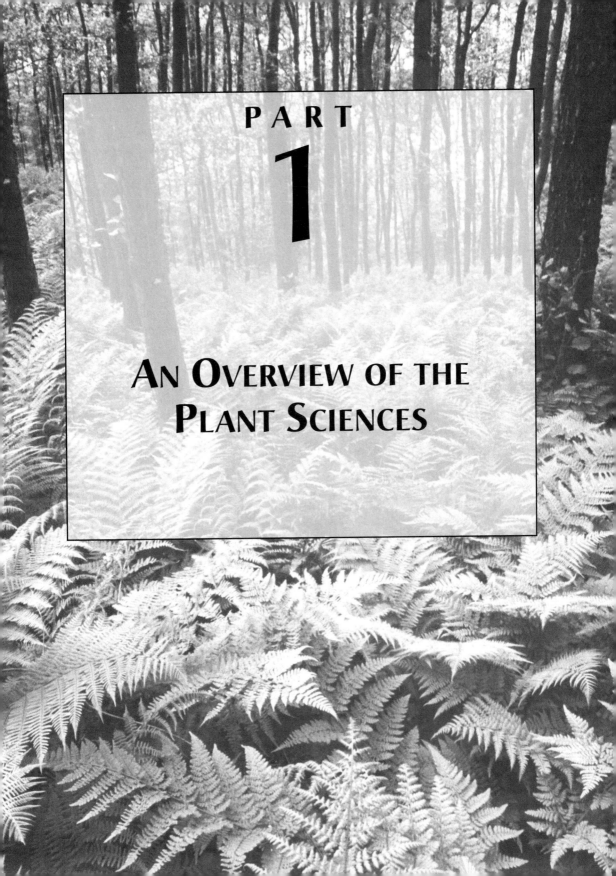

PART

1

AN OVERVIEW OF THE PLANT SCIENCES

1

Introduction to the Plant Sciences

Plant science is a specialized area of study of botany that emphasizes the use of plants in agricultural applications. This study area of botany is also referred to as the applied phases of plant study, the plant sciences, or economic botany. The plant sciences can be divided into more specialized fields that can be broadly categorized by commodity (such as agronomy, horticulture, and forestry) or cross commodity (such as plant pathology, entomology, integrative pest management) orientation. This introductory chapter provides an overview of the historical importance of plants in the development of societies, industries, and science. These subjects will be covered more in-depth in Chapters 2 through 4.

Plants in Society

Plants are intimately connected with the development of societies and many of the fields of science. Early humans undoubtedly observed that plants (a major source of food and other raw products) were greatly affected by the environment. Seasonal cropping cycles, serendipitous successful plantings, and beneficial uses of certain plants were noted (as paintings, drawings, and written and oral stories) by these early ancestors and passed on to future generations. As a result, humans became less nomadic, which allowed communities and societies to evolve.

Plants in Industries

The success of most of today's agricultural plant production businesses depends on growing crops to acceptable quality in economical technological ways in the available environment so that the plants can be sold for a profit (Fig. 1.1). Agriculturalists have long known the importance

3

FIGURE 1.1   A roadside market can be successful if appropriate crops are selected for the growing environment and if these crops can be sold at a profit.
Courtesy of Dr. Mike Orzolek, Penn State University.

of the environment on plant growth, and early farmers were some of the first to manipulate or modify the growing environment to provide a more conducive environment for plant growth. Protective shelters were built, water needs of the plant were provided for, and sunlight was modified (by shading or by training and pruning plant material) to provide more optimal growing environments to enhance plant growth and production.

Plants in Science

As growing plants (and animals) became more specialized and agriculture continued to evolve, the understanding of the role of the environment in successful food and fiber production became more of a science. The plant-growing environment was determined to be a myriad of environmental factors that directly influenced the physiology of the plant and how well it grew. Botany, or the study of plants, developed as science and investigations on the taxonomy, anatomy, morphology, physiology, and ecology of plants were institutionalized as fields of advanced study at institutions of higher learning.

The areas of plant study that placed major emphasis on the use of plants by humans or as agricultural products (or crops) became known as the plant sciences. The plant sciences also became institutionalized with the establishment of the land grant universities (with their emphasis on the agricultural and mechanical sciences), and federal and state funding for the establishment of the various state experiment stations and extension services. The plant sciences evolved to specialized crop commodity fields such as agronomy, horticulture, and forestry, and specialized cross commodity fields such as plant pathology, entomology, and integrated pest management.

Summary

Plant science is a field of botany that emphasizes the use of plants by humans, often in the context of agricultural production. Domesticating plants often coincided with or was the precursor for the evolution of societies and communities. As plant growing became more successful, specialty disciplines such as agronomy, horticulture, forestry, pathology, and entomology were developed and institutionalized.

Review Questions

1. What are some of the ways our early ancestors passed on information on plants and plant culture?
2. How can the plant-growing environment be manipulated?
3. Define plant sciences.

Selected References

Janick, J. 1979. *Horticultural science*. 3rd ed. San Francisco: W. H. Freeman.
Norstog, K., and R. W. Long. 1976. *Plant biology*. Philadelphia: W. B. Saunders.
Stern, K. R. 1991. *Introductory plant biology*. 5th ed. Dubuque, IA: Wm. C. Brown.

Selected Internet Sites

www.aces.uiuc.edu/~sare/history.htm/   A Brief History of Agriculture, University of Illinois College of A.C.E.S.
www.agr.ca/backe.htm/   Agriculture and Agri-Food Canada: History, Agriculture and Agri-Food Canada.
www.usda.gov/history2/back.htm History of American Agriculture 1776–1990, USDA Economic Research Service.

2

Plants and Society

The human race came into existence about two million years ago, and it is believed that our earliest ancestors were food gatherers that spent most of their time hunting animals, catching fish, and collecting edible plants. Ancient cave art often depicted important plants that were used for food. Plants also appear to have been used by primitive humans for uses other than food, as flowers for ornamental purposes and plant fiber in cloth were included in some of the burials of 60,000 years ago.

It is suspected that humans were in the Americas as early as 15,000 years ago. The first Americans probably crossed the Bering Strait from Asia during the most recent Ice Age, following large game during the game's migration. Eventually the humans settled in the Americas where large numbers of game were present.

This chapter provides a summary of the history of agriculture with a discussion of important events that paralleled the evolution of societies. One of the most important events in the establishment of agriculture was the selection and domestication of important plants and the development of agricultural crops. Events that led to the development of agricultural crops are presented, including a discussion of contemporary crop improvement programs.

History of Agriculture

Evidence indicates that agriculture and the plant sciences originated during the Neolithic Age (around 8000 BC) in the semiarid mountainous regions near the river valleys of Mesopotamia, in present-day Iraq. About that time, men and women began collecting grain plants and keeping domesticated animals such as sheep and dogs.

Eventually, through cultivation of plants and domestication of animals, humans were able to produce sufficient food to free them from the constant search for their next meal. With the development of a dependable food source, humans began to live together and villages and towns came into existence. Wheat and barley seeds dated at about 6750 BC have been found in evacuations in Iraq. Early centers of crop development include the Near East, Southeast Asia, Meso-America, and possibly Western Sudan. The further domestication of plants and animals paralleled the development of villages, towns, and cities along the Nile around 3500 BC in present-day Egypt. The Egyptians are credited with developing technologies for food storage, such as pickling and drying, and the agricultural technologies of drainage, irrigation, and land preparation.

The Romans, around 500 BC to 500 AD, developed efficient agricultural production systems that became the cornerstone of their strength. During the collapse of the Roman Empire (around 500 AD) and before the Renaissance, monasteries became important reservoirs for agricultural skills perfected by the Romans. Monasteries often had cultivated fields of grain and vegetables as well as orchards. The monks in the monasteries maintained important collections of herbal and medicinal plants. Many of the plants kept in cultivation in the monasteries would become important later during the Renaissance.

Because the earliest uses of many agricultural plants were medicinal, herbals are some of the earliest and most important plant science manuals. Their origin traces to the Greek interest in cataloging and describing plants. As originators of the study of botany, the Greeks produced writings that listed common plants and often supplied their medicinal usage. For much of the Middle Ages, there was little distinction between medicine and botany, as plants were used to cure ills.

By the second half of the 1500s botany had established itself as a science. The drawings of the plants became more accurate, and, with the exploration of new worlds, many new plants were discovered and cataloged. It was fashionable during that period of history for the aristocracy throughout Europe to maintain large gardens full of exotic plants and to have extensive collections of herbals in their libraries.

The First Cultivated Plants

The tribes in the Old World that engaged in hunting and in gathering wild edible plants made attempts to domesticate dogs, goats, and possibly sheep as early as 8000 BC. Centers of origin of major cultivated crops are found in both the Old and New worlds (Table 2.1). A plant's center of origin is the geographical area where a species is believed to have evolved through natural selection from its ancestors. This is also the plant's center of diversity where a pool of genes exists for use by plant breeders in crop improvement programs.

Earliest evidence of plant cultivation in the Americas appears to be about 10,000 years ago in South America, where potatoes and a variety of other crops were domesticated. Agriculture probably began in Mexico about 5000 BC. Early cultivated plants included maize, avocado, squash, and chili peppers. Research also indicates that agriculture arose very early in the Andean valleys of Peru. It seems probable that agriculture originated independently in at least two regions in the New World—Mexico and Peru—and the earliest site may have been Peru.

TABLE 2.1   *Centers of Origin of Important Cultivated Plants and Commercially Important Crops that Originated from These Centers*

Old World

China—The mountainous and adjacent lowlands of central and western China represent the earliest and largest centers of origin of cultivated plants and world agriculture. Important crops that originated from this center include millet, buckwheat, soybean, bamboo, pear, cherry, citrus, cinnamon, tea, radish, eggplant, several legumes, Brassicas, onions, and various cucurbits.

India (India and Burma)—Important crops that originated from this center include orange, other citrus, mango, black pepper, many legumes, and gourds.

Indo-Malaysia (Indochina, Malaysia, Java, Borneo, Sumatra, and Philippines)—Important crops that originated from this center include giant bamboo, ginger, banana, coconut, and nutmeg.

Central Asia—Important crops that originated from this center include common wheat, pistachio, apricot, pears, apple, garden pea, broad bean, carrot, radish, garlic, spinach, and mustard.

Near East (Asia Minor, Iran, Transcaucasia, and Turkmenistan Highlands)—Important vegetable crops that originated from this center include poppy, fig, pomegranate, cherry, hazelnut, cantaloupe, cabbage, lettuce, and muskmelons.

Mediterranean—Important crops that originated from this center include olive, peppermint, lavender, thyme, sage, rosemary, beet, parsley, leek, chive, celery, parsnip, and rhubarb.

Abyssinia (Ethiopia and Somaliland)—Important crops that originated from this center include wheats, barley, sesame, coffee, okra, and garden cress.

New World

Southern Mexico and Central America (South Mexico, Guatemala, El Salvador, Honduras, Nicaragua, and Costa Rica)—Important crops that originated from this center include chayote, upland cotton, papaya, agave, cacao, corn, common bean, lima bean, sweet potato, pepper, and cherry tomato.

South America (Peru, Ecuador, and Bolivia)—Important crops that originated from this center include lima bean, pepper, coca, tobacco, potato, tomato, and pumpkin.

Chiloé (South Chile)—An important crop that originated from this center is the white potato.

Brazil–Paraguay—Important crops that originated from the semiarid region include peanut and pineapple; important crops from the tropical Amazon region include manioc and the rubber tree.

Sources: Adapted from G. Acquaah, *Horticulture: Principles and Practices* (Upper Saddle River, NJ: Prentice Hall, 1999); J. Janick, *Horticultural Science,* 3rd ed. (San Francisco: W. H. Freeman, 1979).

Maize (corn) appears to have been introduced into North America from South America about 1,700 years ago and was a relatively minor crop until about 800 years ago. Different varieties of maize with different cob shapes and sizes appeared in different regions. Europeans recorded the existence of short- and long-season corn varieties, but it is not known when or how Native Americans developed these varieties. Also about this time, the common bean was introduced.

Development of Agricultural Crops

All of the modern crops (plants used in agriculture) that we produce today had their earliest beginnings as wild plants. Early plant gatherers applied self-serving criteria to decide which plants to gather. Of utmost importance was size. As a result of continually choosing the largest fruit of the wild plants through the years, many of the crops that underwent domestication have bigger fruits than their wild ancestors had.

Another criterion used by plant gatherers was taste. Many wild seeds are bitter tasting, yet their fruits are sweet and tasty. This was important so seed would not be chewed up and eaten and instead would be expelled. Other criteria for gathering were fleshy or seedless fruit, oily fruits and seeds, and plants with fiber for clothes. Other plants produced fruit that are adapted to being eaten by a particular animal. Strawberries are often eaten by birds, acorns by squirrels, and mangoes by bats. As a result of eating these plants, seeds and other plant parts were dispersed. This was the beginning step in the process of domestication.

Human latrines may have been testing grounds for the first crop breeders, as these may have been primary spots where seeds of ingested fruits would have been deposited. These were also areas where the soils had greater nutrient concentrations than in nonlatrine areas.

Garbage dumps where food scraps were scattered may have also played a role in early plant domestication and growing. Spoiled or rotten fruit would have been placed in these areas and the seeds from these fruits could have germinated. Also seeds of consumed fruits may have been deposited into these areas and provided another source for future plant generations.

Other methods of seed dispersal are unintentional. Dispersal may occur by the plants or their seeds being blown about in the wind or floating in the water. Many wild plants have specialized mechanisms that scatter seeds and generally make them unavailable to humans. Most wild peas have pods that explode (split open) when the seeds within the pod are mature. Because humans collected only the pods that didn't explode, this nonexploding trait was passed on during domestication.

For the survival of plant species, seeds produced by the plant had a scattered germination rate (i.e., they didn't germinate at one time). This resulted in plants not having all their seeds germinating at a time when possibly the climate or some other factor may not have been conducive for early plant growth and development. Thick seed coats contributed to seeds undergoing a scattered germination rate. Many ecologically advantageous traits such as thick seed coats have been reduced or removed during domestication to improve the efficiency of seedling establishment and plant growth in large-scale field plantings.

During the 10,000 or 11,000 years that have passed since the beginning of plant domestication, the plants that humans selected as useful to them have undergone profound changes. Artificial selection of plants during the ages differs considerably from natural selection. During natural selection the plants have properties that preadapt them to a wide variety of environmental conditions, hopefully ensuring continuation of the species. Artificial selection breaks down these stabilized systems, creating gene combinations that possibly could not survive in the wild.

Timeline for Domestication of Important Crops

During the development of agriculture and plant domestication some plants were domesticated earlier or easier than others were. The earliest domesticated plants appear to have been the Near Eastern crops of cereals and legumes (such as wheat, barley, and peas) around 10,000 years ago. These crops may have been domesticated earlier because they came from wild ancestors that had many characteristics that were advantageous for the process of domestication. Some of these advantages include seeds that were edible in the wild and could be readily stored and plants that were easily grown from sowing, grew quickly, were self-pollinated. Overall few genetic changes were required of these crops to go from wild plants to domesticated plants.

Fruit and nut domestication probably began around 4000 BC. The fruit and nut crops typically are not harvested until 3 to 5 years after planting. To grow these types of crops, people needed to be committed to settled life and could no longer be seminomadic. Fruit and nut crops are often grown by cuttings, which has the advantage that the progeny or descendants of the original plants are identical to the original plant. Some fruit trees cannot be grown from cuttings, and it was found to be a waste of effort to grow them from seed. Instead, grafting techniques had to be discovered and perfected before suitable domestication could occur.

Contemporary Crop Improvement Programs

Improvement to crops has been a continuous process since humans began collecting and then growing plants for their use. In the process of selecting and using certain plants, primitive humans passed on these chosen plants and the characteristics that made them be selected to the next set of plants that grew. The criteria for selection may have been fruit size, color, taste, fast rate of growth, resistance to diseases or insects, or simply the lack of toxins to humans. This process of selection was effective, as many of our contemporary cultivated plants no longer resemble their primitive ancestors.

Plant breeding as a science is the systematic improvement of plants and has only been in existence in the last couple of decades. Plant genetics is the study of the mechanisms of heredity of plant traits and is the underlying science of plant breeding. Gregor Mendel did the pioneering work on genetic inheritance in 1865 using garden peas. Today's plant breeding and genetic programs continue to utilize the basic concepts of genetics and controlled plant crosses. Crossing is the transfer of genes (in pollen) from one plant to another.

Some plants are self-pollinated, which means that the flowers of the plant use its own pollen for pollination. Seeds from these plants produce uniform plants very much like the parents. Mass selection and pedigree selection are breeding methods for self-pollinated crops. Mass selection involves selecting the best plants based on appearance (phenotype). Pedigree selection is conducted after creating variability by controlled crossing of two parents.

Cross-pollination is the process of transferring pollen grains from one plant and depositing them on the stigma of the flower of a different one. Seeds from plants that cross-pollinate produce hybrid (nonuniform) plants. Hybrid vigor or heterosis refers to the increase in vigor shown by certain crosses as compared to that of either parent. Many cross-pollinated agricultural crops (such as apple) are also vegetatively propagated. Mass selection, recurrent selection, and other methods of breeding may be applied to cross-pollinated species. Many of today's crop improvement programs also utilize the modern concepts of molecular genetics, tissue culture, and genetic engineering.

Summary

Agriculture began around 8000 BC near the valleys of Mesopotamia when important food plants were collected and sheep and dogs were domesticated. The Egyptians in 3500 BC developed technologies for drainage, irrigation, land preparation, and food storage. The ancient Greeks listed common plants and their medicinal usage in herbals and are credited with establishing the scientific field of botany. Plant cultivation of potatoes and a variety of other crops began in the Americas about 10,000 years ago.

Plant domestication began when early agriculturalists began selecting and cultivating wild plants that had desirable characteristics such as large fruit size, sweet taste, fast growth, and disease and insect resistance. The earliest domesticated plants appear to have been cereals and legumes (such as wheat, barley, and peas). Crop domestication is a continual process, carried on in today's plant breeding and crop improvement programs.

Review Questions

1. What role did the evolution of ancient humans going from primarily food gatherer to food producer have in the development of villages and towns?
2. What is the role of monasteries and herbals in contributing to our contemporary knowledge of plant uses?
3. What are some of the methods of seed dispersal that are used by plants?
4. What were some of the criteria used by early plant gatherers in selecting plants?
5. Why did cereal and legume crops undergo domestication easier and earlier than fruit and nut crops?
6. What is plant breeding?

Selected References

Acquaah, G. 1999. *Horticulture: Principles and practices.* Upper Saddle River, NJ: Prentice Hall.

Janick, J. 1979. History's ancient roots. *HortScience* 14:299–313.

Janick, J. 1979. *Horticultural science.* 3rd ed. San Francisco: W. H. Freeman.

Norstog, K., and R. W. Long. 1976. *Plant biology.* Philadelphia: W. B. Saunders.

Selected Internet Sites

www.aces.uiuc.edu/~sare/history.htm/   A Brief History of Agriculture, University of Illinois College of A.C.E.S.

www.agr.ca/backe.htm/   Agriculture and Agri-Food Canada: History, Agriculture and Agri-Food Canada.

www.usda.gov/history2/back.htm History of American Agriculture 1776–1990, USDA Economic Research Service.

3

Plants as Industries

The share of consumer expenditures for food in the United States is the lowest in the world (only approximately 12% of the disposable income for an average consumer). This is in no small part due to efficient and progressive agricultural industries that have developed through the years. These industries have evolved from the primitive food production systems developed by the colonists and settlers for survival to the highly sophisticated systems of today's producers that supply a wide variety and relatively consistent supply of food for sale nationally and internationally.

This introductory chapter provides an overview of the historical importance of plants in the development of plant-related agricultural industries and the current status of U.S. agriculture and crop production. To provide a better understanding of contemporary plant-related agricultural industries, important agronomic and horticultural industries are presented.

Historical Periods of the United States and the Development of Plant-Related Agricultural Industries

Colonists and Early Settlers

American colonists were largely self-supporting, growing plants for their own use. Early settlers endured a rough pioneer life while adapting to new environments. Small family farms predominated, except for some relatively large plantations that were developed in the southern coastal areas.

The settlers used relatively simple tools such as wagons, plows, harrows, axes, rakes, scythes, forks, and shovels. All seed sowing was done by hand. Most of the early settlers tended to live near forested areas, which provided wood for housing, fencing, and fuel.

In 1819, Jethro Tull helped revolutionize farming by developing and patenting an iron plow with interchangeable parts. By the 1820s and 1830s farmers, blacksmiths, and other innovators introduced various modifications to the plow that provided a sharper and stronger cutting edge with smoother surfaces so that the soil did not stick to either the plowshare or the mold-board. John Deere and Leonard Andrus began manufacturing steel plows in 1837.

Most early settlers planted a mixture of crops both for home consumption and for market. Field corn was often a mainstay because it gave high, reliable yields. Diversification was also needed to supply the various food components for the home table, and vegetables were mainstays of these plantings.

Manufacturing was making its way into the cities and into agriculture by the 1850s. The two-horse straddle row cultivator was patented in 1856. The change from hand power to horses characterized the first American agricultural revolution.

Post–Civil War

Many products that were developed prior to or during the Civil War increased the productivity of each laborer in harvesting, planting, and cultivating fields. In 1800, approximately 75% of the population was directly engaged in agricultural production. By 1850, it was less than 60%, and by 1900 less than 40% of the population was engaged in agricultural production.

During these times the number of farms began to decline and farm size increased. As both farming and manufacturing became more productive, people could choose a broader range of career options in such fields as medicine, law, science, government, and entertainment. The increased productivity of farmworkers led to surpluses of agricultural products and thus lower prices. This affected the livelihood of many farmworkers, and the supply of available labor at times exceeded the demands, resulting in unemployment. By the 1890s agriculture became increasingly more mechanized.

Industrial expansion, which began about 1895, established concentrated population centers or cities. Agricultural improvements made more food available to support a larger nonagricultural population. These large population centers became largely dependent on special producers for their food supply, and, as a result, commercial production of many agricultural crops developed near population centers.

During the 1910s, big open-geared tractors came into use in areas of extensive farming. In 1926, a successful light tractor was developed, and during the 1930s the all-purpose, rubber-tired tractor came into wide use. By the 1930s, 58% of all farms had cars, 34% had telephones, and 13% had electricity.

Pre–World War II

For much of the early 1900s, American agricultural policy was guided by the philosophy that society would best be served by traditional family-size, owner-operated farms. These family farms relied heavily on local labor, supplies, and consumers to sustain their business.

Post–World War II

After 1945, the rapid transition to volume marketing systems began. The improvement in technology in food and fiber handling systems (including refrigeration) developed during wartime, the changing economic structure of American agriculture, and the highway expansion of the 1950s all favored those growers who could supply the market with a large volume over a prolonged period. Small farms were inefficient and many either failed or enlarged to meet new challenges. Growers enlarged through purchases of additional land or through production/marketing cooperatives. They maintained competitiveness by adopting new technology or by stressing high quality in crop production and handling. During these times, much less variety was available than today in terms of number, form, and quality of crop products.

The change from horses to tractors after 1945 and the adoption of a group of technological practices characterized the second American agricultural revolution. By 1954, the number of tractors on farms exceeded the number of horses and mules for the first time. By 1954, 71% of all farms had cars, 49% had telephones, and 93% had electricity.

Relatively Recent Times

The last several decades were highlighted by crop industries that responded to an increasingly diverse population and the products that this population demanded. During the 1970s, no-tillage agriculture was popularized, a trend that continued into the 1980s as more farmers used no-till or low-till methods to curb soil erosion. By 1975, 90% of all farms had phones, 98% had electricity.

In the 1980s, targeted marketing replaced mass marketing. Development of new products occurred at a rapid pace during the 1980s and continued during the 1990s. During the 1990s and early 2000s, more farmers began to use low-input sustainable techniques to decrease chemical applications. Also field production of genetically engineered crops began.

U.S. Agriculture and Crop Production

Total U.S. agricultural output increased at an average annual rate of 1.88% from 1948 to 1996 (Fig. 3.1). In 2002, the total U.S. farm cash receipts for agriculture was in excess of $217 billion (Table 3.1). Crop cash receipts accounted for $97.2 billion of this amount. Feed crops represent the largest crop category in returns.

Exports

With the productivity of U.S. agriculture growing faster than domestic food and fiber demand, U.S. farmers and agricultural firms rely heavily on export markets to sustain prices and revenues. Export revenues accounted for 20% to 30% of U.S. farm income during the last 30 years and are projected to remain at this level until 2010 (Fig. 3.2).

During the early 1990s, U.S. exports of nearly all commodities grew (Fig. 3.3). After U.S. exports peaked in fiscal 1996, export shares of production (in value) for many commodities dropped in the late 1990s, except for nuts, fruits, and vegetables. For some commodities, trade dependency is relatively high. Exports of rice, cotton,

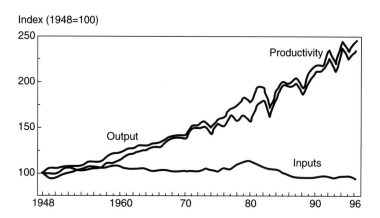

FIGURE 3.1   Growth of agricultural productivity as percentage of 1948 production. *Source:* Economic Research Service, USDA.

TABLE 3.1   *Value Added to the U.S. Economy by the Agricultural Sector*

	1999	2000	2001	2002
	Value ($ billion)			
Final crop output	92.6	94.8	95.1	97.2
Food grains	7.0	6.5	6.4	6.4
Feed crops	19.5	20.5	21.4	23.7
Cotton	4.6	2.9	3.6	3.9
Oil crops	13.4	13.5	13.3	14.7
Tobacco	2.3	2.3	1.9	1.7
Fruits and tree nuts	11.9	12.5	12.0	13.0
Vegetables	15.1	15.6	15.6	16.9
All other crops	18.3	18.5	19.0	19.2
Final animal output	95.3	99.3	106.4	93.2
Services and forestry	25.1	24.4	26.2	26.8
Final agricultural sector output	213.0	218.5	227.7	217.2

Source: Economic Research Service, USDA.

and wheat approached 50% shares of U.S. production in the 1990s. For some fruits and nuts, the shares exported are even larger.

Historically, the bulk commodities (such as wheat, rice, coarse grains, oilseeds, cotton, and tobacco) accounted for most U.S. agricultural exports. However, in the 1990s, as population and incomes worldwide rose, U.S. exports of high-value products (e.g., meats, live animals, oils, fruits, vegetables, and beverages) expanded steadily in response to demand for more food diversity (Table 3.2). In 2000, high-value product exports accounted for a 65% share of total U.S. agricultural exports, while bulk exports accounted for 35%.

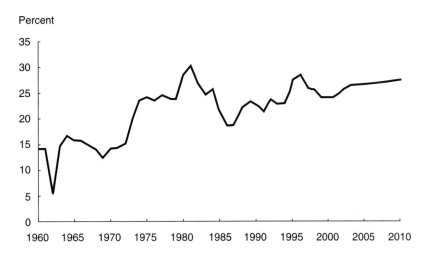

FIGURE 3.2 Value of agricultural exports as a percentage of gross cash income.
Source: USDA Agricultural Baseline Projections to 2010, February 2001. Economic Research Service, USDA.

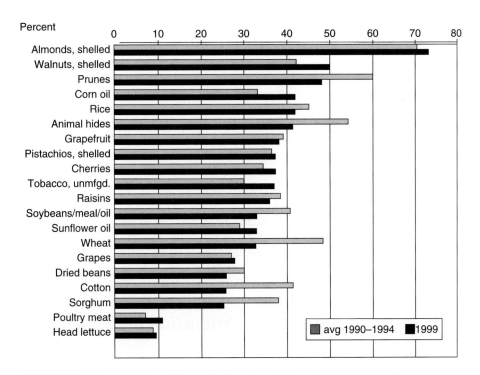

FIGURE 3.3 Export share of U.S. agricultural products.
Source: Economic Research Service, USDA.

TABLE 3.2   *Value of U.S. Agricultural Exports, Ranked by Commodity Groups*

Export Item	Avg. 1990–1996	Avg. 1997–2000
	Value ($ billion)	
Bulk products		
Grains	12.0	9.9
Oilseeds	5.0	6.1
Cotton	2.5	1.9
Tobacco	1.4	1.4
High-value products		
Animals	8.6	10.9
Vegetables	3.0	4.3
Grain	3.5	4.1
Fruits and nuts	3.5	3.9
Other[1]	2.7	3.6
Oilseed	2.4	3.3
Juice, wine, and other beverages	1.4	2.0

[1]Sugar and tropical products, nursery and greenhouse, seeds, essential oils, and miscellaneous vegetable products.

Source: Economic Research Service, USDA.

Imports

U.S. consumers desire expanded food variety, stabilized year-round supplies of fresh fruits and vegetables, and tempered increases in food prices. As a result, U.S. imports have increased steadily as demand for food diversification has expanded. Imports' share of total domestic food consumption was relatively low in 1975 and 1980. However, imports now account for a rising share of total food consumed in American homes. Oils, spices, tree nuts, fruit juices, and some fruits and vegetables are products for which a large share is imported.

U.S. agricultural imports grew throughout the 1990s (Fig. 3.4), despite appreciation of the dollar versus currencies affected by the financial crises. Horticultural products such as fruits, vegetables, nuts, wine, malt beverages, and nursery products were the largest U.S. agricultural imports, at 40% of the total (Table 3.3). Animals are next in importance, followed by the noncompetitive tropical products such as coffee, cocoa, and rubber.

Agronomic Segments of Crop Production Industries

Cereal or Grain Crops

Cereals are grasses grown for their edible seeds. The term *cereal* is used to either describe the grain or the seed itself. *Grain* is a collective term applied to cereals. Important cereals are wheat, oats, barley, rye, maize, rice, millet, and grain sorghum.

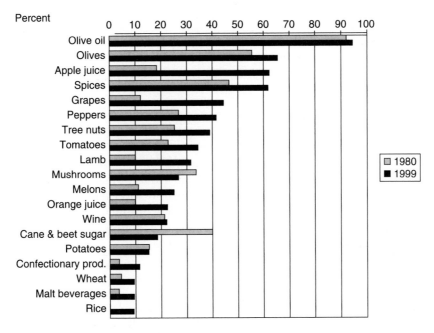

FIGURE 3.4   U.S. imports as a share of food consumption.
Source: Economic Research Service, USDA.

TABLE 3.3   *Value of U.S. Agricultural Imports, Ranked by Commodity Groups*

Import Item	Avg. 1990–1996	Avg. 1997–2000
	Value ($ billion)	
Horticulture	10.9	16.8
Animals	5.7	7.1
Coffee, cocoa, and rubber	4.3	5.8
Grains and feeds	1.8	3.0
Sugar and tobacco	2.1	2.5
Oilseeds	1.3	2.0

Source: Economic Research Service, USDA.

The United States is a major wheat producing country, with output typically exceeded only by China, the European Union, and, sometimes, India. In 2002, wheat ranked third among U.S. field crops in both planted acreage and gross farm receipts, behind corn and soybeans. Presently, almost half of the U.S. wheat crop is exported.

Corn is the most widely produced feed grain in the United States, accounting for more than 90% of total value and production of feed grains. Almost 32 million hectares of land were planted to corn in 2000. Most of the crop is used as the main energy ingredient in livestock feed.

Forage Crops

Forage refers to the vegetative matter, fresh or preserved, utilized as feed for animals. Forage crops include grasses, legumes, crucifers, and other crops cultivated and used for hay, pasture, fodder, silage, or soilage.

Fiber Crops

The fiber crops include cotton, flax, hemp, and kenaf. Cotton is the single most important textile fiber in the world, accounting for over 40% of total world fiber production. Although some 80 countries from around the globe produce cotton, the United States, China, and India together provide over half the world's cotton. The United States, while ranking second to China in production, is the leading exporter, accounting for 25% to 30% of global trade in raw cotton. In 2002, cotton accounted for close to $4 billion gross farm receipts in the United States.

Tobacco

Tobacco is used for cigarettes, cigars, chewing, snuff, and pipes. Different types and kinds of tobacco are used in these tobacco products. With a farm value in 2002 of $1.7 billion, tobacco is one of the top ten U.S. cash crops. The United States is fourth behind China, India, and Brazil in world production, and second behind Brazil in exports.

Oilseed Crops

The major U.S. oilseed crops are soybeans, cottonseed, rapeseed (canola), and sunflower seed. Soybeans are the dominant oilseed in the United States, accounting for about 90% of U.S. oilseed production. Processed soybeans are the largest source of protein feed and vegetable oil in the world. The United States is the world's leading soybean producer and exporter. Farm value of U.S. soybean production in 2001 was $12.5 billion, the second-highest value among U.S.-produced crops, trailing only corn. Soybean and soybean product exports accounted for 43% of U.S. soybean production in 2000.

Horticultural Segments of Crop Production Industries

Vegetable Crops

The United States is one of the world's leading producers and consumers of vegetables and melons. In 2002, the sale of vegetables and melons (including mushrooms) earned farmers $17.7 billion. Annual per capita use of vegetables and melons rose 7% from 1990 to 2002, reaching 200 kg as fresh consumption increased and processed fell. Vegetable (and other food crops) in the United States can be produced for either fresh market or for the processing industry.

Fresh Market Commercial fresh market food crops can be produced on either truck farms or market gardens. The differences between truck farms and market gardens are in where they are located, the number of different types of crops grown, the relative acreage of each crop grown, and how and where the crops are marketed. In addition, small acreage growers may direct market their crops through subscription farming, pick-your-own operations, or farmer's markets (also called curb markets).

Truck farms are often located near transportation systems or highways. They tend to deal with only one or two crops on a substantial acreage for distant marketing, often by trucks.

Market gardens are usually small businesses that tend to be located near population centers and supply a wide variety of home or locally grown produce or plants. Market gardens tend to concentrate on high-profit crops and seek to produce high quality. Their success partially depends on demonstrating that the quality of the crops they produce is better than crops that are shipped in. They grow vegetables for local markets and often use intensive agriculture with high rates of fertilizer and access to irrigation (often city water). Market gardens frequently grow multiple crops per year on the same land and use plastic mulch and row covers to extend the marketing season.

Pick-your-own or U-pick operations require the least grower labor and capital for market facilities. The customers do the harvesting. This method works well for some commodities and in some locations but not for all crops or for all growers. Growers who have pick-your-own operations must be able to effectively deal with the public and must be willing to accept a certain amount of unintentional damage caused by customers.

Farmers' markets (also called curb markets) are similar to roadside markets, but the retailing function is moved closer to the customer. This enables the grower to offset the potential disadvantage of production location.

Processing For most vegetables, growing for processing is distinct from producing for the fresh market. Generally, little diversion between the fresh and processing markets takes place. Most varieties grown for processing are better adapted to mechanical harvesting and often do not have characteristics desirable for fresh market sale. For example, processing tomatoes are smaller and possess different internal attributes than fresh varieties. Most vegetables destined for processing are grown under contractual arrangements between growers and processors, whereas contracting for fresh market sales, although increasing, is still much less common. About 53% of all vegetable and melon production is destined for processing.

Fruit and Nut Crops

The United States is among the top producers and consumers of fruit and tree nuts in the world. Each year, fruit and tree nut production generates about 13% of U.S. farm cash receipts for all agricultural crops. Annual U.S. per capita use of fruit and tree nuts totals nearly 130 kg, fresh-weight equivalent. Oranges, apples, grapes, and bananas are the most popular fruit, and almonds, pecans, and walnuts are the most preferred tree nuts.

Fruit   U.S. fruit production generated $11.2 billion in farm cash receipts in 2000, up 19% from 1995. The most important fruit in terms of farm cash receipts are the following: grapes, $3.1 billion; oranges, $2.1 billion; apples, $1.4 billion; strawberries, $1.0 billion; and avocados, $489 million. The most consumed fruit in the United States in 2000 were oranges (31% of total per capita fruit consumption, all uses), grapes (17%), apples (16%), bananas (10%), and grapefruit (5%).

Most fruits are grown to serve both fresh and processing markets. The fresh market sector accounts for about half the value of U.S. fruit production, with over three-quarters of that generated by noncitrus fruit production. California's production accounts for more than one-fourth the value of all fresh fruits.

Processed fruit products include juice and canned, frozen, and dried fruit, as well as wine. At packinghouses, fruits are inspected and graded for size, shape, and appearance. Some fruits intended for the fresh market are diverted to processors due to quality requirements. Most fruits for processing, on the other hand, are not easily diverted to the fresh market because of the same quality requirements. Most fruit destined for processing are grown under contractual arrangement between growers and processors.

Tree Nuts   U.S. tree nut production increased significantly over the past three decades, from 125 million kg (shelled basis) in 1970 to 890 million kg in 2000. That increase is being driven by increased domestic and foreign demand. U.S. per capita use of all tree nuts was 1.1 kg (shelled basis) in 2000, up from 0.7 kg in 1977. Exports continued to gain a larger share of domestic supplies, increasing from an average of 24% during the 1970s to 41% during the 1990s.

U.S. tree nut production in 2000 generated $1.5 billion in farm cash receipts, with almonds, walnuts, pecans, and pistachios accounting for 97% of U.S. sales. California is the nation's number-one producer of tree nuts. In 2000, 83% U.S. tree nut production was harvested from California orchards, including virtually all almonds, pistachios, and walnuts.

Nursery and Greenhouse Crops

The United States is the world's largest producer and market for nursery and greenhouse crops (together known as the green industry). Grower cash receipts from nursery and greenhouse sales (on sales of plants to retail and distribution businesses) have grown steadily over the last two decades and are increasing at approximately $500 million per year. Floriculture and nursery crops reached $13.8 billion in sales in 2002, up from $13.7 billion in 2001.

According to the Economic Research Service of the U.S. Department of Agriculture, the nursery and greenhouse industry comprises the fastest-growing segment of U.S. agriculture. In terms of economic output, nursery and greenhouse crops represent the third most important sector in U.S. crop agriculture, ranking seventh among all commodities in cash receipts, and among the highest in net farm income. Eighty-five million U.S. households spent $39.6 billion at lawn and garden retail outlets in 2002, according to the National Gardening Association and Harris Interactive, and more than 24.7 million households spent $28.9 billion on professional landscape, lawn, and tree care services. In total, Americans spent $68.5 billion improving their homes in 2002.

Niche Crops

Niche markets are those markets that specialize in nontraditional crops. Common niche markets for plants include specialty crops, such as unusual or exotic plants and organically grown plants. Larger ethnic populations and growth in their cultural expression have increased the demand for food product diversity. Also, a broader portion of the population is experimenting with foods once considered ethnic or regional.

Home Gardens

Home garden plants are grown in small quantities in sections of the property of many homeowners. Home gardening has been designated as the number-one outdoor leisure activity. In previous years people gardened to save money. Today many people garden for fresher tasting vegetables, better quality food, better nutrition, and improved health.

Summary

U.S. agriculture had its beginnings in the primitive food production systems developed by the colonists and settlers for survival. Early settlers planted a mixture of crops both for home consumption and for market. The two-horse straddle row cultivator was patented in 1856. The change from hand power to horses characterized the first American agricultural revolution. Industrial expansion, which began about 1895, established concentrated population centers or cities. Agricultural improvements made more food available to support a larger nonagricultural population. These large population centers became largely dependent on special producers for their food supply, and, as a result, commercial production of many agricultural crops developed near population centers.

For much of the early 1900s, American agricultural policy was guided by the philosophy that society would best be served by traditional family-size, owner-operated farms. After 1945, the rapid transition to volume marketing systems began. The improvement in technology in food and fiber handling systems (including refrigeration) developed during wartime, the changing economic structure of American agriculture, and the highway expansion of the 1950s all favored those growers who could supply the market with a large volume over a prolonged period.

The change from horses to tractors after 1945 and the adoption of a group of technological practices characterized the second American agricultural revolution. The highly sophisticated systems of today's crop producers supply a wide variety and relatively consistent supply of food for sale nationally and internationally. In the 1980s, targeted marketing replaced mass marketing.

Total U.S. agricultural output increased at an average annual rate of 1.88% from 1948 to 1999. With the productivity of U.S. agriculture growing faster than domestic food and fiber demand, U.S. farmers and agricultural firms rely heavily on export markets to sustain prices and revenues. U.S. consumers desire expanded food variety, stabilized year-round supplies of fresh fruits and vegetables, and tempered increases in food prices. As a result, U.S. imports have increased steadily as demand for food diversification has expanded.

Agronomic crop production industries include cereal or grain crops, forage crops, fiber crops, tobacco, and oilseed crops. Horticultural crop production disciplines include vegetable crops, fruit and nut crops, green industry (nursery and greenhouse) crops, and niche crops. In addition, home gardening is a popular outdoor leisure activity.

Review Questions

1. What are some of the reasons that consumer expenditures for food in the United States is the lowest in the world?
2. What were the U.S. plant producing industries like during the following historical periods?
   a. American colonists
   b. Post–Civil War
   c. Pre–World War II
   d. Post–World War II
   e. Relatively recent times
3. What characterized the first and second American agricultural revolutions?
4. How might labor affect which crops are grown by a commercial producer?
5. Why are exports important to crop producers in the United States?
6. What helped spark the recent increases in imports of food products?
7. What is the importance of the green industries to U.S. agriculture?

Selected References

Brewer, T., J. Harper, and G. Greaser. 1994. *Fruit and vegetable marketing for small-scale and part-time growers.* Penn State Coop. Ext. Serv. (Agricultural Alternatives).

Cook, R. L. 1992. The dynamic U.S. fresh produce industry: An overview. In *Postharvest technology of horticultural crops,* ed. A. A. Kader, pp. 3–13, Oakland,

CA: University of California, Division of Agriculture and Natural Resources Publication 3311.

The Packer. 1997. *Fresh trends: A profile of the fresh produce consumer.*

Snyder, R. G. 1996. Greenhouse vegetables—Introduction and U.S. industry overview. *Proc. Natl. Ag. Plastics Congress* 26:247–252.

U.S. Department of Agriculture. 2003. *Floriculture and nursery crops yearbook.* USDA—ERS Publication FLO-2003.

VanSickle, J. 1998. 1998 vegetable outlook. *American Vegetable Grower,* January, 20–21.

Selected Internet Sites

www.aces.uiuc.edu/~sare/history.htm/   A Brief History of Agriculture, University of Illinois College of A.C.E.S.

www.econ.ag.gov/   Economic Research Service, USDA.

www.jan.manlib.cornell.edu/data-sets/specialty/89011/   Vegetable Yearbook, USDA Economics and Statistical System, Cornell University.

www.usda.gov/nass   National Agricultural Statistical Service, USDA.

www.5aday.com/   The Produce for Better Health Foundation.

http://www.ers.usda.gov/Briefing/AgTrade/usagriculturaltrade.htm   Ag Trade Briefings, Economic Research Service, USDA.

4

The Sciences of Plants

The Earth has more than 400,000 species of plant life. These plants, powered by light from the sun, carbon dioxide from the air, and nutrients from the soil, pass on energy to the life forms that consume them. Plants also provide humans with raw material for clothing, structures, and aesthetic pleasures.

This chapter begins by defining botany and describing some of the more "basic" and "applied" botanical disciplines. Plant classification schemes that provide a framework for discussion of various plants are presented along with some of their uses.

Botany

Botany, one of the oldest branches of biology, is the scientific study of plants. It was established as a science in early classical times when Aristotle and his pupils designed a systematic approach to studying and classifying plant species. Botany includes the study of the chemical and physical natures of the materials and processes of plant cells, organization of cells into tissues and tissues into organs, history of plant life, and relationship of plants to all phases of their environment. These study areas are often referred to as the basic or pure phases of plant study. The more specialized basic fields of botany can be categorized into form and structure disciplines, and growth and development disciplines.

Botany also includes those aspects of plant study that place major emphasis on industrial and agricultural applications of plants. These study areas of botany are often referred to as the applied phases of plant study, the plant sciences, or economic botany. The more specialized applied fields of botany can be broadly categorized by commodity or cross-commodity. Plant scientists study plants and their growth in soils, helping producers of food, feed, and fiber crops to continue to feed a growing population while conserving natural resources and maintaining the environment.

The "Basic" Botanical Sciences

Examples of Form and Structure Disciplines

- **Plant taxonomy,** also called plant systematics, is concerned with the classification of the members of the plant kingdom. Its object is to identify the members by name and description and to arrange them according to their natural relationships into species, genera, families, and orders.
- **Plant morphology** is the study of the form and structure of plants. Its object is to describe the structure of the plant body and to trace underlying similarities in form among various plant groups.
- **Plant anatomy** is the study of the internal structure of a plant, such as cell and tissue arrangements.
- **Plant cytology** is the study of plant cell structure, function, and life history. It is involved with evolution and compartmentation of plant cells; with structure, function, and development of plant cell components (endomembrane system, vacuoles, nuclei, mitochondria, plastids, cytoskeleton, and cell wall); and with ontogenetic development of plant cells and its regulation (cell division, growth, morphogenesis, cell senescence, and programmed cell death).

Examples of Growth and Development Disciplines

- **Plant genetics** is the study of plant heredity, genes, and gene function. Genetic engineering is the artificial manipulation or transfer of genes from one organism to another. Genes have been directly transferred to plants either using crown gall bacterium (*Agrobacterium tumefaciens*) as a vector or by bombarding genes into plants using a particle gun.
- **Plant physiology** is the study of the mechanisms and processes in plants and the interpretation of plant behavior in terms of physical and chemical laws. It includes studies on the internal processes (such as assimilation, photosynthesis, translocation, or transpiration) that are involved in vital functions and the influence that one or more environmental factors (e.g., humidity, water, light, mineral nutrients, and/or temperature) have on these functions and processes.
- **Plant ecology** is the study of the underlying order of plant species and vegetation. It is usually centered on the relationships and interactions of species within communities (collective organisms within a location) and the manner in which populations of a species are adapted to a characteristic range of environmental factors. Vegetation consists of all plant species in a region (the flora), and plant ecology is concerned with the pattern of how all those species are spatially and temporally distributed. If the region is large, the vegetation will consist of several plant communities. The life form of its dominant plants and the architecture of its canopy layers characterize each vegetation type.
- **Plant ecophysiology** is a science that seeks to describe the physiological mechanisms that underlie ecological observations (Fig. 4.1). Ecophysiologists address the ecological questions about controls over the growth, reproduction,

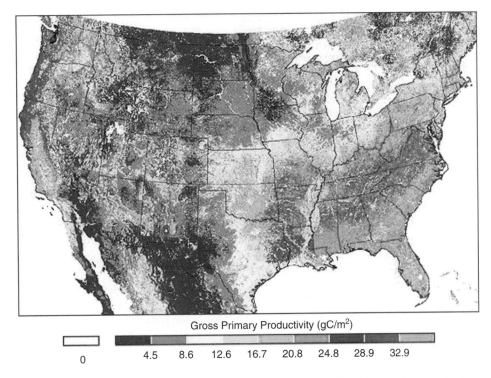

Gross Primary Productivity (gC/m^2)

0   4.5   8.6   12.6   16.7   20.8   24.8   28.9   32.9

FIGURE 4.1   Ecophysiologists may study the gross primary productivity or photosynthesis of a region. This composite image over the continental United States, acquired and compiled by NASA during the period March 26 to April 10, 2000, shows regions where plants were more or less productive.
Source: Earth Observatory, NASA.

survival, abundance, and geographical distribution of plants as these processes are affected by the interactions between plants and their physical, chemical, and biotic environment.
- **Plant geography** is the study of the geographical distribution of plants and the factors that determine this distribution.

The "Applied" Botanical Sciences or Plant Sciences

The "applied" botanical sciences or the plant sciences are the aspects of plant study that place major emphasis on the use of plants by humans. These are often referred to as agricultural plant production or plant sciences within an agricultural context or as economic botany within a more sociological context.

Within the plant sciences, specialized disciplines are based on the commodity (or type of plant produced) disciplines such as agronomy, horticulture, and forestry or cross-commodity (that emphasize growth and development subjects) disciplines such as plant propagation, plant breeding, plant pathology, entomology, agricultural mechanization, integrated pest management, and crop ecology.

Examples of Commodity-Based Disciplines of Plant Science

- **Agronomy** is the branch of agriculture that studies the principles and practice of crop production and field management. It pertains to field crops, such as cereals and fodder. The terms *agronomy* and *crop science* are often used interchangeably. It is also known as *scientific agriculture.*
- **Horticulture** is the branch of plant agriculture concerned with the intense cultivation of garden crops produced for food, medicine, enjoyment, and recreation. In general, the monetary return on investment per unit of production (often per plant or hectare) is also generally higher for horticultural crops than for many of the other agricultural crops. Also many horticultural crops are utilized by the consumer or purchaser fresh with minimal postharvest processing (often referred to as fresh market) or as living materials (such as in landscapes); whereas agronomic and forestry products are generally highly processed after harvest and the plant material is generally considered as nonliving or nongrowing before being used by the consumer (as grain, fiber, timber).

  The three general divisions of horticulture are olericulture or vegetable production, pomology or fruit production, and ornamental horticulture. Ornamental horticulture includes floriculture or the culture of cut flowers, potted flowers, and foliage plants; nursery plants; and landscape horticulture (design and construction).
- **Forestry** is the study of forest management with the goals of conservation and the production of timber and is associated with nonfood tree crops and their products.

Examples of Cross-Commodity Disciplines of Plant Science

- **Plant propagation** is the study or practice of producing numerous plants from an individual plant (vegetative reproduction) or plants (sexual reproduction) to preserve unique characteristics.
- **Plant breeding** is the study of the genetic improvement of crop plants for the benefit of society. It encompasses a broad range of activities, from molecular studies of crop plant genomes to field evaluations in multiple locations.
- **Plant pathology** is the study of the cause, nature, prevalence, severity, and control of plant diseases. It includes investigating the cause, nature, prevalence, and severity of parasitic, nonparasitic, and viral diseases that attack plants; conducting experiments in and establishing methods for preventing and controlling such diseases; and/or studying the relationship of such diseases to the practices involved in planting, cultivating, transporting, and storing plants and plant products.
- **Entomology** is the study of insects. Insects are the predominant species on earth, representing the greatest biodiversity with over 1 million known

species. Not surprisingly, insects significantly affect crop production, whether it's a positive effect through pollination of our food plants or a negative effect as competitors with our food supply.

- **Agricultural mechanization** is the study of the mechanical, structural, natural resource, processing, and electronic technologies applied in agriculture systems.
- **Agroecology** is the study of the interactions among the many biological, environmental, and management factors that make up and influence agriculture. Further, it is the study of material and energy flow within and across agricultural fields, from the level of the individual soil organism to the global scale. Important interactions within this complex web include those among soil, plants, animals, humans, landscapes, and the atmosphere.
- **Crop ecology** is the part of agroecology that specifically addresses crop production. The relationship between soil quality and crop health is an important feature of crop ecology. The major objectives of crop ecological management are to enhance soil quality, manage pests and diseases with minimum environmental impact, and recycle nutrients and residues effectively and efficiently.
- **Integrated pest management,** or IPM, is the sustainable approach used by crop producers to manage pests by combining biological, cultural, physical, and chemical tools in a way that minimizes economic, health, and environmental risks. It utilizes approaches to pest management that combines a wide variety of crop production practices with careful monitoring of pests and their natural enemies. IPM uses improved decision making to reduce reliance on purchased inputs while maintaining crop yield and quality.

Plant Classification Schemes

Because of the large number of different plants that are found around the globe, it is often useful to classify these plants into groups or classes so as to better understand and discuss them. Placing individual plants into groups or classes with other crops that share some characteristics reduces the rather large number of crops into a smaller number of groups and results in logical associations. As with many techniques that are used to simplify, no one classification or grouping is perfect and some classifications are more useful than others depending on situations and needs.

Botanical

The botanical classification scheme is based on similarity or dissimilarity in morphological structures, often with flower structure as the main criteria for determining relationships. The successive levels of morphological relationships are a result of evolution. The successive groupings of plants in the botanical classification (from broadest grouping to most specific) are kingdom, division, subdivision, phylum, subphylum, class, subclass, order, family, genus, and species.

Most common agricultural crops belong to the division Anthophyta. The division Anthophyta is generally broken down into two classes: monocotyledons and dicotyledons. The classes are further divided into families (with names that end in "aceae"), which are composed of individual related plant species.

The broadest grouping that crops are typically discussed is family. The genus and species make up the scientific name. Scientific names are accepted worldwide and serve as positive identification, regardless of language. Plants recognized as a single crop, even if they have different scientific names, are said to be of one kind.

Because management systems may be governed by botanical similarities, knowledge of botanical classifications of plants is useful for producers. Also the climatic requirements of a particular family or genus are usually similar. The use of the crop for economic purposes within families is similar, and disease and insect controls are quite often similar for the related genera.

Use

Classification by use (sometimes referred to as agronomic classification) provides a grower or handler with broad plant groupings that imply specific cultural or handling techniques. For example, leafy crops are very perishable and require rapid chilling after harvest to preserve quality. Root crops are similar in how they are affected by soil fertility, water management, and soil texture. While classification by use is used, some crops do have more than one use.

Life Cycle

All plants can be classified according to the time required to complete their life cycle. Annual plants complete their life cycle during a single growing season. Biennial plants require two seasons to complete their life cycle, and perennial plants grow for more than two years.

Many of the common agricultural crops are annuals. Examples of annual crops include corn, soybeans, cotton, and beans. Other crops are biennials but are grown as annuals. These include many of the cole crops such as broccoli, cauliflower, and cabbage and root crops such as celery and parsnips. Many biennials tend to be sensitive to temperature regulation of flowering. Other crops are perennials and can remain in production for many years. Examples of these include fruit trees, ornamental shrubs, globe artichoke, asparagus, and rhubarb.

Other Classifications

Other classification schemes of agricultural crops refer to their sensitivity to environmental factors. Included among these are those that are grouped according to sensitivity to temperature, soil pH, tolerance to nutrient levels, preference for soil moisture levels, and sensitivity to chilling damage. Other classification schemes are based on morphological features such as edible plant part, seed size, and depth of rooting.

Summary

Botany is the scientific study of plants. It includes the study of the chemical and physical natures of the materials and processes of plant cells, organization of cells into tissues and tissues into organs, history of plant life, and relationship of plants to all phases of their environment. These study areas are often referred to as the basic or pure phases of plant study. Examples include plant taxonomy, plant morphology, plant anatomy, plant cytology, plant genetics, plant physiology, plant ecology, plant ecophysiology, and plant geography.

Botany also includes those aspects of plant study that place major emphasis on the agricultural applications of plants. These study areas of botany are often referred to as the applied phases of plant study, the plant sciences, or economic botany. The more specialized fields of the plant sciences include commodity-based disciplines, such as agronomy, horticulture, and forestry, and cross-commodity disciplines, such as plant propagation, plant breeding, plant pathology, entomology, agricultural mechanization, agroecology, crop ecology, and integrated pest management.

Agricultural plants are often placed into groups or classes with other crops that share some characteristics. This reduces the rather large number of crops into a smaller number of groups and results in logical associations. Some of the more common groupings are those based on botany, use, and life cycle.

Review Questions

1. Define botany.
2. List some examples of the basic and applied botanical science.
3. Why is it important to classify crops into groups or classes?
4. What is the botanical classification scheme based on?
5. What are the advantages to classifying crops according to plant part?
6. What are some of the characteristics of warm-season crops and cool-season crops?

Selected References

Asian Vegetable Research and Development Center. 1990. *Vegetable production training manual.* Shanhua, Taiwan: Asian Vegetable Research and Development Center. Reprinted 1992.

Janick, J. 1979. *Horticultural science*, 3rd ed. San Francisco: W. H. Freeman.

Martin, J. H., W. H. Leonard, and D. L. Stamp. 1976. *Principles of field crop production*, 3rd ed. New York: Macmillan.

Maynard, D. N., and G. J. Hochmuth. 1997. *Knott's handbook for vegetable growers*, 4th ed. New York: John Wiley.

Nonnecke, I. L. 1989. *Vegetable production*. New York: Van Nostrand Reinhold.

Yamaguchi, M. 1983. *World vegetables*. New York: Van Nostrand Reinhold.

Selected Internet Sites

http://www.nbii.gov/disciplines/botany/index.html   National Biological Information Infrastructure home page. The National Biological Information Infrastructure is a broad-based, collaborative program amongst federal, state, international, non-government, academic, and private industry partners.

www.yenra.com/seeds/catalogs/   Resource list of online Internet seed catalog sites.

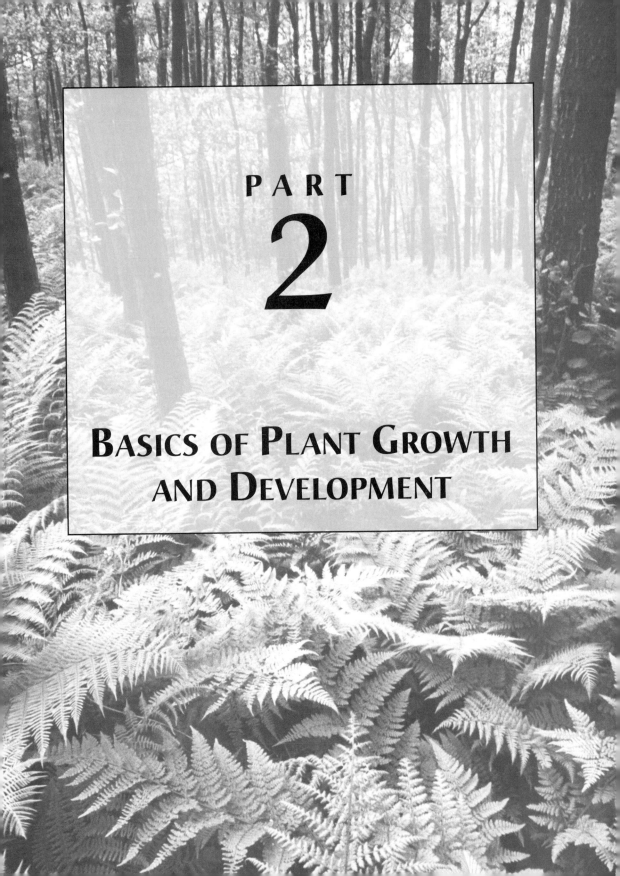

PART

2

BASICS OF PLANT GROWTH AND DEVELOPMENT

5

Introduction to Plant Growth and Development

A generalized young plant body consists of two distinct sections: the aerial (above ground) section and rhizosphere (below ground) section. The aerial section of the plant body consists mainly of a shoot or shoots. Stems, leaves, buds, and sometimes reproductive structures collectively form the shoot. The rhizosphere section of the plant body consists primarily of roots. Roots serve to hold the plant in place, facilitate water and nutrient uptake, and store carbohydrates. This introductory chapter provides some definitions to important terms that will be used in the following chapters.

Plant Organs

Stems

Stems are the supporters and producers of leaves and flowers. A primary function of the stem is to distribute the leaves in space so that the leaves can intercept sunlight for photosynthesis. Stems can also store food and provide for the movement of water and nutrients to and from the leaves.

In most crops the stems are above ground and are called aerial stems. These above-ground stems may be rigid (erect) or flexible and form climbing stems or vines. The stem is made up of distinct areas where leaves and/or buds are attached. These areas are called nodes. The area between two successive nodes is called an internode. Stems usually possess a terminal bud at the tip and axillary buds in the axils of each leaf. The axillary buds may produce stems or flowers.

Some stems undergo modifications and appear as thorns, prickles, or tendrils. Some stems also occur as stolons, producing at intervals erect stems and adventitious roots as a modification of the horizontal or prostrate stems. The rhizomes of some plants, such as potatoes, are thick and fleshy or partially thickened and called tubers. Some plants are propagated by short, stout, erect underground stems in which food is stored. Such a stem is broader than it is long and is know as a corm. Some bulbs, such as onions, have layers of thickened leaves (bud scales) surrounding a short, erect stem.

Leaves

Leaves are appendages of the stem and primarily carry on photosynthesis. Some leaves are not photosynthetic, though, and even photosynthetic leaves have other functions such as storage of food and water, reproduction, root formation, climbing, protection, or flower formation. Leaves of many agricultural plants (such as spinach, onion, and cabbage) are useful because they store water, salts, food, minerals, and vitamins in their leaves.

Flowers

Flowers provide plants with the necessary apparatus to carry on sexual reproduction. Sexual reproduction is the foundation for evolution and crop improvement through genetic changes. The immature flowers or flower parts are the edible and marketable plant parts in several agricultural plants. These include broccoli, cauliflower, and Brussels sprouts. Edible flowers, such as squash flowers, are also becoming more popular.

Roots

Most roots function to anchor plants and to absorb water and nutrients from the soil. In some cases they also serve as food storage reservoirs. The two basic types of roots are fibrous roots and taproots. In the taproot system, the primary root is enlarged and the secondary roots are few and slender. The type of root system and size varies with different species of plants and also with environmental factors such as moisture and soil composition. The enlarged taproots of carrots and beets and the enlarged lateral roots of sweet potatoes are examples of root plant parts that are consumed as the edible portion.

Plant Growth

Plant growth is an irreversible increase in mass (weight) or size (volume) of the plant due to the division and enlargement of cells. The term can be applied to an organism as a whole or to any of its parts (Fig. 5.1). Growth is not uniformly distributed throughout the plant but is restricted to certain zones containing cells in a meristem. Meristems are located near the root and shoot tips (apices), in the vascular cambium, near the nodes of monocots, and in certain parts of young leaves. The root and shoot apical meristems develop during embryo development as the seed forms, but the vascular cambium and meristematic areas of leaves are not distinguishable until after germination.

Stem Growth

Stem growth occurs at several sites within a stem. Primary growth results in an increase in stem length and occurs as cells divide and elongate at the stem apical

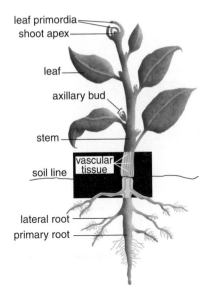

FIGURE 5.1   The parts of a vascular plant.
Source: Oregon State University Extension Service Master Gardener Handbook, Chapter 1,
Botany Basics by Ann Marie VanDerZanden. 1999.

meristem. Many stems further increase their length by retaining meristematic tissue just above each node. Secondary growth is an additional aspect of plant expansion for many stems. In woody plants, a vascular cambium develops between the xylem and the phloem. This layer forms more vascular tissue (xylem and phloem) and thereby causes lateral expansion of the stem.

Leaf Growth

The earliest sign of leaf development usually consists of divisions in one of the three outermost layers of cells near the surface of the shoot apex. Subsequent leaf development is highly variable, as suggested by the wide variety of leaf shapes observed in the plant kingdom. When the leaf is a millimeter or so long, meristematic activity begins throughout its length. An increase in width of the leaf base in angiosperms results from meristems along each margin of the leaf axis. Most cell divisions cease well before the leaf is full grown. The remainder of leaf expansion is caused solely by the growth of preformed cells with specific functions (Fig. 5.2).

Flower Growth

Flowers often develop after roots, stems, and leaves are produced by the plant. Fruits then generally form after appropriate pollination (the transfer of pollen from a stamen to a stigma of a flower) and seeds subsequently are produced thus completing the life cycle and providing a mechanism for perpetuating the species.

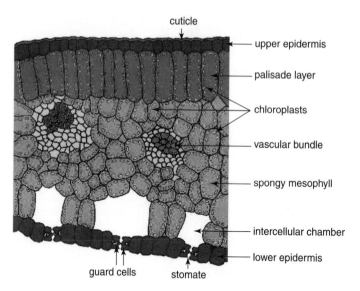

FIGURE 5.2   Cross section of a leaf with organelles indicated.
Source: Oregon State University Extension Service Master Gardener Handbook,
Chapter 1, Botany Basics by Ann Marie VanDerZanden. 1999.

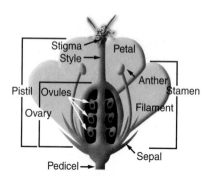

FIGURE 5.3   The parts of a perfect flower.
Source: Oregon State University Extension Service Master Gardener Handbook, Chapter 1,
Botany Basics by Ann Marie VanDerZanden. 1999.

Many plants produce perfect flowers that contain female and male parts (Fig. 5.3). Dioecious plants have imperfect flowers with male and female flowers on different individuals. Monoecious plants form male and female flowers at different positions along a single stem.

The opening of flowers with appropriate plant parts available for pollination is referred to as anthesis. After anthesis and pollination, the petals eventually wither and die. After pollination, the zygote and embryo sac develop into the seed. Development of fruits usually depends on germination of pollen grains on the stigma or on this plus subsequent fertilization. Developing seeds are usually essential for normal fruit development.

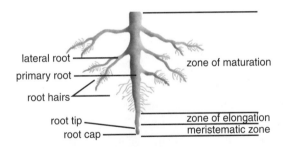

FIGURE 5.4   The root tip is responsible for primary growth of roots.
Source: Oregon State University Extension Service Master Gardener Handbook, Chapter 1,
Botany Basics by Ann Marie VanDerZanden. 1999.

Production of fruits lacking seeds is referred to as parthenocarpic fruit development. In some plants species, parthenocarpic fruit may occur naturally or may be induced by treating the unpollinated flowers with various plant hormones. Parthenocarpy may result from ovary development without pollination (citrus, banana, pineapple), from fruit growth stimulated by pollination but without fertilization (certain orchids), or from fertilization followed by abortion of embryos (grapes, peaches, cherries). Normal production of parthenocarpic fruits is common among fruits that produce many immature ovules, such as bananas, melons, figs, and pineapples. Many species, especially members of the Solanaceae and Cucurbitaceae families, will develop parthenocarpic fruit in response to applied auxins. In other species (grapes, apples, pears, cherries, apricots, peaches), exogenous application of a gibberellin (a plant hormone) is more effective.

Root Growth

Root growth occurs in two major ways. Primary growth results from cell production at the tip of each root (root apical meristem) and the subsequent elongation of these cells (Fig. 5.4). This elongation precedes the formation of root hairs and lateral roots. An additional tissue layer, the root cap, is produced by the root apical meristem. The root cap surrounds and protects the root apical meristem as the root tip grows through the soil. Some roots exhibit an additional pattern of growth called secondary growth in which a vascular cambium develops between the xylem and the phloem and produces more vascular tissue (Fig. 5.5).

Determinate versus Indeterminate Growth

Some plants or plant parts exhibit determinate growth in that they attain a certain size, stop growing, and then die. Other plants or plant parts exhibit indeterminate growth and continue to be active for many years. Many plants go through a regular sequence of growth rates, with an initial period of rapid growth followed by minimal if any increase in growth and eventually death after cessation of growth and breakdown of tissues.

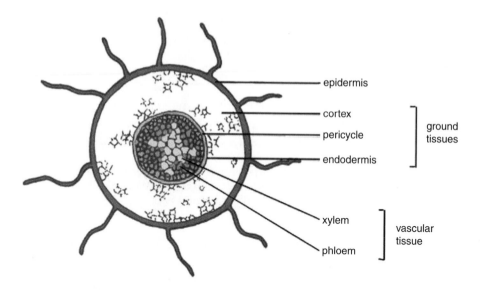

FIGURE 5.5   Secondary growth of some roots occurs when a vascular cambium develops between the xylem and phloem and produces more vascular tissue.
Source: Oregon State University Extension Service Master Gardener Handbook, Chapter 1, Botany Basics by Ann Marie VanDerZanden. 1999.

Measurement of Growth

The principle methods of measurement of plant growth involve determining either increases in volume (size) or weight (mass). Volume increases are often approximated by measuring expansion, such as length, height, width, diameter, or area. Weight increases can be determined by harvesting the entire plant or that plant part of interest and weighing it rapidly before too much water evaporates from it. This is called fresh weight. Because fresh weight can be variable (due to loss of some moisture from the tissue after harvest but before weighing or moisture status of the plant at harvest), dry weight is often preferred as a measurement for plant growth. The dry weight is commonly obtained by drying the freshly harvested plant material for 24 to 48 hours at 70°C to 80°C.

Development

The process by which cells become specialized is called differentiation. Differentiation occurs as various cells mature, become larger, and assume different forms adapted to specific functions such as conduction, support, or secretion of special substances. The term *development* is applied to the organized and systematic formation of plant organs from differentiated cells. Another term for this is *morphogenesis*, which is derived from the Greek *morpho* meaning "form" and *genesis* meaning "origin" or "beginning."

Most plant species undergo distinctive life phases after germination. These include a seedling phase, juvenile phase, reproductive phase, and senescence phase.

Seedling Phase

Plants generally grow most rapidly during the seedling phase. Due to their small root systems, seedlings are especially vulnerable to desiccation from minor occurrences of soil drying. When seeding or planting densities are high, seedlings also experience strong competition for light. Most plant mortality occurs in the seedling phase through the interactive effects of environmental stress, competition, pathogens, and herbivory, so rapid growth to acquire resources during this phase is beneficial. Often seed size is the major determinant of early seedling size and growth rate.

Juvenile Phase

During the juvenile phase plants begin to accumulate reserves (nutrients and carbohydrates) to buffer the plant against any future unfavorable environmental conditions. Annual plants (annuals) allocate little of their required resources to storage during the juvenile phase, whereas perennial plants (perennials) are characterized by storage of nutrients and carbohydrates. The greater resource allocation to storage rather than leaf area partly accounts for the slower growth rate of perennials. The stored reserves, however, allow perennials to start growth early in a seasonal climate and to survive conditions that are unfavorable for photosynthesis or nutrient acquisition.

Reproductive Phase

Plants often go through an abrupt hormonally triggered shift to the reproductive phase, where some shoot meristems produce reproductive rather than vegetative organs. Plants use some internal developmental clue to initiate flowering. For some plants, this may be either day length and/or temperature. After flowers are produced, pollination may occur with the assistance of pollinators such as insects, birds, or bats. Pollination is required for the development of a fruit with seeds. Parthenogenic fruits develop without seeds and often without prior pollination.

Allocation to reproduction differs among species. In general, annuals and other short-lived species allocate a larger proportion of annual production to reproduction than do long-lived perennials. Because of the large investments of carbon and nutrients to produce seeds, allocation to female function is generally considered the most costly component of reproduction. After flowering, phloem-mobile nutrients (such as nitrogen) are exported from the senescing leaves and roots to the developing fruits.

Vegetative reproduction is a method of producing physiologically independent individuals without going through reproduction and establishment. Plants that undergo vegetative reproduction often evolved in environments where flowering is infrequent and seedling establishment is rare. The taxonomic process by which vegetative reproduction occurs differs among species. These mechanisms include production of new tillers (a new shoot and associate roots) in grasses and sedges, initiation of new shoots from the root system (root suckering) in some shrubs and trees, production of new shoots at the base of the parental shoot (stump sprouting) in other shrubs and trees, initiation of new shoots from below-ground stems as in many Mediterranean shrubs, and rooting of lower limbs of trees that become covered by soil organic matter (layering) in many conifers.

Senescence Phase

Senescence in plants is a hormonally controlled developmental process. Plant senescence is affected by environmental factors (e.g., irradiance, photoperiod, and nitrogen supply) and plant growth regulators. The plant hormones ethylene and abscisic acid promote senescence while cytokinins and/or gibberellins slow or reverse senescence. An early visible symptom of leaf senescence is leaf yellowing due to loss of chlorophyll. Chloroplast proteins are hydrolyzed by proteolytic enzymes, and free acids are exported via the phloem.

Summary

Plant growth is the irreversible increase in mass or size of the plant. Meristems are areas of the plant where growth typically occurs. Some plants may exhibit determinate growth (where they attain a certain size, stop growing, and then die), whereas others exhibit indeterminate growth and continue to grow for many years. Plant growth is often measured by determining increases in volume or weight of the plant.

Differentiation is the process by which cells become specialized and assume specific function. Plants undergo distinctive life phases along with differentiation. These include the seedling phase, juvenile phase, reproductive phase, and senescent phase.

Review Questions

1. How does growth differ from development?
2. What are some ways that plant growth can be measured?
3. During which plant growth phase is growth the most rapid?

Selected References

Asian Vegetable Research and Development Center. 1990. *Vegetable production training manual.* Shanhua, Taiwan: Asian Vegetable Research and Development Center. Reprinted 1992.

Cavigelli, M. A., S. R. Deming, L. K. Probyn, and R. R. Harwood, eds. 1998. *Michigan field crop ecology: Managing biological processes for productivity and environmental quality.* Michigan State Univ. Extension Bulletin E-2646.

Hartmann, H. T., A. M. Kofranek, V. E. Rubatzky, and W. J. Flocker. 1988. *Plant science: Growth, development, and utilization of cultivated plants,* 2nd ed. Upper Saddle River, NJ: Prentice Hall.

Lambers, H. F., S. Stuart III, and T. L. Pons. 1998. *Plant physiological ecology.* New York: Springer-Verlag.

Odum, E. P. 1971. *Fundamentals of ecology,* 3rd ed. Philadelphia: W. B. Saunders.

Salisbury, F. B. and C. W. Ross. 1978. *Plant physiology,* 2nd ed. Belmont, CA: Wadsworth.

Stern, K. R. 1991. *Introductory plant biology,* 5th ed. Dubuque, IA: Wm. C. Brown.

Selected Internet Sites

http://www.caf.wvu.edu/~forage/growth.htm   Plant Growth and Development as the Basis of Forage Management publication by Edward B. Rayburn, West Virginia University Extension Service, 1993.

http://extension.oregonstate.edu/mg/botany/growth.html   Plant Growth and Development Module, based on Chapter 1, Botany Basics in the Oregon State University Extension Service Master Gardener Handbook by Ann Marie VanDerZanden. 1999.

6

An Overview of Photosynthesis and Respiration

Because photosynthesis is the only process of biological importance that can absorb and chemically store energy from the sun (Fig. 6.1), and respiration is the only process whereby energy stored by photosynthesis as carbohydrates is released in a controlled manner, the influence of the various environmental factors on these two processes is a major determining factor on how well a plant will grow and produce in a particular environmental location. Gaseous exchanges (CO_2, O_2, and H_2O vapor) are important components of photosynthesis and respiration. Annually some 10^{19} kcal of solar energy is used to convert CO_2 into biomass (carbohydrates) by photosynthetic organisms in the biosphere (the zone of the earth that contains living organisms, extending from the crust into the surrounding atmosphere) (Fig. 6.2). The energy in these organic compounds can be used immediately by the plant through the processes of respiration and tissue building or stored in the plant in biochemical forms (such as those in fossil fuels) that can last for hundreds of millions of years.

This chapter is designed to provide some basic understandings of photosynthesis and respiration. More detailed influences of the various environmental factors on these two processes are presented in later chapters devoted to the specific environmental factor.

Photosynthesis

Overview

The overall process of photosynthesis involves the reduction (addition of electrons to the molecule) of CO_2 to form carbohydrates and the oxidation (removal of electrons from the molecule) of H_2O with the release of O_2. For each mole of carbon fixed there is a required input of energy

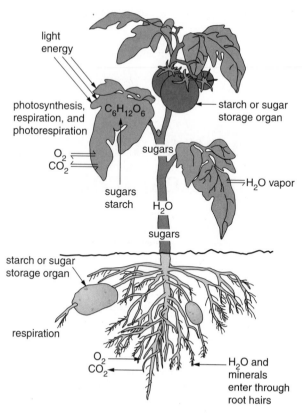

light
energy

photosynthesis,
respiration, and
photorespiration

$C_6H_{12}O_6$

starch or sugar
storage organ

O_2
CO_2

sugars

H_2O vapor

sugars
starch H_2O

sugars

starch or sugar
storage organ

respiration

O_2
CO_2

H_2O and
minerals
enter through
root hairs

FIGURE 6.1   Photosynthesis, respiration, leaf water exchange, and translocation of photosynthates in a plant.
Source: Oregon State University Extension Service Master Gardener Handbook, Chapter 1, Botany Basics by Ann Marie VanDerZanden. 1999.

from sunlight of at least 4.8×10^5 J. The photosynthesis reaction can be summarized as

$$6H_2O + 6CO_2 + \text{light energy and chlorophyll} \rightarrow C_6H_{12}O_6 + 6O_2.$$

Chloroplasts   Chloroplasts are the primary organelles in which photosynthesis occurs. They are small football-shaped organelles consisting of an outer and inner membrane that encloses a complex network of thylakoid membranes surrounded by a gel-like inner matrix called the stroma (Fig. 6.3). The thylakoids are stacked to form grana.

The leaf is a specialized plant organ that enables plants to intercept light energy necessary for photosynthesis (Fig. 6.4). The light is captured by a large array of chloroplasts that are in close proximity to air and not too far from vascular tissue

September 2000

Chlorophyll *a* Concentration (mg/m^3)

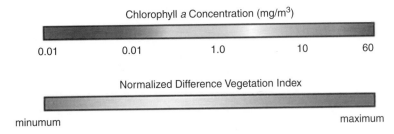

0.01   0.01   1.0   10   60

Normalized Difference Vegetation Index

minumum   maximum

FIGURE 6.2   A view of the global biosphere and relative amounts of photosynthetic organisms (vegetation and phytoplankton) that through the process of photosynthesis consume atmospheric carbon dioxide. The Normalized Difference Vegetation Index measures the amount and health of plants on land, while chlorophyll *a* measurements indicate the amount of phytoplankton in the ocean. Land vegetation and phytoplankton both consume atmospheric carbon dioxide.
Source: Earth Observatory, NASA.

(xylem and phloem), which supplies water and exports the products of photosynthesis. Most chloroplasts are located in the mesophyll cells of the leaves, which could contain 50 or more chloroplasts per cell.

The uptake of CO_2 occurs through leaf stomata. Each stomate consists of two guard cells that can rapidly change their aperture and regulate the amount of CO_2 that may enter the leaf. Once inside, CO_2 diffuses from intercellular air spaces to the sites of carboxylation in the chloroplast or in the cytosol. The demand for CO_2 is affected by the rate of processing CO_2 in the chloroplast, which is regulated by the structure and biochemistry of the chloroplast, by environmental factors such as light, and/or by factors that affect plant demand for carbohydrates.

(a)

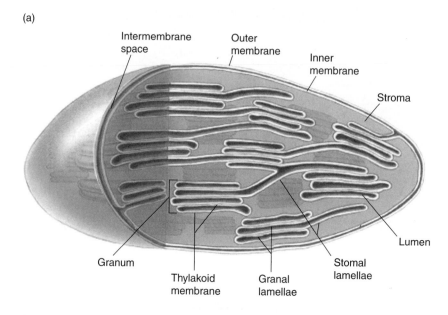

(b)

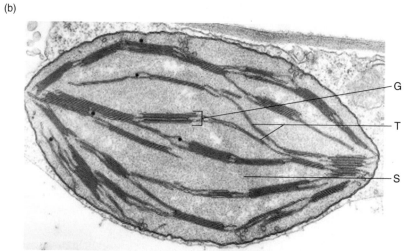

FIGURE 6.3 Chloroplast and thylakoid structure. Shown in this (a) illustration and (b) electron micrograph cross section of a spinach leaf chloroplast are the grana (G), thylakoid membrane (T), and the stroma (S).

Source: Horton, H. R., L. A. Moran, R. S. Ochs, D. J. Rawn, and K. G. Scrimgeour. *Principles of Biochemistry*, 3rd ed. (Upper Saddle River, NJ: Prentice Hall, 2002). Used with permission: Pearson Education, Inc., Upper Saddle River, NJ.

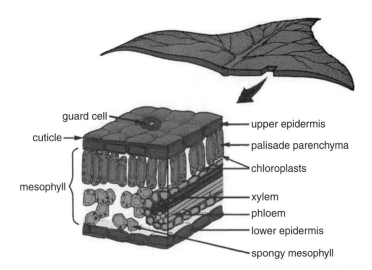

FIGURE 6.4   Cross section of a typical leaf and location of chloroplasts.
Source: Oregon State University Extension Service Master Gardener Handbook, Chapter 1,
Botany Basics by Ann Marie VanDerZanden. 1999.

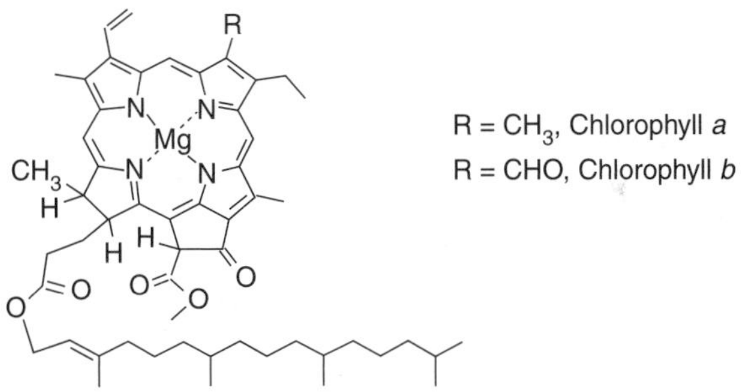

R = CH$_3$, Chlorophyll *a*
R = CHO, Chlorophyll *b*

FIGURE 6.5   Chemical structure of chlorophyll *a* and chlorophyll *b*.

Chlorophyll *a* and *b*   Photosynthesis occurs primarily using the pigments of chlorophyll *a* and chlorophyll *b* that are mainly in the thylakoids. A pigment is any substance that absorbs light, and each pigment generally has its own characteristic absorption spectra and is often colored in appearance. While having similar chemical structures (Fig. 6.5), chlorophyll *a* is bluish green in color, whereas chlorophyll *b* is yellowish green. The content of chlorophyll *a* in green plants is two to three times that of chlorophyll *b*. When a molecule of chlorophyll *b* absorbs light, it transfers the energy to a molecule of chlorophyll *a*. Consequently, chlorophyll *b* makes it possible for photosynthesis to occur over a broader spectrum of light than would be possible with only chlorophyll *a*.

A comparison of the action spectrum (the relative effectiveness of the different wavelengths of light at generating chemical reactions) of a particular process with

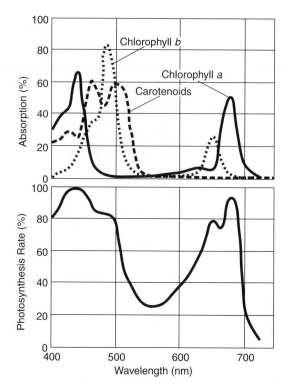

FIGURE 6.6   Representative action spectrum and absorption spectrum of photosynthesis (bottom graph) and absorption spectra of various pigments important in photosynthesis (top graph).
Source: "Concepts in Photobiology: Photosynthesis and Photomorphogenesis," Edited by G.S. Singhal, G. Renger, S.K. Sopory, K-D Irrgang and Govindjee, Narosa Publishers/New Delhi; and Kluwer Academic/Dordrecht, pp. 11–51. Used with permission: Kluwer Academic.

absorption spectra of potentially involved pigments can implicate pigment(s) involvement in that process. The action spectrum of photosynthesis and absorption spectra of various pigments suggests that there are other accessory pigments involved in photosynthesis besides the chlorophylls (Fig. 6.6). Other plant pigments involved in photosynthesis include carotenoids, phycobilins, and several other types of chlorophylls.

Major Stages of Photosynthesis

Photosynthesis occurs in two major stages or sets of reactions referred to as the light reactions and the CO_2 fixation reactions (often previously referred to as the dark reactions). The light reactions of photosynthesis occur in or on the thylakoid membranes and are light requiring. These reactions trap light energy and cleave water molecules into hydrogen and oxygen and serve as electron and proton transfer reactions. The CO_2 fixation reactions of photosynthesis occur in the stroma and are non-light-requiring. Electrons and hydrogen atoms are added to CO_2 during these

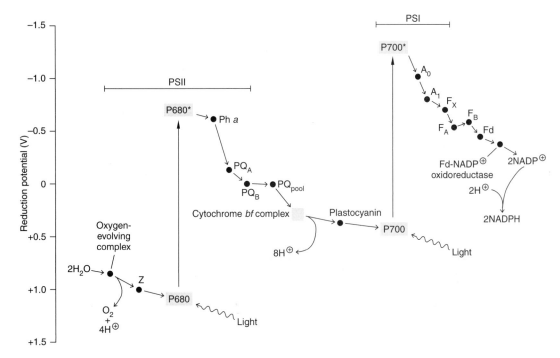

FIGURE 6.7   The Z scheme of photosynthesis illustrates the reduction potentials and electron flow during photosynthesis. Light energy absorbed by the special-pair pigments, P_{680} and P_{700}, drives electron flow uphill. Abbreviations: Z, electron donor to P_{680}; Pha, pheophytin a, electron acceptor of P_{680}; PQ_A, plastoquinone tightly bound to PSII; PQ_B, pool made up of PQ and PQH2; Ao, chlorophyll a, the primary electron acceptor of PKI; A1, phylloquinone; Fx, Fb and Fa, iron sulfur clusters; and Fd, ferredoxin. NADP+ is reduced by a hydride ion (H-) donated by the FADH2 prosthetic group of ferredoxin-NADP+ oxidoreductase.
Source: Horton, H. R., L. A. Moran, R. S. Ochs, D. J. Rawn, and K. G. Scrimgeour. *Principles of Biochemistry*, 3rd ed. (Upper Saddle River, NJ: Prentice Hall, 2002). Used with permission: Pearson Education, Inc., Upper Saddle River, NJ.

reactions, resulting in the formation of sugar. The CO_2 fixation reactions require reducing power in the form of reduced nicotinamide adenine dinucleotide phosphate (NADPH) and chemical energy in the form of adenine triphosphate (ATP). ATP is the universal cellular energy carrier and is formed from adenine diphosphate (ADP) and inorganic phosphate. The NADPH and the ATP required by the CO_2 fixation reactions are supplied by the light reactions.

Light Reactions   Most chloroplasts contain two types of photosynthetic units (photosystems) (Fig. 6.7). These are designated as photosystem I and photosystem II. These two photosystems connected by an electron carrier chain are commonly referred to as the "Z scheme" of photosynthesis.

The photosystems are packed into thylakoids and contain pigments, proteins, other compounds, and different ions (including manganese, iron, calcium, and chlorine). Their function is to transfer electrons excited by the absorption of light energy down a chain of associated molecules in the thylakoid membrane called an electron transport chain. This results in the uptake of hydrogen ions from the stroma into the lumen (or interior) of the thylakoid membrane, and eventually the creation of ATP (through photophosphorylation) and the reduction of nicotinamide adenine dinucleotide (NADP) to NADPH. Photosystem I is the only photosystem to absorb wavelengths above 680 nm, but both photosystems can absorb shorter wavelengths. The reaction center for photosystem II is P_{680} and for photosystem I is P_{700}. The letter P stands for pigment and the numbers 700 and 680 refer to peaks in the absorption spectra of light as in nm for photosystem I and photosystem II (700 and 680 nm, respectively).

Photolysis is the manganese-dependent process of splitting water (forming a molecule of oxygen and four protons) during the early stages of photosynthesis. When a photon of light strikes a P_{680} molecule of photosystem II, an electron is excited to a higher energy level. This excited electron is picked up by a pheophytin a (Pha) and is then passed to two plastoquinone acceptors, PQ_A and PQ_B. Electrons extracted from a water molecule replace the electrons lost by the P_{680} molecules.

The electrons from the plastoquinones of photosystem II are transferred to the Cytochrome b_6 complex of the electron transport system. The Cytochrome b_6 complex transfers electrons to plastocyanin, which in turns reduces P_{700}. While electrons pass along the electron transport system, protons are being moved through a coupling factor and photophosphorylation occurs.

During photophosphorylation, ATP molecules are formed from ADP and phosphate (using ATP synthase) as electrons are passed along the electron transport system through series of oxidation-reduction reactions. When a photon of light is absorbed by a P_{700} molecule in photosystem I, an electron is excited and is transferred to a primary acceptor molecule (probably a chlorophyll) and then onto iron-sulfur acceptor molecules (Fx, Fa, and Fb). The electron is then transferred to an iron-sulfur acceptor molecule, ferredoxin (Fd), which in turn releases it to $NADP^+$, which is reduced to NADPH. The ATP that is produced as a result is called noncyclic photophosphorylation.

Photosystem I can work independently of photosystem II. This occurs when the electrons excited from P_{700} are passed to the electron transport chain between the two photosystems (instead of to ferredoxin and $NADP^+$) and back into the reaction center of photosystem I. ATP formed as a result of this cyclic electron flow is called cyclic photophosphorylation. A probable summary equation for the light reactions of photosynthesis is $2H_2O + 2NADP^+ + 3ADP^{-2} + 3H_2PO4^- + 8 - 12$ photons $\rightarrow$ $O_2 + 2NADPH + 2H^+ + 3ATP^{-3} + 3 H_2O$.

CO_2 Fixation Reactions   The CO_2 fixation reactions of photosynthesis normally occur during the daylight and are affected by temperature. These reactions occur in the stroma of the chloroplasts as long as products of the light reactions are present. The 3-carbon (C3) pathway, the 4-carbon (C4) pathway, and the crassulacean acid metabolism (CAM) pathway are three known mechanisms through which CO_2 is converted to carbohydrates during the CO_2 fixation reactions. The most widespread pathway in the plant kingdom is the 3-carbon pathway (also called the Calvin-Benson cycle or the reductive pentose phosphate pathway).

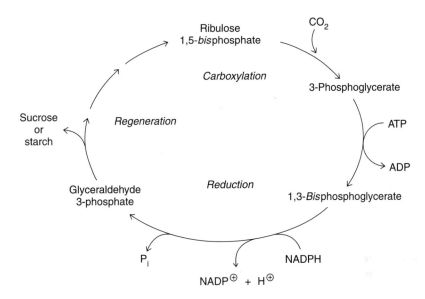

FIGURE 6.8 The 3-carbon (C3) pathway or the Calvin-Benson cycle (also called the reductive pentose phosphate cycle). The 3-carbon pathway has three stages: CO_2 fixation (by rubisco), shown in red; carbon reduction to (CH$_2$O), shown in blue; and regeneration of the CO_2 acceptor molecule (ribulose 1,5-bisphosphate), shown in black.
Source: Horton, H. R., L. A. Moran, R. S. Ochs, D. J. Rawn, and K. G. Scrimgeour. *Principles of Biochemistry*, 3rd ed. (Upper Saddle River, NJ: Prentice Hall, 2002). Used with permission: Pearson Education, Inc., Upper Saddle River, NJ.

The 3-carbon (C3) pathway Six molecules of CO_2 combine with six 5-carbon molecules of ribulose 1,5-bisphosphate (RuBP), with the aid of RuBP carboxylases (also called rubisco), during the 3-carbon (C3) pathway or Calvin-Benson cycle (also called the reductive pentose phosphate cycle) (Fig. 6.8). The resulting six 6-carbon unstable complexes are immediately split into twelve 3-carbon phosphoglycerate (PGA) molecules. The NADPH and ATP from the light reactions supply energy to convert the PGA to twelve molecules of glyceraldehyde 3-phosphate (GA3P). Ten of the twelve GA3P molecules are restructured and become six molecules of RuBP. The two remaining GA3P molecules are used in carbohydrate (sucrose or starch) formation or in other pathways resulting in the production of lipids or amino acids.

The 4-carbon (C4) pathway Some tropical plants produce a 4-carbon compound, oxaloacetic acid (OAA), instead of the 3-carbon PGA during the initial stages of the CO_2 fixation reactions of the 4-carbon (C4) pathway. OAA is formed as a result of the combining of the phosphoenolpyruvate (PEP), a 3-carbon compound, with carbon dioxide in mesophyll cells (Fig. 6.9). OAA may then be converted to 4-carbon organic

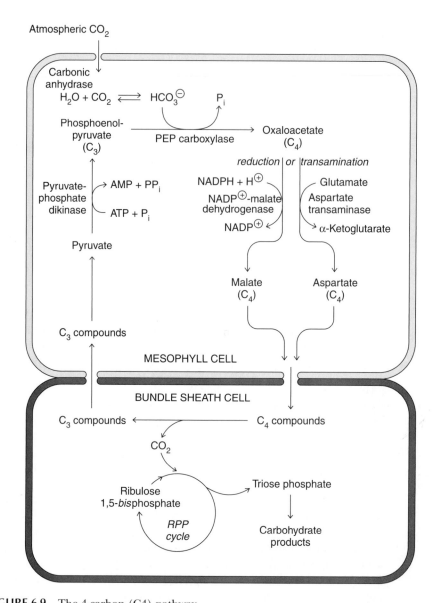

FIGURE 6.9   The 4-carbon (C4) pathway.
Source: Horton, H. R., L. A. Moran, R. S. Ochs, D. J. Rawn, and K. G. Scrimgeour. *Principles of Biochemistry*, 3rd ed. (Upper Saddle River, NJ: Prentice Hall, 2002). Used with permission: Pearson Education, Inc., Upper Saddle River, NJ.

acids of aspartic, malic, or other acids (this appears to be species dependent). These organic acids diffuse to the bundle sheath cells surrounding the vascular bundles of leaves where they are transported into the chloroplast and converted to pyruvate and CO_2. Additional PEP molecules are formed when the pyruvate returns to the mesophyll cells and interacts with ATP. The CO_2 released in the chloroplast of the bundle sheath cells combines with RuBP and is converted to PGA and related molecules in the 3-carbon pathway. Plants having the 4-carbon pathway are referred to as C4 plants, whereas plants only having the 3-carbon pathway are referred to as C3 plants.

C4 plants generally have higher rates of net photosynthesis (gross photosynthesis rate minus the respiration rate) than C3 plants. C4 plants engage in both C3 and C4 photosynthesis, whereas C3 plants lack the C4 pathway. C4 plants require slightly more energy than C3 plants, but this is offset by other features such as the absence of photorespiration (light dependent respiration) in C4 plants. Photorespiration lowers the apparent efficiency of CO_2 assimilation in C3 plants because the rate of photorespiration increases with temperature faster than gross photosynthesis. As a consequence, many C3 plants are nonproductive at high temperatures, whereas C4 plants (such as the tropical grasses) increase in productivity at higher temperatures.

The leaf anatomy of C4 plants differs from that of C3 plants. C4 plants are characterized by their Kranz anatomy, which is a sheath of thick-walled cells that surround the vascular bundle. In some C4 species, the cells of the bundle sheath contain large chloroplasts with mainly stroma thylakoids and very little stacking of grana.

Crassulacean acid metabolism   Crassulacean acid metabolism (CAM) is used by plants in at least 18 families, including Cactaceae, Orchidaceae, Bromeliaceae, Liliaceae, and Euphorbiaceae. CAM is similar to 4-carbon metabolism in that 4-carbon compounds (OAA and malate) are produced during the CO_2 fixation reactions. Stomata are open in CAM plants during the night to allow diffusion of CO_2 into the plant (stomata are generally only open during the daylight for C3 plants). PEP carboxylase converts PEP and CO_2 to malate, which is stored in the cell vacuole (Fig. 6.10). Stomata of CAM plants close during the day thereby preventing H_2O loss and further uptake of CO_2. The malate diffuses out of the vacuole and is converted back to CO_2 for use in the Calvin-Benson cycle. A much larger amount of CO_2 can be converted to carbohydrate each day under conditions of both limited water supply and high light intensity with CAM plants than otherwise would be possible with non-CAM plants.

Many of the CAM plants are succulents that grow in regions of high light intensity and dry conditions. A few succulents do not have CAM photosynthesis, and several nonsucculents do. Species with CAM usually lack a well-defined palisade layer of cells, and most of the leaf or stem cells are spongy mesophyll. The CAM plant of greatest economic importance is the pineapple, *Ananas sativus.*

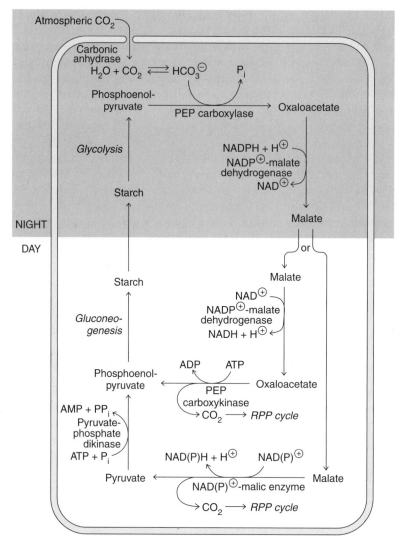

FIGURE 6.10   Crassulacean acid metabolism (CAM).
Source: Horton, H. R., L. A. Moran, R. S. Ochs, D. J. Rawn, and K. G. Scrimgeour. *Principles of Biochemistry,* 3rd ed. (Upper Saddle River, NJ: Prentice Hall, 2002). Used with permission: Pearson Education, Inc., Upper Saddle River, NJ.

Respiration

Overview of Respiration

All active cells respire continuously, often absorbing O_2 and releasing CO_2 in nearly equal volumes. This is initiated in the cytoplasm and completed in the mitochondria. It is during respiration that living organisms utilize the energy-containing compounds produced during photosynthesis. The small, enzyme-controlled steps of respiration release the usable energy of these energy-containing compounds and store the released energy in ATP molecules, which permits the available energy to be used more efficiently and the process to be controlled more precisely. The overall process is an oxidative-reduction reaction in which carbohydrates or other assimilates are oxidized to CO_2, and the O_2 absorbed is reduced to form H_2O. Starch, fructans, sucrose, fats, organic acids, and proteins can serve as respiratory substrates.

Respiration is often viewed as the reverse of photosynthesis. Much of the energy released during respiration (approx. 686 kcal/mol of glucose) is heat. Far more important than the heat released is the energy trapped in compounds that can be used later for many essential processes of life, such as those involved in growth and ion accumulation. ATP is the most important of these compounds; NADH and NADPH are also important because of their ability to transfer electrons. The common equation for the respiration of glucose is written as

$$C_6H_{12}O_6 + 6O_2 \rightarrow 6\ CO_2 + 6\ H_2O + \text{energy}.$$

Respiration is not a single reaction but a series of reactions, each catalyzed by a specific enzyme. Usually only some of the respiratory substrates are fully oxidized to CO_2 and H_2O, and the rest are used in anabolic or tissue-building processes. The energy trapped during oxidation can be used to synthesize the larger molecules required for growth. Whether most of the carbon atoms respired are converted to CO_2 or to any of the larger molecules depends on the kind of cell involved, its position in the plant, and whether the plant is rapidly growing. When plants are growing rapidly, most of the disappearing sugars are metabolized into molecule synthesis reactions and never appear as CO_2.

Energy demand, substrate availability, and oxygen supply control the types and rates of plant respiration. At low levels of oxygen, respiration cannot proceed by the oxygen-requiring (aerobic) pathways, and instead proceeds through the process of fermentation, with ethanol and lactate as end products. The yield of ATP by fermentation is considerably less than that of aerobic respiration.

Major Phases of Respiration

The process of aerobic cellular respiration occurs in four phases: breakdown of storage forms of carbohydrates into glucose, glycolysis, the Krebs cycle (also called the tricarboxylic acid cycle or citric acid cycle), and the electron transport chain.

Breakdown of Storage Forms of Carbohydrates The first phase in cellular respiration is the breakdown of the storage forms of carbohydrates, most commonly starch, into glucose. Starch accumulated in the chloroplast during photosynthesis is an important reserve in the leaves. Also, starches that have been formed in amyloplasts of storage organs after translocation of sucrose or certain other sugars are also principal respiratory substrates for those organs at certain stages of their development. For example, potato tubers are rich in starch-containing amyloplasts, and most of this disappears during early plant development. Parenchyma cells in both roots and stems in perennial species commonly store starch. The starch stored during the growing season in these plants is maintained during the winter months and used for new growth the following spring.

The enzymes α-amylase, β-amylase, and starch phosphorylase catalyze most of the steps in the degradation of starch to glucose. Only α-amylase can metabolize starch granules, and β-amylase and starch phosphorylase typically metabolize the first products released by α-amylase reactions.

Both amylases are hydrolytic enzymes that require the uptake of one H_2O for each bond cleaved. The amylases are widespread in various tissues but are most active in germinating seeds. α-amylase is located primarily inside the chloroplast near the starch grains that it metabolizes. The starch phosphorylase is a phospholytic enzyme that breaks down starch by incorporating phosphate.

Sucrose is the major sugar translocated in plants (and is the form of carbon most tissues import) and is broken down into glucose and fructose. Glucose or fructose is phosphorylated twice, producing two molecules of glyceraldehydes-3-phosphate. This series of reactions requires two molecules of ATP per glucose.

Glycolysis   The next major phase of respiration is glycolysis (Fig. 6.11), which is a group of reactions that occurs in the cytoplasm and is non-oxygen-requiring in which glucose, glucose-1-P, or fructose is converted to pyruvate. Important functions of glycolysis are the formation of molecules that can be removed from the pathway to synthesize several other constituents of which the plant is composed and the production of ATP. There is a total of four ATPs per hexose (6 carbon sugar) used and a net production of two ATPs per hexose. If glucose-1-P, glucose-6-P, or fructose-6-P is used, the net ATP is three per hexose phosphate.

An additional function of glycolysis is the production of NADH, which is formed by reduction of NAD^+ during the oxidation of glyceraldehyde-3-phosphate to 1,3-bisphosphateglycerate. NADH may enter the mitochondria where it may be oxidized during the electron transport reactions and the energy converted into the terminal phosphate bond of two ATPs, or the NADH can be used as a source of electrons in numerous reactions.

Glycolysis can be summarized as glucose + $2NAD^+$ + $2ADP^{-2}$ + 2 H_2PO4^{4-} → 2 pyruvate + 2NADH + $2H^+$ + $2ATP^{-3}$ + 2 H_2O. Two ATP molecules supply the energy needed to start the process. Four ATP molecules are produced from energy released during the formation of pyruvate, for a net gain of two ATP molecules. The hydrogen ions and electrons released during the process are picked up and temporarily held by the hydrogen acceptor nicotinamide adenine dinucleotide (NAD). The pyruvate formed in glycolysis moves into the mitochondria, where it is oxidized to CO_2 by the reactions of the Krebs cycle.

The Krebs Cycle   The initial steps leading to the Krebs cycle involve the loss of CO_2 from pyruvate and the combination of the remaining 2-carbon acetate unit with a sulfur-containing compound, coenzyme A (CoA), to form acetyl CoA. Acetyl CoA then enters the Krebs cycle. Small amounts of energy and hydrogen are successively removed from a series of organic molecules during the Krebs cycle and transferred to compounds such as NADH, ATP, and reduced flavin adenine dinucleotide ($FADH_2$). Carbon dioxide is produced as a by-product while this cycle is proceeding.

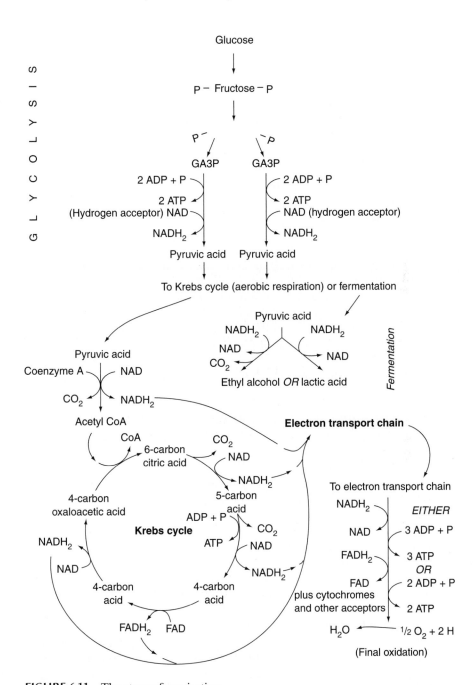

FIGURE 6.11   The steps of respiration.
Source: Stern, K. R. 1991. *Introductory Plant Biology,* 5th ed. (Dubuque, IA: Wm. C. Brown, 1991). Used with permission: The McGraw-Hill Companies.

Because two pyruvates are produced in glycolysis from each glucose, the overall reaction from the Krebs cycle is

$$2 \text{ pyruvate} + 8NAD^+ + 2FAD + 2ADP^{-2} + 2H_2PO^{4-} + 4H_2O \rightarrow$$
$$6 CO_2 + 2ATP^{-3} + 8NADH + 8H^+ + 2FADH_2.$$

Electron Transport Chain   ATP is produced when the NADH and $FADH_2$ produced in the Krebs cycle or in glycolysis are oxidized during oxidative phosphorylation. The electrons are transferred via several intermediate compounds before H_2O is produced. These electron carriers collectively are the electron transport chain or cytochrome system of the mitochondria.

The electron transport chain consists of acceptor molecules (include iron-containing proteins called cytochromes) that are located on the inner membrane of the mitochondria. Energy is released in small increments and ATP is formed from ADP and P at several parts along the chain. Water is formed as the hydrogen ions and electrons combine with oxygen. As a result of the electron transport chain, the recoverable energy of glucose is released and stored in ATP molecules. This stored energy is then available for use in the synthesis of other molecules, for growth, active transport, and a host of other metabolic processes.

Aerobic respiration produces a net gain of 36 ATP molecules from 1 glucose molecule, using a net total of 6 molecules of water. For each mole (180 grams) of glucose respired, 686 kcal of energy is released, with about 39% of that stored as ATP molecules and the remainder released as heat. The summary reaction for the electron transport system is

$$10NADH + 10H^+ + 2FADH_2 + 32ADP^{-2} + 32 H_2PO_4^- + 6O_2 \rightarrow 10NAD^+ + 2FAD$$
$$+ 32ATP^{-3} + 42 H_2O.$$

The net reaction of respiration is

$$\text{glucose } (C_6H_{12}O_6) + 6O_2 + 36ADP^{-2} + 36H_2PO_4^- \rightarrow 6 CO_2 + 36ATP^{-3}$$
$$+ 42 H_2O.$$

Fermentation Metabolism

When oxygen is not available, the Krebs cycle and the electron transport chain cannot function. Plants metabolize pyruvate in the absence of O_2 by undergoing some form of fermentation metabolism. In alcohol fermentation, pyruvate is broken down, producing ethanol and CO_2 and NADH in the process. In lactic acid fermentation, NADH is used during the reduction of pyruvate to lactate.

Summary

Photosynthesis transforms CO_2 and H_2O in the presence of light into carbohydrates. This occurs in the chloroplasts, which contain the photosynthetic pigments and the photosystems. Plant pigments involved in photosynthesis include chlorophyll *a*, chlorophyll *b*, and the accessory pigments (which include carotenoids, phycobilins, and other types of chlorophylls).

The two major stages or sets of reactions of photosynthesis are the light reactions and the CO_2 fixation reactions. The light reactions trap light and cleave water molecules into hydrogen and oxygen and serve as electron and proton transfer reactions. During the CO_2 fixation reactions, electrons and hydrogen atoms are added to CO_2 resulting in the formation of sugar. The three known mechanisms through which CO_2 is converted into carbohydrates during the CO_2 fixation reactions are the 3-carbon (C3) pathway, the 4-carbon (C4) pathway, and the crassulacean acid metabolism (CAM) pathway. The C3 pathway utilizes a 3-carbon intermediate (phosphoglycerate), whereas the C4 pathway utilizes a 4-carbon intermediate (oxaloacetic acid) during the initial stages of the CO_2 fixation reactions. Many C3 plants have lower rates of net photosynthesis than C4 plants and are often nonproductive at high temperatures, whereas C4 plants increase in productivity at higher temperatures. Many CAM plants are succulents that are found in desert-type conditions. CAM is similar to C4 metabolism in that a 4-carbon compound is produced, but the stomata in CAM plants typically only open at night. A much larger amount of CO_2 can be converted to carbohydrates under conditions of both limited water and high light with CAM plants than is possible with non-CAM plants.

Respiration is a series of reactions in which carbohydrates and other assimilates are oxidized to CO_2 and the O_2 absorbed is reduced to form H_2O. Some of the respiration substrates (such as starch, sugars, fats, organic acids, and proteins) are fully oxidized to CO_2 and H_2O, and the rest are used in tissue-building processes. The process of aerobic cellular respiration consists of the following phases: breakdown of storage forms of carbohydrates into glucose, glycolysis, the Krebs cycle, and the electron transport chain.

Review Questions

1. What are the end products of photosynthesis and why are they important?
2. What are chloroplasts and what is their importance to photosynthesis?
3. Describe some difference between chlorophyll *a* and chlorophyll *b*.
4. What are some differences between the light reactions and CO_2 fixation reactions of photosynthesis?
5. Describe the process of photolysis.
6. What is the importance of the action spectrum of photosynthesis and what additional information would you need to implicate pigments in this process?
7. Draw or describe the Z scheme of photosynthesis (label photosystems, important reaction centers, where water enters, and where energy might be produced).
8. What kinds of plants are generally associated with the following:
CAM metabolism
C4 photosynthesis
C3 photosynthesis
9. What is the ecological advantage of certain plants undergoing CAM metabolism?

10. How do C4 plants differ from C3 plants?
11. In which environments are most of the crassulacean acid metabolism plants found?
12. What are the end products of respiration and why are they important?
13. What are the four broad steps that comprise the process of respiration?
14. What are the important products formed during glycolysis and the Krebs cycle?

Selected References

Horton, H. R., L. A. Moran, R. S. Ochs, D. J. Rawn, and K. G. Scrimgeour. 2002. *Principles of biochemistry*. 3rd ed. Upper Saddle River, NJ: Prentice Hall.

Lambers, H. F., S. Stuart III, and T. L. Pons. 1998. *Plant physiological ecology*. New York: Springer-Verlag.

Lehninger, A. L. 1975. *Biochemistry*. 2nd ed. New York: Worth.

Stern, K. R. 1991. *Introductory plant biology*. 5th ed. Dubuque, IA: Wm. C. Brown.

Salisbury F. B., and C. W. Ross. 1992. *Plant physiology*. 4th ed. Belmont, CA: Wadsworth.

Taiz, L., and E. Zeiger. 1998. *Plant physiology*. 2nd ed. Sunderland, MA: Sinauer Assoc.

Whitmarsh, J., and Govindjee. 1995. Photosynthesis. *Encyclopedia of Applied Physics* 13:513–532.

7

Plant Hormones

Plant hormones are important in the integration of plant development and the response of plants to the external physical environment. The phenotype of the plant is controlled by the interactions between a plant's genetic make up or genome and its environment. Environmental factors induce changes in hormone metabolism and distribution within the plant.

This chapter is intended to provide an overview of the major classes of plant hormones. The discussion begins by defining plant hormones and plant growth regulators, highlighting similarities and differences. A summary of two suggested mechanisms by which plant hormones may regulate gene expression is followed by descriptions of the various classes of plant hormones. Within the individual classes of plant hormones, information is presented to describe the chemical classes of compounds to which they belong and their roles in plant development. Examples of commercial uses of plant regulators in agricultural plant production are also presented.

Definitions

A plant hormone is an organic substance other than a nutrient that is active in very small amounts (< 1 mM and often < 1 μM), formed in certain parts of plants, and usually (though not always) translocated to another part of the plant where it evokes specific biochemical, physiological, and/or morphological responses. Growth-regulating, inorganic ions (such as K^+ and Ca^{2+}), synthesized substances (such as 2,4-D), and organic plant compounds (such as sucrose) that are generally needed in high concentrations to cause a plant growth response are not considered plant hormones but plant growth regulators. A plant growth regulator is an organic compound that in small amounts (< 1 mM) promotes, inhibits, or qualitatively modifies growth and development. Plant nutrients are generally not classified as plant growth regulators. All plant hormones are growth regulators, but not all growth regulators are plant hormones. Hundreds of synthetic compounds qualify as growth regulators but are not plant hormones.

General Mechanisms of Action

The mechanisms by which hormones regulate gene expression are poorly understood. Steroid hormones appear to pass through the plasma membrane into the cytoplasm where they bind with a receptor molecule to form a hormone-receptor complex (Fig. 7.1). The complex may then enter the nucleus and affect mRNA synthesis, which could result in a physiological response. Peptide hormones appear to bind to target cell receptor proteins. The receptor protein will subsequently undergo a conformational change ultimately resulting in modification of enzyme activity, altered metabolic processes, and differentiation. It is also likely that certain

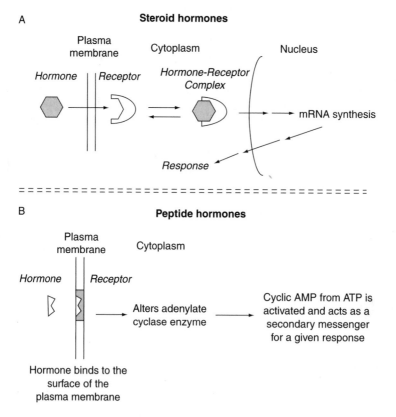

FIGURE 7.1   Diagrams suggesting methods of action for (a) steroid hormones and (b) peptide hormones.
Source: Arteca, R. N. 1996. *Plant Growth Substance Principles and Applications.* Chapman and Hall. Used with permission: Kluwer Publ.

enzymes affect gene regulation after initial hormone binding. Hormones may indirectly control gene expression through these enzymes and messengers at a number of control sites such as at transcription, mRNA processing, mRNA stability, translation, and posttranslation.

Classes of Plant Hormones

The plant hormones include the auxins, gibberellins, cytokinins, abscisic acid, and ethylene. More recently, other compounds that affect plant growth are beginning to gain acceptance as plant hormones. These include brassinosteroids, salicylates, jasmonates, and polyamines.

Auxins

Auxins were the first plant hormones to be discovered. The term *auxin* is derived from the Greek *auxein*, which means "to grow." An auxin is characterized by its capacity to induce elongation in shoot cells. Physiologically, all auxins chemically resemble indole-3-acetic acid (IAA) (Fig. 7.2), the only known naturally occurring auxin.

Auxins are synthesized by living protoplasm from the amino acid tryptophan that is found in both plant and animal cells. Production of auxins occurs primarily in apical meristems, buds, young leaves, and other active young parts of plants. Plants control the amount of IAA present in tissues at a particular time by controlling the synthesis of the hormone. Another control mechanism involves the conjugates, which are molecules that resemble the hormone but are inactive. Degradation of the hormone is the final method of controlling auxins.

Auxins are involved in numerous aspects of plant growth and development including flower initiation, sex determination, growth rate, fruit growth and ripening, rooting, aging, dormancy, and apical dominance. Roots appear to be more sensitive to auxins than are stems. The best-known synthetic auxins are 2,4-dichlorophenoxyacetic acid (2,4-D), 2,4,5-trichlorophenoxyacetic acid (2,4,5-T), 2-methyl-4-chlorophenoxyacetic acid (MCPA), indolebutyric acid (IBA), and naphthaleneacetic acid (NAA).

Gibberellins

Gibberellins are compounds that stimulate cell division and/or cell elongation. More than 110 different gibberellins have been isolated from seeds or from algae.

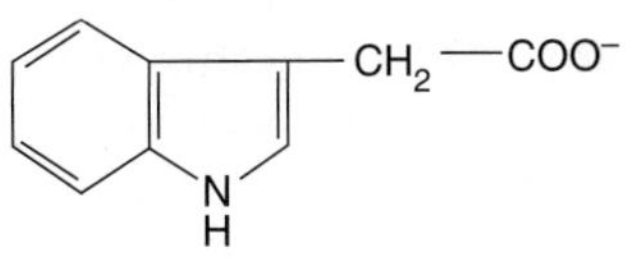

FIGURE 7.2   Chemical structure of indole-3-acetic acid (IAA).

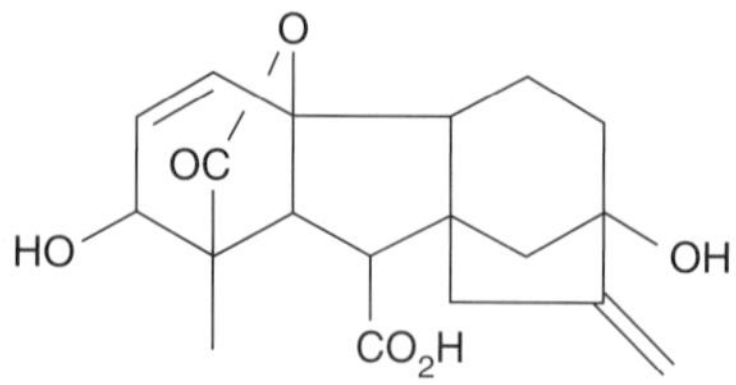

FIGURE 7.3   Chemical structure of GA$_3$, one of the most common gibberellic acids.

FIGURE 7.4   Use of an application of a gibberellin inhibitor to reduce subsequent height increases (stem elongation) of potted plants. Representative plants for the following treatments are: left—gibberellin inhibitor application; center—gibberellin inhibitor application followed by a gibberellin application (that negated the inhibitor action); and right—nonchemical application (control).
Courtesy of Dr. Jay Holcomb, Penn State University.

All gibberellins are acidic compounds and are called gibberellic acids (GAs) (Fig. 7.3). Different subscripts are used to distinguish the GAs. Acetyl coenzyme A, which is important in respiration, functions as a precursor to gibberellins.

Most dicots and a few monocots grow faster when gibberellins are applied to their stems (Fig. 7.4). Conifer trees generally show little or no response to gibberellins. In some plants, GA is involved in flower initiation and sex expression. Senescence of plant parts and dormancy in seeds and buds are also affected by GAs. In general, gibberellins are involved in almost all the same regulatory processes in plant development as auxins.

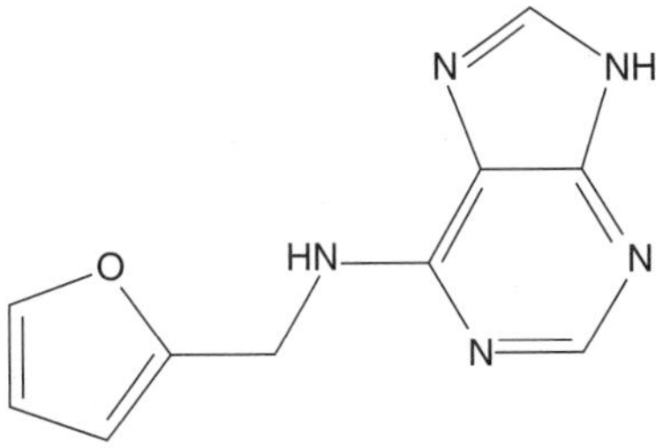

FIGURE 7.5 Chemical structure of kinetin, the first isolated and most common cytokinin.

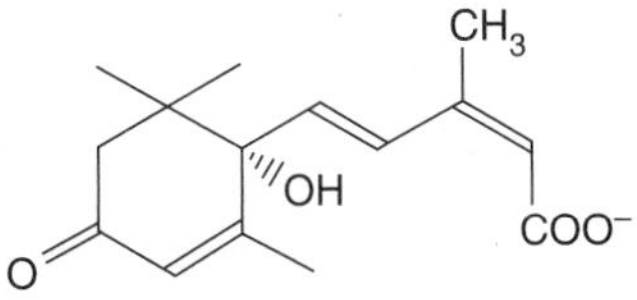

FIGURE 7.6 Chemical structure of abscisic acid (ABA).

Cytokinins

Cytokinins are adenine-resembling molecules that stimulate cell division. The name was probably derived from its property of causing cell division or cytokinesis. Cytokinin concentrations are typically highest in the meristematic regions and areas of continuous growth potential such as roots, young leaves, developing fruits, and seeds. They are thought to be synthesized in the roots and translocated via the xylem to shoots. Cytokinins are involved in cell enlargement, tissue differentiation, chloroplast development, cotyledon growth, delay of leaf aging, and in many of the growth processes also regulated by auxins and gibberellins.

Kinetin (6-furfurylaminopurine) was the first isolated cytokinin and is one of the most common (Fig. 7.5). Though kinetin is a natural compound, it is not formed in plants and is usually considered a synthetic cytokinin (meaning it is synthesized somewhere other than in the plant). Although there are more than 200 natural and synthetic cytokinins, the most common form of naturally occurring cytokinin in plants is zeatin.

Abscisic Acid

Abscisic acid (ABA) is a sesquiterpenoid (15-carbon) compound (Fig. 7.6) that inhibits plant growth, often counteracting the growth-promoting effects of auxins and gibberellins. ABA is synthesized via the mevalonic pathway with transport occurring in both the xylem and phloem. ABA production by plants is accentuated by stresses such as water loss and freezing temperatures. Though ABA functions in mainly inhibitory roles, it also has growth-promoting functions.

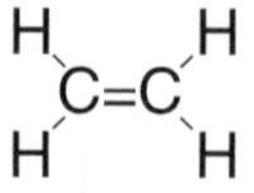

$$H \quad \quad H$$
$$\underset{H}{\overset{}{\diagdown}}C=C\underset{\diagup}{\overset{\diagup}{}}\underset{H}{\diagdown}$$

FIGURE 7.7   Chemical structure of ethylene and its precursors ACC and methionine.

ABA is synthesized in plastids and is found in many plant materials, including fleshy fruit (where it apparently prevents seeds from germinating while they are still on the plant). ABA almost universally inhibits cell growth and moves readily throughout the plant. The actual role of ABA in promoting abscission is not well understood. Other important roles for ABA in plants are the regulation of stomatal closure during periods of water stress and the induction of dormancy.

Ethylene

Ethylene (C_2H_4) is a hydrocarbon gas (Fig. 7.7) that is a product of plant metabolism and is produced by healthy as well as senescent and diseased tissue. It is primarily produced in actively growing meristems in plants, in ripening and senescing fruit, in senescing flowers, and in germinating seeds. Plants respond to many biological and physical stresses by increasing ethylene production, often called stress ethylene, that may be involved in wound healing and disease resistance.

Ethylene exerts regulatory control or influence over plant growth. The many physiological effects attributed to ethylene include stimulation of ripening of fleshy fruits (Fig. 7.8), stimulation of leaf abscission, inhibition of root growth, stimulation of adventitious root formation, inhibition of lateral bud development, epinasty of leaves, flower fading, flower initiation in bromeliads, and root geotropic responses.

Other Compounds Exhibiting Plant Hormone Characteristics

Brassinosteroids   Approximately 60 steroidal compounds are referred to as brassinosteroids. They are named after brassinolide (Fig. 7.9), which was isolated from the rape plant and was the first identified brassinosteroid. Brassinosteroids appear to be widely distributed in the plant kingdom, with some of their effects on plant growth and development including enhanced resistance to chilling, disease, herbicides, and salt stress; increased crop yields, stem elongation, and seed germination; decreased fruit abortion and drop in apples; inhibition of root growth and development; and promotion of ethylene biosynthesis and epinasty.

Salicylates   Salicylates are a class of compounds having activity similar to salicylic acid (ortho-hydroxybenzoic acid) (Fig. 7.10). Willow (*Salix sp.*) bark and leaves were used by the ancient Greeks and Native American Indians for minor pains and fevers and were found to contain salicylic acid. Salicylates, suspected to be widely distributed within the plant kingdom, have only recently been recognized as potential growth regulators. Salicylic acid is synthesized from the amino acid

FIGURE 7.8 Ripening generator that releases ethylene to ripen mature green tomatoes in storage.

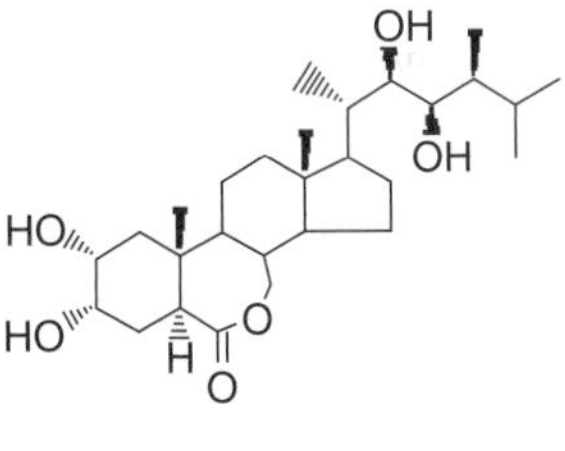

FIGURE 7.9 Chemical structure of brassinolide, the first identified brassinosteroid.

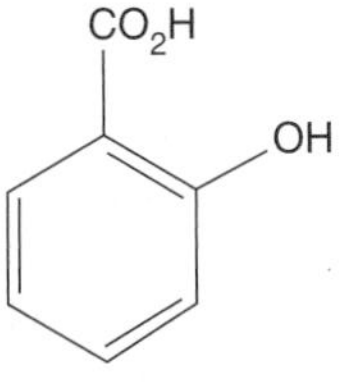

FIGURE 7.10 Chemical structure of salicylic acid, which belongs to the salicylates class of compounds.

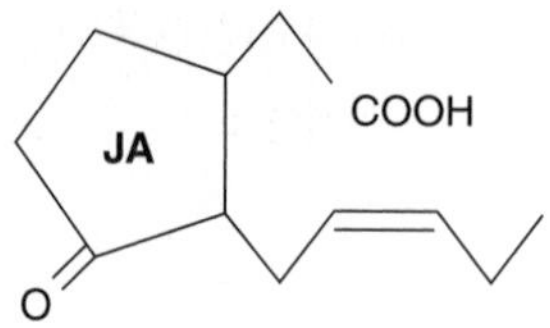

FIGURE 7.11   Chemical structure of jasmonic acid, which belongs to the jasmonates class of compounds.

phenylalanine, and it has numerous reported effects including thermogenesis (temperature-regulated development) in Arum flowers, enhanced plant pathogen resistance, enhanced longevity of flowers, and inhibition of ethylene biosynthesis and seed germination.

Jasmonates   Jasmonates are a specific class of cyclopentanone compounds with activity similar to (-)-jasmonic acid (Fig. 7.11) and/or its methyl ester. They were first isolated from the jasmine plant in which the methyl ester is an important product in the perfume industry. Jasmonic acid is synthesized from the fatty acid linoleic acid.

Jasmonates have been isolated from many plants and may be ubiquitous. A number of effects on plant growth and development have been attributed to jasmonates, including inhibition of growth and germination, promotion of senescence, abscission, tuber formation, fruit ripening, pigment formation, and tendril coiling; and they appear to have important roles in plant defense by inducing proteinase synthesis.

Polyamines   Polyamines are widespread in all cells and exert regulatory control over plant growth and development at very low concentrations. Examples of polyamines include putrescine, spermidine, and spermine (Fig. 7.12). Polyamines have a wide range of effects on plants and appear to be essential in growth and cell division.

Practical Uses of Growth Regulators in the Plant Sciences

Rooting

Propagation of plants that are genetically similar to a stock plant is necessary with many agricultural plant species. To produce genetically identical plants, clonal vegetative propagation is required and is accomplished by the rooting of cuttings (adventitious root formation), root cuttings and division, layering, grafting and budding, and similar techniques. Auxins (such as indole-3-acetic acid, indole-3-butyric acid or IBA, Hormodin, 1-naphthaleneacetic acid or NAA) are used for enhancing the rooting of plant cuttings. The basal end of cuttings are inserted into an auxin-containing formulation prior to placing in media for rooting. A fungicide is often included in the auxin rooting formulation to help reduce rooting of stem and root tissue.

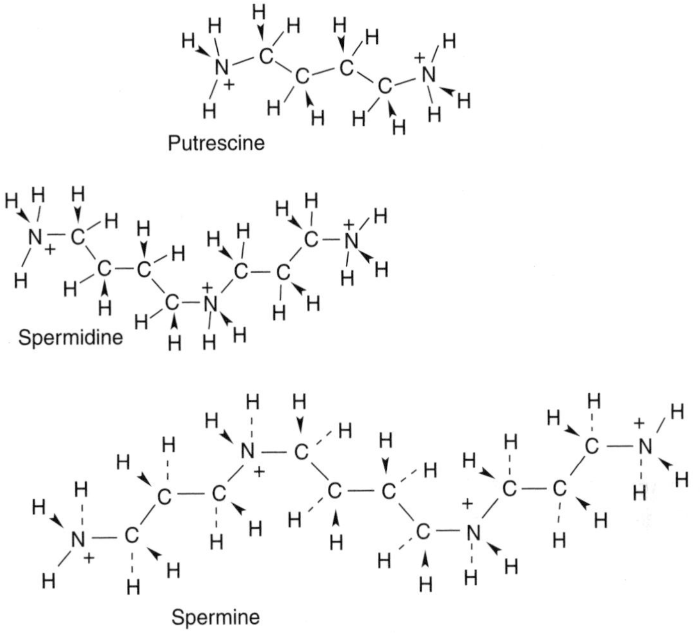

FIGURE 7.12   Chemical structures of the polyamines putrescine, spermidine, and spermine.

Height Control

Most of the plant growth regulators used in greenhouse crop production are used to regulate (often retard) the growth of bedding plants, poinsettias, and other containerized crops. Typical growth retardants are daminozide (B-Nine), chlomequat (Cycocel), ancymidol (A-Rest), paclobutrazol (Bonzi), chlorphonium (Phosfon), and uniconazole-P (Sumagic). These chemicals reduce plant height by inhibiting the production of gibberellins primarily in stem, petiole, and flower stalk tissue. The benefits of using these growth retardants in plant production include improved plant appearance by maintaining plant size and shape in proportion with the pot. They can also increase the stress tolerance of plants during shipping, handling, and retail marketing. The growth retardants are applied prior to rapid shoot growth. Excessive rates of plant retardants can cause persistent growth reductions.

Herbicides

Herbicides are plant growth regulators that are used to destroy weed species. For example, 2,4-dichlorophenoxyacetic acid (2,4-D is available under a large selection of trade names, most often formulated as an inorganic salt, amine, or ester) is a common herbicide that is used for postemergent selective broad-leaved weed control in wheat, corn, sorghum, other small grains, rice, and sugarcane. Glyphosate (Roundup) is a

ROSEWARNE
LEARNING CENTRE

foliar-applied translocated herbicide used for control of many annual and perennial grasses and broadleaf weeds plus many tree and woody brush species. It is applied to undesirable vegetation by a variety of delivery methods including by boom equipment, hand-held and high-volume rollers, and wipers, and, in some states for forestry, aerial application equipment.

Atrazine (AAtrex, Atranex) is triazine herbicide that is widely used as a selective herbicide for control of broadleaf and grassy weeds in corn, sorghum, sugarcane, macadamia orchards, and turf grass sod. It is used also in some areas for selective weed control in conifer reforestation and Christmas tree plantations as well as for nonselective control of vegetation in chemical fallow. Chloramben (Amiben) is used for pre-emergence weed control that is applied at planting of soybeans, dry beans, peanuts, sunflowers, corn, sweet potatoes, lima beans, seedling asparagus, squash, and pumpkins.

Branching and Shoot Growth

Some plant growth regulators are used to enhance branching. For floricultural crops, these include Florel, Atrimmec, and Off-Shoot-O. They generally inhibit the growth of the terminal shoots or enhance the growth of the lateral buds, thereby increasing the development of the lateral branches. They can be used to replace the manual pinching of many crops. Often this increased branching also will reduce the overall height of the plant. The ethylene released inside the plant by Florel also inhibits internode elongation, keeping plants more compact than untreated plants. The plant must have sufficient growth to allow for sites of lateral development.

A gibberellin inhibitor, prohexadione-calcium (Apogee), is applied to apple trees to reduce shoot growth. Controlling excessive shoot growth on apple trees can reduce pruning costs, increase color of apples when light is limiting, facilitate easier spraying, and aid in control of some disease and insect pests. By decreasing the level of GA in the plants, Apogee inhibits the shoot's ability to elongate. Applications usually occur when the longest shoots are between 1 to 3 inches long (oftentimes this is about at the petal fall period).

Flowering

Plant growth regulators can be used to enhance flowering or to remove flowers. To improve flowering of some woody ornamentals typically forced in the greenhouse (such as azalea), GibGro, which contains gibberellic acid, can be used to substitute for all or part of the chilling requirement. In addition to overcoming dormancy, these compounds can improve flowering and/or bloom size of camellia, geranium, cyclamen, statice, and calla lily.

Flower removal is especially desirable for stock plants maintained for cuttings of vegetatively propagated ornamentals such as geraniums, fuchsia, begonias, or lantana. Ethephon (Florel) is the primary compound used for flower removal. Once ethephon is absorbed by the plant it is converted to ethylene, the primary hormone responsible for flower senescence. Flower removal results in more energy going into vegetative growth, increasing the number of laterals available for cuttings on stock plants and promoting increased branching of plugs and finished plants. Early flower removal also allows synchronization of flowering for a more dramatic appearance and/or for initiation of flowering on a specific marketing date. Because initiation and development of flowers requires time, Florel is not generally used on crops within 6 to 8 weeks of marketing.

Fruiting

Chemical thinning sprays to reduce fruit number is one of the most important spray applications an apple grower typically makes during the growing season of established trees. Accel is a growth regulator whose active thinning ingredient is 6-benzyladenine (6-BA). 6-BA is a cytokinin that enhances cell division. The actual mechanism of action is not known.

Ethephon (Ethrel) is often used when other thinners have been use and insufficient thinning has occurred. Ethephon has been shown to effectively remove Delicious apples up to 20 mm in size. It is usually applied 10 to 20 days after full bloom.

Napthaleneacetic acid (NAA, Fruittone, Fruit Fix) has been used as fruit thinners for many years. NAA stimulates ethylene production by fruit tissues, which in turn slows the development of the youngest and weakest fruits more than the older fruit in a cluster. The result is that the weaker fruit cannot compete for resources and they abscise. Late applications can cause small fruit called pygmies to remain on the tree until harvest.

Ripening

Ethephon (Ethrel or Cepha) is an ethylene-releasing compound applied to plants as an agricultural spray for increasing ethylene concentration in plant tissue. Ethylene stimulates ripening in fruits that have reached a certain minimum stage of maturity. Ethylene-releasing compounds are used for stimulation of fruit ripening, degreening of citrus fruit, and abscission induction prior to mechanical harvesting of fruit.

Stimulated ripening may be desirable where apple fruit is needed for early fresh market, for increasing the sweetness of early cider, or for getting an early start on harvesting a large crop. In certain cultivars it may be possible to harvest in one picking rather than two or three pickings. Use of ethephon advances development of all the maturity-dependent changes: fruit becomes softer, an abscission zone develops between fruit stem and spur, starch in the fruit is converted to sugars, internal production of ethylene increases, the rate of respiratory heat production increases, and for some fruit the ability to develop a ripened color in the fruit skin increases.

Fruit Abscission

Some plant growth regulators are used to retard abscission of fruit. NAA is an excellent fruit abscission control for many cultivars of apples. It becomes active within 4 to 5 days after application with fruit drop reduced for about 10 days. When ethephon is used for enhancing fruit ripening, NAA is applied 4 to 5 days before applying ethephon as a safeguard against premature fruit drop.

Amino-ethoxyvinylglycine, AVG (Retain), works by retarding ethylene production and delaying fruit maturation. The delay is manifested by delays in loss of fruit firmness, starch conversion to sugar, watercore development, and increases in fruit-soluble solids. Retain may be helpful in holding fruit on the tree longer, thus allowing better color development.

Enhance Postharvest Life

Handling and storage practices for fruits and vegetables are designed with the goal of maintaining quality and delaying overripening and senescence. Ethylene is the natural plant product that coordinates ripening processes (e.g., softening, color change, conversion of starch to sugars, loss of acidity) in fruits and vegetables. 1-methylcyclopropene (1-MCP or SmartFresh) has been shown to specifically but reversibly suppress ethylene responses and extend the postharvest shelf life and quality of numerous fruits and vegetables including apple, tomato, and avocado fruits. Harvested fruits are exposed to the volatile active ingredient of SmartFresh in enclosed areas, such as storage rooms, greenhouses, coolers, shipping containers, enclosed truck trailers, or controlled-atmosphere food-storage facilities.

1-MCP works by attaching to a site (receptor) in fruit tissues that normally binds to ethylene. Binding of ethylene to these sites is how plant tissues perceive that ethylene is present in the environment. If ethylene binding is prevented, ethylene no longer promotes ripening and senescence. This causes fruits to ripen and soften more slowly, therefore maintaining their high-quality, edible condition for longer periods. Even some fruits and vegetables that do not go through a ripening phase (e.g., broccoli, lettuce, carrots) may benefit from 1-MCP exposure.

1-MCP has been shown to greatly delay softening and red color development of tomato fruit compared to untreated fruit. In this regard, 1-MCP has potential to increase overall fruit quality by allowing fruit to ripen on the vine for an extended period. Consequently, the use of 1-MCP is compatible with vine-ripe harvesting practices resulting in fruit that can tolerate the rigors of shipping and handling better than nontreated fruit.

Summary

A plant hormone is an organic substance other than a nutrient that is active in small amounts, formed in one part of the plant and translocated to another part where it evokes a response. A plant growth regulator is an organic compound other than a nutrient that in small concentrations modifies growth and development. Although all plant hormones are growth regulators, not all growth regulators are plant hormones.

The exact mechanisms by which plant hormones affect growth and development through gene expression are not well understood, but two mechanisms (depending on hormone type) are suggested. Steroid hormones bind with receptor molecules in the cytoplasm of the cells to form a hormone-receptor complex. This complex subsequently affects mRNA synthesis resulting in specific physiological responses. Peptide hormones appear to bind on receptor proteins on target cells, which subsequently undergo conformational modifications resulting in changes in physiological processes.

The plant hormones include the auxins, gibberellins, cytokinins, abscisic acid, and ethylene. Auxins resemble indole-3-acetic acid and characteristically induce elongation in shoot cells. Gibberellins are compounds that stimulate cell division or cell elongation, or both. Cytokinins are adenine-resembling molecules that stimulate cell division. Abscisic acid is a sesquiterpenoid compound that functions as a growth inhibitor, and ethylene is a hydrocarbon gas that is a natural product of plant metabolism and is involved in several plant processes including ripening, abscission, and root growth and formation. Other classes of plant compounds that are gaining acceptance as plant hormones include brassinosteroids, salicylates, jasmonates, and polyamines.

Review Questions

1. Define the term *plant hormone.*
2. How does a plant hormone differ from a plant growth regulator?
3. What was the first group of plant hormones discovered?
4. Which classes of plant hormones are synthesized via the mevalonic acid pathway?
5. What is the role of abscisic acid in plants?
6. What is stress ethylene?
7. What are some of the suspected roles of salicylic acid in plants?

Selected References

Arteca, R. N. 1995. *Plant growth substances: Principles and applications.* New York: Chapman & Hall.

Huber, D., J. Jeong, and M. Ritenour. 2003. Use of 1-methylcyclopropene (1-MCP) on tomato and avocado fruits: Potential for enhanced shelf life and quality retention. Florida Cooperative Extension Service, Institute of Food and Agricultural Sciences, University of Florida Publication HS-914.

Latimer, J. G. 2001. Selecting and using plant growth regulators in floricultural crops. Virginia Coop. Ext. Serv. Publ. No. 430-102.

Moore, T. C. 1979. *Biochemistry and physiology of plant hormones.* New York: Springer-Verlag.

Salisbury, F. B., and C. W. Ross. 1978. *Plant physiology,* 2nd ed. Belmont, CA: Wadsworth.

Selected Internet Sites

www.chm.bris.ac.uk/motm/brassinolide/brassinolideh.htm   Brassinolide page created by Martin A. Iglesias-Arteaga at the University of Havana.

http://hcs.osu.edu/plants.html   Plant dictionary created and maintained by Tim Rhodus, Ohio State University.

http://tfpg.cas.psu.edu/   Penn State College of Agricultural Sciences Tree Fruit Production Guide.

8

Some Ecological Principles in Plant Growth and Production

The word *ecology* is derived from the Greek *oikos,* which means "house" or "place to live." Ecology therefore could be considered as the study of organisms "at home" or in their environment. The environment is the summation of all biotic and abiotic factors that surround and potentially influence an organism. This chapter provides some background information on basic principles of ecology and plant ecology.

The chapter begins with explanations of ecological concepts of organism groupings and interactions between organisms. Although ecological principles are most often discussed in concept in natural ecosystems, a discussion of farms as ecosystems is presented to blend the ecological concepts with commercial plant production systems.

Important Ecological Concepts

Organism Groupings

In the study of ecology, organisms are usually placed into groupings. These groupings can be small such as at the specie level (from which our scientific taxonomic systems have developed) or large such as in biomes. The usefulness of the chosen grouping depends on the level of complexity at which an observed event is occurring (e.g., disappearance of a specie of plants versus the destruction of a forest ecosystem that may have many interacting species).

Following are some of the groupings of organisms used in ecology.

Specie   A specie consists of a group of morphologically and ecologically similar populations that may or may not be inbreeding. Living organisms fall into more or less well-defined groups, which are commonly called species. Traditional taxonomy has developed according to the existence of discrete species.

Population   Populations are the individuals of a species growing in a particular area (e.g., a field of planted tomatoes). A subgrouping of population is a breeding population. Individuals in a breeding population have the opportunity to reproduce.

Community   A community is made up of individuals that just happen to exist in the same location. Plant communities can be applied to vegetation types of any size or durability.

Ecosystem   An ecosystem is the sum of the plant community, animal community, and environment in a particular region or habitat. It is a geographic location on the earth's surface where energy and nutrients are captured and transformed by plants, animals, and microbes. Biotic or living organisms comprise the biological component of an ecosystem. The biological components are dependent on the physical components of the ecosystem. Topography, geology, and climate of the area determine the physical environment.

Biomes   A biome is a group of ecosystems with similar biological features, though the same species are often not involved. Examples of biomes include temperate forests, grasslands, savannas, and barren regions such as deserts (Fig. 8.1).

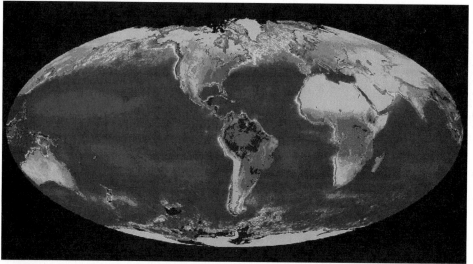

KEY:
Tropical forests, very productive temperate forests
Temperate forests and moist savanna
Dry savanna, mixed forests, grassland
Coniferous forests, grasslands
Semi-arid steppes and tundra
Barren regions (deserts, ice)

FIGURE 8.1   Locations of the major biomes of the world.
Source: Goddard Flight Center, NASA.

Significant Interactions among Organisms

Plant species are influenced by interactions with other organisms (including other plants) that may modify the genetic potential of each species (its physiological optimum and range) to yield the observed plant growth and development and ecological community. These interactions can be beneficial or harmful or show no effect on the interacting organisms. During evolution many plants evolved defenses against potentially harmful interactions (such as from herbivores and pathogens). These defenses can be physical (such as thorns and spines, thick cuticles, lignified and suberized tissue) or chemical (such as allelochemicals or toxins).

Significant interactions between organisms can be broadly grouped into the following.

Commensalism   Commensalism is an interaction that stimulates one organism but has no effect on the other. An example of commensalism is the growth of epiphytes on host trees.

Mutualism   Mutualism is an obligate interaction, where the absence of the interaction depresses both partners. Closely related organisms often do not seem to form such an interaction. Mutualism occurs between plants and some other organism. The most widespread mutualistic relationship is between plants and insect pollinators. Some plants are also involved in mutualistic interactions (or symbiosis) with nitrogen-fixing bacteria and mycorrhizal fungi.

Competition   Competition results in mutually adverse effects to organisms that utilize a common resource in short supply. The most intense competition exists where there are similarities between plants in their environmental requirements. Light is the primary resource for which plants compete, followed by water and nutrients.

Antagonism or Amensalism   Antagonism or amensalism is an interaction that depresses one organism while the other remains stable. Antagonisms are common between plants and other kinds of organisms, although it can be plants. Allelopathy, herbivory, and parasitism are types of antagonistic interactions. Allelopathy occurs when one plant produces chemicals that inhibit the growth of other plants. For example, the black walnut tree secretes toxins from its roots that inhibit the growth of herbaceous plants in its vicinity. Herbivory is the consumption of all, or part, of a plant by a consumer. This occurs when animals ranging from microscopic arthropods to elephants consume plants. Parasitism involves mostly viral, bacterial, and fungal pathogens (though there are a few parasitic angiosperms).

FIGURE 8.2   Farm ecosystems are parts of landscapes that can affect other ecosystems downwind and downstream.

Farm as Ecosystems

Farms are parts of landscapes (Fig. 8.2) that can affect other ecosystems downwind and downstream and in turn are influenced by forces and events in other ecosystems that may be upwind and upstream. Farms are human-managed ecosystems generally designed to produce as much harvestable and/or marketable biomass (yield) as environmental conditions will allow. Contrasted with natural ecosystems, farms are typically populated with relatively few species. Often, though not always, the number of species on a farm varies inversely with management intensity.

Farmland (crops and pasture land) covers approximately 37% of the land surface of the earth and 47% of the United States (Table 8.1). Natural inputs contribute tremendously to a farm's productivity, as do supplemental inputs such as fertilizers and pesticides. As a result farms tend to be more productive than the natural ecosystems they replace. This increased productivity often places increased demands on the field's environmental resources. Farms are usually managed to maximize crop yield while trying to minimize the environmental degradation they may cause.

TABLE 8.1   *Land Use Patterns for the United States and the World*

Use	Area (million ha)[*]	
---	World	U.S.A.
Crops	1,441	188
Pasture	3,357	239
Forest	3,897	287
Urban	?	99
Unmanaged	4,345	202
Total	13,041	1,015

Source: Cavigelli, M. A., S. R. Deming, L. K. Probyn, and R. R. Harwood, eds. 1998. *Michigan field crop ecology: Managing biological processes for productivity and environmental quality.* Michigan State University Extension Bulletin E-2646.

[*]1 hectare (ha) is about 2.5 acres.

Summary

Ecologists are interested in the understanding between organisms with their environment. Organisms are often placed into groupings for ecological study. The smallest grouping is the specie and the largest is the biome. The usefulness of the chosen grouping depends on the level of complexity at which an observed event is occurring. Numerous types of interactions also can occur between plants and other organisms. These may be beneficial (such as mutualistic interactions of certain plants with nitrogen-fixing bacteria) or detrimental (such as plants in competition with other organism for limited resources).

Although discussions of ecological principles are commonly centered on natural ecosystems, many ecological principles are utilized when crop producers employ integrated pest management into their systems. Farm ecosystems differ from natural ecosystems in that farms are usually populated with relatively few species and supplemental inputs (such as fertilizer and pesticides) contribute greatly to their productivity.

Review Questions

1. What is the general goal of a farm as it applies to ecosystems?
2. How does a community differ from a population?
3. What are some examples of antagonistic relationships that plants may have with other organisms?
4. How does a commensalistic interaction (commensalism) between organisms differ from a mutualistic interaction (mutualism)?

Selected References

Barbour, M. G., J. H. Burk, and W. D. Pitts. 1980. *Terrestrial plant ecology.* Reading, MA: Benjamin/Cummings.

Cavigelli, M. A., S. R. Deming, L. K. Probyn, and R. R. Harwood, eds. 1998. *Michigan field crop ecology: Managing biological processes for productivity and environmental quality.* Michigan State University Extension Bulletin E-2646.

Lambers, H., F. S. Stuart III, and T. L. Pons. 1998. *Plant physiological ecology.* New York: Springer-Verlag.

Odum, E. P. 1971. *Fundamentals of ecology.* 3rd ed. Philadelphia: W. B. Saunders.

Pugnaire, F. I., and F. Valladares, eds. 1999. *Handbook of functional ecology.* New York: Marcel Dekker.

Sinclair, T. R., and F. P. Gardener, eds. 1998. *Principles of ecology in plant production.* New York: CAB International.

Selected Internet Sites

hortipm.tamu.edu/   Hort IPM Web site, Texas A&M University.

www.ippc.orst.edu/cicp/Vegetable/veg.htm   Internet resources on IPM on vegetables.

www.nrcs.usda.gov/technical/ECS/   Technical information on ecological sciences including agricultural ecology, aquatic ecology, ecological climatology, forestry and agroforestry, range and grazing land ecology, understanding ecosystem, and wildlife management, Natural Resources Conservation Service, USDA.

www.nysipm.cornell.edu   Integrated pest management information, Cornell University.

www.peak.org/~mageet/tkm/ecolenv.htm   Internet resources related to the science of ecology and the state of the environment, Peak Organization.

www.westminster.edu/staff/athrock/ECOLOGY/Botlinks.htm   Internet resources for botany and plant ecology, Westminster College, New Wilmington, PA.

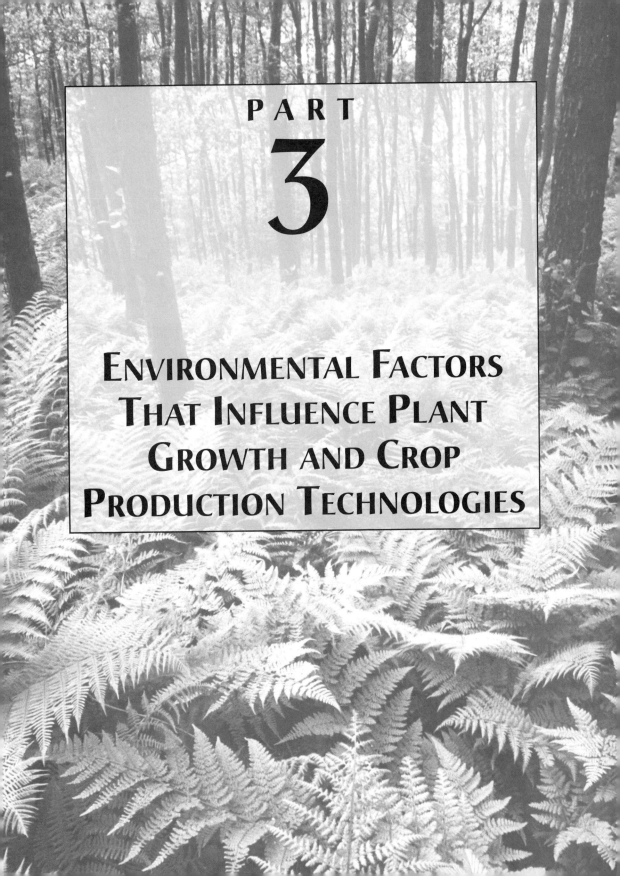

PART
3

ENVIRONMENTAL FACTORS THAT INFLUENCE PLANT GROWTH AND CROP PRODUCTION TECHNOLOGIES

9

Introduction to the Role of the Environment in Plant Growth and Development

The growth and development of plants depends on the environment in which they are growing. Through the years and successive generations, each plant has evolved to prefer certain environmental parameters for optimum plant growth. To attain the highest potential yield a crop must be grown in an environment that meets these parameters. A crop is generally considered well matched with its climate or growing conditions if it can be grown with minimal environmental adjustments. Growing plants in unfavorable environmental conditions can stress the plants resulting in reduced growth and lower yields. For many horticultural crops the environment can be artificially modified, such as in greenhouses, to minimize stresses and more suitably meet the specific crop's requirements.

This introductory chapter defines environment, weather and climate, and plant stress. Some of the reasons and methods for studying how the environment affects plant growth are also discussed.

Environment

The environment is the sum of all the external forces or factors that affect the life of an organism. These factors interact not only with the organism but also among themselves. The components of the environment can be divided into abiotic (physical) and biotic (biological) factors. Abiotic factors include the physical environmental conditions of irradiance (light), temperature, water, air, and mechanical disturbances. The biotic factors include biological organisms such as animals, insects, diseases, and other plants.

Weather and Climate

Weather and climate are major factors of the field environment. Weather is the composite of the temperature, rainfall, light intensity and duration, wind direction and velocity, and relative humidity of a specific location for a set amount of time. It is the immediate day-to-day, local combination of the various environmental factors. For a specific location, weather factors assume a certain pattern, changing day by day, week by week, month by month, and season to season. The same pattern generally repeats itself year by year.

Climate, sometimes called macroclimate, is the weather pattern for a particular location over several years. Landmasses, prevailing wind patterns, bodies of water, altitude, and the latitude affect the climate of the region. The soil of an area is greatly dependent on the climate. Crops and landscape plants that thrive under one set of climatic conditions may not grow well under another. This is the reasoning behind the development of the plant hardiness zones found on the USDA plant hardiness map (discussed further in Chapter 12). As a rule, macroclimates are not easily changed.

The microclimate is the little weather variations that exist in a location or field. Frost pockets (areas of a field with the highest probabilities of experiencing frost) fit into this category. Microclimates are often easily changed or modified. For example, shade trees can be added or removed to affect light and temperature, and windbreaks can be used to block or redirect wind currents. Adverse microclimate situations can often be avoided with proper plant selection, such as choosing plants that are more tolerant of cold temperatures for use in areas of frost pockets.

Although climate determines which crops can be grown optimally in a particular location, the rate of growth and development of these crops largely depends on the weather. Cloudiness, amount of rainfall, and wind movement, for example, all influence how well a crop will grow at a particular time. Weather also determines when some farm operations such as cultivating, fertilizing, harvesting, and irrigating can occur.

Topographical features such as mountains, hills, large bodies of water, and deserts all modify and create special micro or regional deviations from the climate for the area. For example, cool-season crops can be grown in warm semitropical or tropical areas by using the cooler, higher elevations for crop production to escape the excessive summer heat of the lowlands (Fig. 9.1).

Plant Stress

The term *stress* in biological sciences is often used for any environmental condition that is potentially unfavorable to living organisms for normal growth and development. When stressed, plants may show a physical response (e.g., bending of a stem) or a chemical response (e.g., shift in metabolism). Stressing plants almost always has

FIGURE 9.1   Lettuce, normally a cool-season crop, growing in the cooler, higher elevations in Panama.

the suggestion of causing possible injury, and the plant may be severely injured or die if the stress is sufficiently severe. The extent of plant injury due to a stress often depends on the duration of time that the plant is exposed to the stress, but this is not universally true (such as a killing temperature, which may be fatal with a brief exposure). Stress resistance is the ability of the plant to survive the unfavorable factor and even to grow in its presence. The amount of stress resistance exhibited by a plant depends on a variety of plant characteristics such as age, metabolism, genetic make up, and structure.

Reasons to Study How the Environment Affects Plant Growth

In natural ecosystems, knowledge of plant responses to the environment can be used to better understand why plants evolved in certain areas with unique environmental characteristics. Also, the presence or absence of certain plants can be indicative of changing environmental factors or climates. For agricultural systems, the probability of growing a crop that can result in a beneficial economic return is greatest by growing it in suitable environmental conditions. This is accomplished by either finding a location and growing season with the preferred environmental conditions to grow the crop or modifying the existing environmental conditions to the crop's needs or environmental specifications. Alternately, other crops may be chosen for that location and season that may have a better chance of survival in that environmental situation.

Methods for Studying How the Environment Affects Plant Growth and Development

The effects of the various environmental factors on crop growth can be studied either in the field under natural or cultivated conditions, or in the controlled environments of the lab, growth chambers, or greenhouses. Both approaches have advantages. The field method provides typical amounts of environmental components such as light and certain soil characteristics that are present in many agricultural systems, and thus is more typical of the conditions the plant normally experiences (Fig. 9.2). In the field situation, many parameters may be measured and statistical methods employed to deduce which factors are influencing plant growth and development and in what ways.

Certain techniques and environmental factors are relatively difficult to work with under field conditions and are best evaluated in controlled environment conditions. For example, it is almost impossible to control the field environment so that the effects of contrasting amounts of a given component can be studied. Lab, growth chamber, or greenhouse research often emphasizes the control of the environment (Fig. 9.3). A typical experiment in a controlled environment would maintain all environmental parameters but one (e.g., air temperature) constant. The selected parameter chosen for study would vary in the experiment and therefore would be implicated for the observed plant response, as suggested by statistical analysis of the data.

FIGURE 9.2   Example of field research evaluating growth and production of various geranium cultivars in a location. Courtesy of Dr. Richard Craig, Penn State University.

FIGURE 9.3   Example of controlled environment research where plants can be placed within individual airtight chambers in a controlled environment room and treated with various concentrations of a gas.

Summary

Important environmental factors affecting crop growth and production include abiotic (e.g., irradiance, temperature, water, air, and mechanical disturbances) and biotic (e.g., animals, diseases, insects, and other plants) influences. Major factors of the field environment include weather and climate. Weather is the immediate day-to-day, local combinations of the various environmental factors, whereas climate is the weather pattern for a particular location over several years. Microclimates are the little weather variations that exist from one area of a location to another and can be affected by agricultural production practices.

Plant growth and differentiation can be adversely affected by any environmental factor. When a factor is unfavorable for plant growth and development, it is considered a plant stress. Often (though not always) the duration of time that a plant is exposed to the stress determines the extent of the injury. Resistance to a particular stress depends on such plant characteristics as age, metabolism, genetic make up, and structure.

Review Questions

1. What requirements must be met for a crop to be grown to its maximum yield?
2. How do weather and climate differ and what effect do they exert on a field-grown crop?
3. What are some responses that plants exhibit when stressed and what is meant by stress resistance?
4. What are some reasons we should study how the environment affects plant growth in both natural and agricultural systems? (Give two reasons for each system.)
5. What are the two general methods for studying how the environment affects plant growth and what are their advantages and disadvantages?

Selected References

Asian Vegetable Research and Development Center. 1990. *Vegetable production training manual.* Shanhua, Taiwan: Asian Vegetable Research and Development Center. Reprinted 1992.

Cavigelli, M. A., S. R. Deming, L. K. Probyn, and R. R. Harwood, eds. 1998. *Michigan field crop ecology: Managing biological processes for productivity and environmental quality.* Michigan State University Extension Bulletin E-2646.

Hartmann, H. T., A. M. Kofranek, V. E. Rubatzky, and W. J. Flocker. 1988. *Plant science: Growth, development, and utilization of cultivated plants.* 2nd ed. Upper Saddle River, NJ: Prentice Hall.

Lambers, H. F., S. Stuart III, and T. L. Pons. 1998. *Plant physiological ecology.* New York: Springer-Verlag.

Odum, E. P. 1971. *Fundamentals of ecology.* 3rd ed. Philadelphia: W. B. Saunders.

Salisbury, F. B., and C. W. Ross. 1978. *Plant physiology.* 2nd ed. Belmont, CA: Wadsworth.

Stern, K. R. 1991. *Introductory plant biology.* 5th ed. Dubuque, IA: Wm. C. Brown.

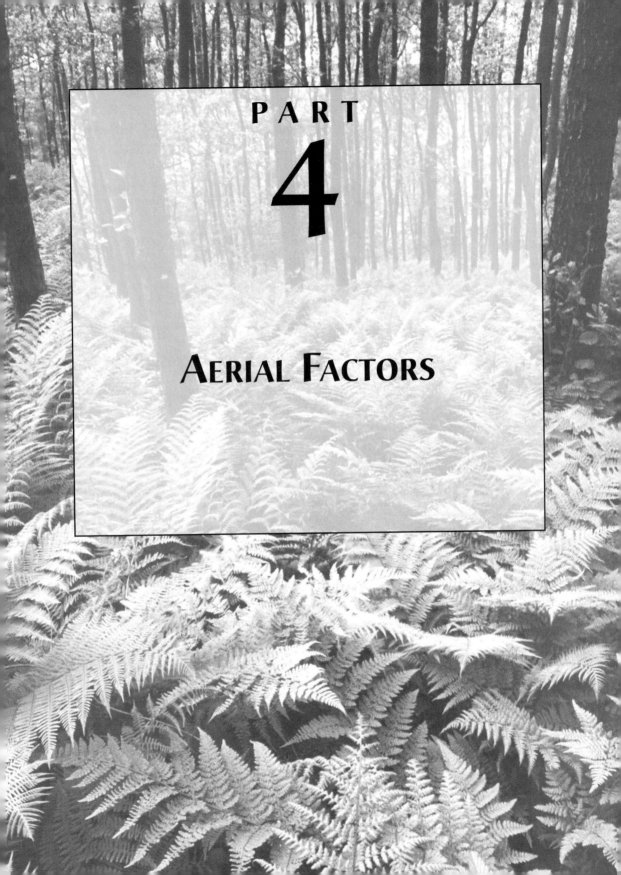

PART

4

AERIAL FACTORS

10

Overview of the Aerial Environment

The aerial environment is that portion of the ecosphere that exists predominantly above the soil surface in relative proximity to the growing plant and that has the potential to affect the physiology and growth of the plant (Fig. 10.1). This includes irradiance (light), temperature, atmosphere (including weather), and organisms (plants, animals, and pathogens) that primarily live above the soil surface.

This introductory chapter describes the ecosphere and provides a brief discussion of the components of the aerial environment, which will be covered more in-depth in Chapters 11 to 15.

The Ecosphere and the Aerial Environment

The ecosphere or biosphere is a shell around the earth in which life exists (Fig. 10.2). This is a thin layer of air, water, and soil on or near the earth's surface having an approximate thickness of only 12 to 20 km. Because the ecosphere is a closed system, which means that no matter is leaving or entering, the chemical components (such as mineral nutrients, gases, and water) necessary for life must be continually cycled and recycled.

The ecosphere can be divided into three major spherical layers. The atmosphere (air) is a thin layer no more than 12 km high. The hydrosphere (water) is the limited supply of water in rivers, glaciers, lakes, oceans, and underground deposits. The lithosphere (soil) is a thin crust of soil, minerals, and rocks extending only a few thousand meters into the earth's interior.

FIGURE 10.1 The aerial environment of a plant is exemplified by the above-ground growth of this celery planting and the environment that surrounds it.

FIGURE 10.2 The earth's ecosphere as seen from the moon is a thin shell around the earth in which life exists.
Source: Earth Observatory, NASA.

Irradiance and Temperature

During the thermonuclear reactions within the sun, energy is liberated as an almost continuous spectrum of radiation. Radiation is electromagnetic energy, which can be described or quantified as waves (often as wavelength) or discrete packets (often as quanta or photon energy). A portion of radiation is in the visible wavelengths and is often referred to as light. Radiation travels at the speed of light (3×10^8 ms^{-1}) in a vacuum and is slightly slower in the atmosphere. Solar irradiance is the main energy source for almost all forms of life. Important to plant growth are the wavelengths between 400 to 700 nm (often referred to as photosynthetically active radiation, PAR) that drive the carbohydrate-producing reactions of photosynthesis and the wavelengths of red, far-red, and blue that affect the plant development reactions of photomorphogenesis.

Temperature measures the average kinetic energy of molecules. The amount of radiant energy emitted from a source (such as the solar energy) is a strong function of temperature of the energy source. The sun is estimated to have a surface temperature of approximately 5900K. Radiation emanating from a high-energy source (such as the sun) can be absorbed by matter of a lower energy state and at a lower temperature (such as the earth or organisms or objects on the earth) and increase the temperature of that matter.

The Earth's Atmosphere

The earth's atmosphere surrounds all terrestrial organisms even to the roots of plants and organisms in the soil. The factors of weather (e.g., precipitation, winds, clouds, humidity, and temperature) also occur in the lower altitudes of the atmosphere. Although the atmosphere primarily consists of relatively inert nitrogen, it supplies organisms with the oxygen necessary for respiration and with carbon dioxide for photosynthesis. The atmosphere is also the reservoir from which nitrogen slowly cycles via certain bacteria into living parts of the ecosystem. Other atmospheric gases include water vapor, argon, and other minor constituents.

Another component of the atmosphere is air pollution. Air pollutants are typically emitted into the atmosphere from a source and transported to the affected organism. Air pollution can be both natural and anthropogenic (human-generated) (Fig. 10.3). Natural sources evolved with the development of the ecosphere and include volcanoes, forest fires, dust storms, decaying marshes, salt spray from oceans, methane from animals, and hydrocarbon emissions from some trees. Anthropogenic sources predominantly evolved with the development of the industrial revolution and include emissions from industrial and manufacturing plants and combustion engines.

Gravity, although not strictly an aerial environmental factor, exerts an orientation effect on above-ground organisms and other earth surface constituents (such as rocks, animals, water). Plants and animals use gravity for stability and as a reference to determine the direction in which to grow and/or move. Gravity is also responsible for maintaining the spherical shells of the atmosphere and for providing the pressure to prefer to hold organisms and other relatively heavy components (soil, rocks, and water) on the surface of the earth and not suspended in the atmosphere.

FIGURE 10.3   Power plants, cattle, and cars are some of the major contributors of air pollutants and greenhouse gases such as carbon dioxide and methane.
Source: Earth Observatory, NASA.

Above-Ground Living Organisms

The above-ground living organisms of the aerial environment include plants, animals, and pathogens that interact with the plant above the soil surface. These interactions may be beneficial to plants (such as from pollinators for subsequent fruiting) or harmful (such as from diseases and insects) and influence whether the crop can be successfully produced and marketed. Often the harmful effects exerted by other living organisms to the plant are due to competition for physical components of the environment (such as mineral nutrients, water, and light) or as the result of some type of mechanical disturbance (such as insect damage or animal grazing).

Autotrophs and heterotrophs are two broad categories that living organisms can be divided into based on how they derive their nourishment. Autotrophs or autotrophic organisms, often called producers and are predominantly plants, capture light energy and use simple inorganic substances (mineral nutrients) through the process of photosynthesis to develop carbohydrates and other complex substances. The heterotrophs or heterotrophic organisms acquire their nourishment from others through digestion and/or decomposition, and the resulting rearrangement of complex materials. The heterotrophs are made up of macroconsumers and microconsumers. Macroconsumers are primarily animals that ingest other organisms or particulate matter, whereas microconsumers are primarily bacteria and fungi that break down the complex compounds of dead cells, absorbing some of the decomposition products and releasing inorganic nutrients that are then usable by the producers.

Summary

The plant's aerial environment, or that part of the environment that surrounds the plant above the soil surface, consists of factors that can affect plant growth and development, including irradiance, temperature, atmosphere, and other primarily above-ground living organisms. Irradiance and temperature occur as the result of energy liberated from thermonuclear reactions within the sun. The earth's atmosphere provides plants with oxygen necessary for respiration and with carbon dioxide for photosynthesis. The atmosphere also contains air pollutants that can be injurious to plants. The above-ground living organisms that interact with plants can be beneficial (such as pollinators) or harmful (such as from diseases and insects). Living organisms are made up of autotrophs, which are producers that carry out photosynthesis for the energy needs (primarily plants), and heterotrophs, which are consumers that derive their nourishment from other organisms through digestion and/or decomposition.

Review Questions

1. What is the ecosphere?
2. Define the aerial environment.
3. What wavelengths of light are important for plants?
4. What are some general effects that gravity exerts on plants?
5. How do autotrophs differ from heterotrophs?

Selected References

Applied Science Associates. 1976. *Diagnosing vegetation injury caused by air pollutants.* Environmental Protection Agency Handbook. Washington, DC: Applied Science Associates.

Etherington, J. R. 1982. *Environment and plant ecology.* 2nd ed. New York: John Wiley.

Odum, E. P. 1971. *Fundamentals of ecology.* 3rd ed. Philadelphia: W. B. Saunders.

Selected Internet Sites

www.epa.gov/   Home site of the Environmental Protection Agency (EPA).

www.epa.gov/oar/   Government site providing information on air pollution, clean air, and air quality information. Good links for acid rain, ozone depletion, and climate change. Office of Air and Radiation, EPA.

http://www.eia.doe.gov/   Official energy statistics of the U.S. government. Energy Information Administration, DOE.

http://www.nrcs.usda.gov/technical/airquality.html   Air quality information and report of Agricultural Air Quality Task Force. Natural Resources Conservation Service, USDA.

www.wcc.nrcs.usda.gov/wcc.html   Government weather site containing observations, forecasts, maps and models, weather safety, and education. National Water and Climate Center, USDA.

www.nws.noaa.gov/   Government weather site containing observations, forecasts, maps and models, weather safety, and education. National Weather Service, National Oceanic and Atmospheric Administration.

11

Irradiance

The electromagnetic spectrum is a graphical representation of the known distributions of electromagnetic energies arranged according to wavelengths, frequencies, or photon energies (Fig. 11.1). Irradiance, often called light or visible radiation (the terms are often used interchangeably), is energy from the visible and neighboring wavelengths portion of the electromagnetic spectrum and is defined as the radiant flux density on a given surface. The entire spectrum extends over 20 orders of magnitude with the visible portion being a part of only one order of magnitude.

This chapter describes the characteristics of sunlight and radiation, discusses the effect of geographical location and season on the plant light environment, and provides an overview of radiation measurements and instrumentation. Selected effects of light quantity and quality on plant growth and development are also presented, and the chapter concludes with discussions on how plant spacing, weeds, row orientation, supplemental lights and filters, pruning and training practices, and reflective mulches can modify the plant light environment and plant development.

Background Information

Characteristics of Sunlight

The sun is the ultimate source of irradiance (solar radiation) in the solar system, though irradiance can also emanate from other objects. The sun's mass is converted into energy through a series of continuous thermonuclear fusion reactions occurring within the sun. Fusion is a nuclear reaction in which nuclei combine to form more massive nuclei with the simultaneous release of energy. Hydrogen is converted to helium in the core (or central portion) of the sun providing energy to sustain a core temperature of 10 to 15 million degrees C. As a result, the sun has an extremely high surface temperature and radiates a great amount of energy into space.

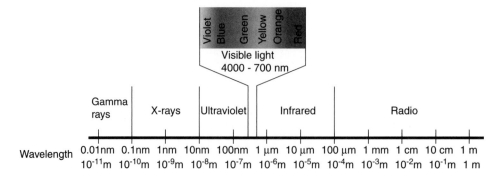

FIGURE 11.1 The electromagnetic spectrum.
Source: Surface Radiation Research, NOAA.

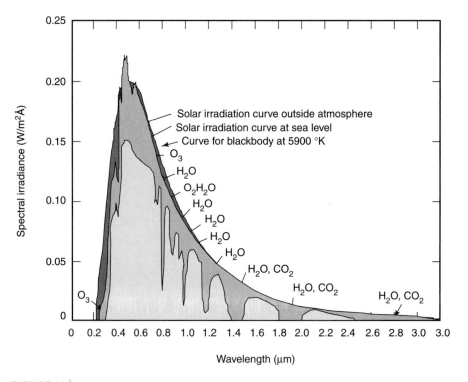

FIGURE 11.2 Solar spectrum and atmospheric absorptions. The sun (solar irradiation curve outside atmosphere) has a distribution that is equivalent to that of a blackbody with a surface temperature of 5900°K. Solar irradiation curve at sea level is reduced due to selective absorption of molecules in the atmosphere.
Source: Goddard Space Flight Center, NASA.

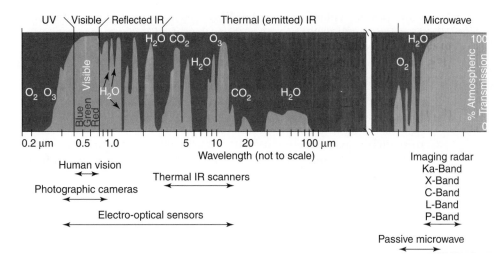

FIGURE 11.3   Gases and particles in earth's atmosphere selectively absorb and reflect certain wavelengths of radiant energy while allowing others to pass through relatively unhindered. The graph above shows the percentage of radiant energy that is allowed to pass through the atmosphere for each wavelength of the electromagnetic spectrum. Source: Earth Observatory, NASA.

The sun can be considered a "black body," which means that it is a perfect absorber and emitter of energy. The spectral energy distribution from a body depends on its surface temperature. Every object at a temperature of $> 0°$ K (approx. $-273°$ C) emits some radiation at all wavelengths; however, the amount of energy radiated at each wavelength depends on temperature. For example, when a body is heated to $500°$ C, it emits primarily infrared radiation (heat); and when it is heated to $> 500°$ C, it emits visible radiation in addition to heat. The sun's spectral distribution is equivalent to that of a black body with a surface temperature of $5900°$ K, though the actual spectral distribution received on the earth's surface is influenced by absorption of atmospheric gases and location on the earth (Fig. 11.2). Because the maximum emission of solar energy is at 490 nm, many biological processes have evolved physiological mechanisms to use this portion of the radiant energy spectrum.

Short-wave radiation from the sun that is not reflected by the atmosphere is absorbed at the earth's surface and transformed into outgoing long-wave radiation (heat) (Fig. 11.3). Subsequently, the earth then becomes a radiating body at a temperature that averages $14°$ C ($290°$ K). Radiation from the sun is termed solar radiation, whereas that from the earth or atmosphere is termed terrestrial radiation.

Modern physics assumes that the electromagnetic spectrum has both wave and particle characteristics. Usually the propagation of energy is referred to in wave terms (such as wavelength, frequency, and wave number), whereas its absorption is often referred to in particle terms (such as quanta). *Wavelength* refers to the length of one "wave" (peak to peak or trough to trough), whereas *frequency* refers to the number of waves passing a plane per second. *Wave number* refers to the number of wave crests in 1 cm length of wave. The units for wavelength of radiation are generally given as microns (μm or 10^{-6} m) or, more commonly, nanometers (nm or 10^{-9} m). The bending of the light wave path when it passes from one material or medium into another and exits as bands of different colors of light, such as observed with white light passing through a prism or through atmospheric water droplets (resulting in a rainbow), illustrates the wave properties of light.

Radiation also consists of small particles that are emitted by a source and travel through transparent materials. When light is absorbed by matter, it behaves as a stream of discreet indivisible packets or quanta. The particle properties of light are demonstrated by the photoelectric effect. The photoelectric effect occurs when zinc exposed to ultraviolet light releases electrons and becomes positively charged. These electrons create an electrical current that can be measured using a photoelectric sensor.

The quantum theory states that the quantum energy of a wavelength is directly proportional to its frequency and inversely proportional to the length of the wavelength, or

$$E_\lambda = h\upsilon = h/\lambda$$

where E_λ = quantum energy of a wavelength λ
$\quad$ h = Planck's constant (6.626×10^{-34} J s)
$\quad$ υ = frequency of oscillation of λ
$\quad$ c = speed of light in a vacuum (3.00×10^8 m s^{-1})
$\quad$ λ = wavelength

Thus, shorter wavelengths have more energy than longer wavelengths (Table 11.1). For example, a quantum of orange light (representative wavelength of 620 nm) has about twice the absolute energy of a quantum of infrared (representative wavelength of 1400 nm).

Sunlight (solar radiation) is one of the most important and variable components of the plant environment. Unfiltered sunlight entering the earth's ionosphere has a brightness or intensity between the wavelengths of 225 and 3200 nm of 1.39 kW/m^2 (this is also referred to as the solar constant). The ozone layer in the stratosphere (the atmospheric zone extending variably from 12 to 50 km above the earth's

TABLE 11.1 *Definitions and Characteristics of the Various Wavelength Regions of Light*

Color	Approximate Wavelength Range (nm)	Representative Wavelength (nm)	Energy (kJ mol^{-1})
Ultraviolet	below 400	254	471
Violet	401 to 425	410	292
Blue	426 to 490	460	260
Green	491 to 560	520	230
Yellow	561 to 585	570	210
Orange	586 to 640	620	193
Red	641 to 740	680	176
Infrared	above 741	1400	85

surface) absorbs a proportion of the ultraviolet radiation while water vapor, carbon dioxide, and oxygen in the troposphere (the atmospheric zone from the earth's surface to the lower level of the stratosphere where weather occurs) absorb the wavelengths of 1100 to 3200 nm. In addition, clouds and particulates in the air reflect (Fig. 11.4), scatter, or absorb the remaining sun's radiation. As a result, only 47% of the radiation emanating from the sun reaches the earth's surface, some of which is reradiated from the earth into space.

The scientific community became concerned in the early 1970s that chlorofluorocarbons (CFCs) used by industrial nations in the production of a variety of commercial products (e.g., refrigerants, aerosol sprays) could potentially reduce levels of the ozone layer (Fig. 11.5). Organisms on the earth's surface would be exposed to increased UV light if the ozone layer was reduced. In October 1985, a British team of scientists discovered a significant reduction in stratospheric ozone (about 40% less than it had been the previous year) over Halley Bay, Antarctica. Soon after, NASA researchers reported that it too had detected a dramatic loss of stratospheric ozone over all of Antarctica. Subsequently, this ozone depleted zone was referred to as the "ozone hole."

In the years following the discovery of the ozone hole, NASA satellites recorded increasing depletion of the ozone layer with each passing year (Fig. 11.6). In response, 43 nations signed the Montreal Protocol in 1987 in which they agreed to reduce the use of CFCs by 50% by the year 2000. By the end of 1998, production of CFCs had fallen by 95% in industrialized countries. Since 2000, other countries have signed newer versions of the Montreal Protocol and CFC production continues to decrease worldwide.

The term *visible light* is generally used to describe a relatively small region of the continuous electromagnetic spectrum to which the average light-adapted human eye is sensitive. The range of visible light, determined by the photochemical properties of the human visual pigments, extends from about 380 nm to about 770 nm wavelengths and is represented by the International Commission on Illumination (abbreviated as CIE from its French title Commission Internationale de l'Eclairage) photopic curve (Fig. 11.7). In biological science, the term *light* is generally used to describe a wider region of the radiant energy spectrum whose wavelengths possess sufficient energy to alter the outer electronic energy levels of atoms, but not to ionize those atoms.

Photochemical Reactions

The ability of irradiance to affect biological organisms (such as plants) is a result of its effect on photochemical reactions. The fundamental nature of photochemical reactions indicates that light must be absorbed by a specific photoreceptive molecule (photoreceptor) whose chemical reactivity is thereby changed. Einstein's law of photochemical equivalence states that in the primary photochemical reaction, one quantum must be absorbed for every atom or molecule reacting for photoactivation to occur. One gram of molecule of an absorbing substance requires one Einstein (μmol) of light for the reaction to proceed. By virtue of its changed chemical reactivity, the photoreceptor then initiates a sequence of metabolic processes that lead ultimately to observed developmental and photosynthetic changes. The absorption of quanta can lead to a change in the energy levels of the outer electrons and thus to a photochemical reaction.

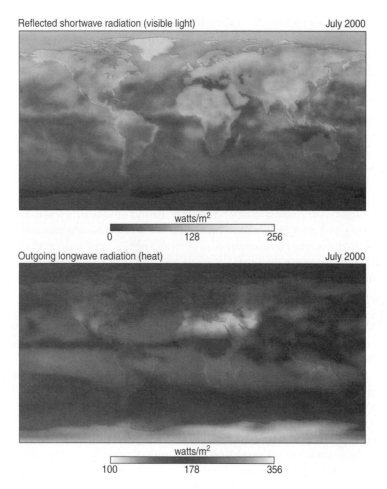

Reflected shortwave radiation (visible light)   July 2000

watts/m^2

0   128   256

Outgoing longwave radiation (heat)   July 2000

watts/m^2

100   178   356

FIGURE 11.4 The average amount of sunlight (in watts per square meter) that was reflected from the earth back into space during a day in July 2000 (top). White pixels show where more sunlight (or reflected short-wave radiation) is escaping the top of the atmosphere. Green pixels show intermediate values, and blue pixels show the lowest values. Clouds and snow-covered surfaces are highly reflective, whereas the ocean strongly absorbs sunlight (Antarctica appears dark because it is night time there in the month of July). The average amount of heat (in watts per square meter) that was emitted from the earth back into space during a day in July 2000 (bottom). Yellow pixels show where more heat (or outgoing long-wave radiation) is escaping the top of the atmosphere. Purple and blue pixels show intermediate values and white pixels show the lowest values. Desert regions and other areas experiencing heat waves (e.g., southwestern United States) emit a lot of heat, whereas the cold cloud tops of the thunderheads along the equator and the snow- and ice-covered continent of Antarctica emit very little heat.
Source: Earth Observatory, NASA.

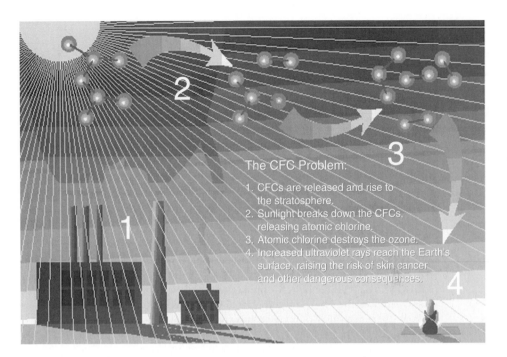

FIGURE 11.5 Diagram illustrating how chlorofluorocarbons (CFCs) are harmful to the ozone layer. CFC molecules are broken down in the stratosphere to ozone destroying atomic chlorine.
Source: Food and Drug Administration, USDA.

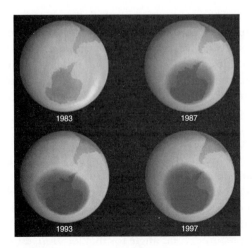

FIGURE 11.6 The progression of the ozone "hole" in four different years. With each passing year since the early 1980s, ozone concentrations over the South Pole have grown less during the months of September and October.
Source: Earth Observatory, NASA.

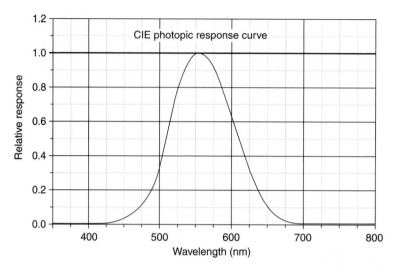

FIGURE 11.7   Visible light for humans is a narrow portion of the electromagnetic spectrum and is represented by the International Commission on Illumination (abbreviated as CIE from its French title Commission Internationale de l'Eclairage) photopic curve.

Variations in the Light Environment

The light environment can vary as a result of geographical location (latitude and physiography) and seasonal effects (day length and intensity).

Solar Zenith Angle and Latitude   The sun's rays are essentially parallel to each other when they strike the earth. This is due to the great distance between the earth and the sun. Because the earth is a sphere, the parallel rays of the sun strike the earth at different angles at different latitudes, resulting in seasons. The tilt of the earth's axis to the plane of its orbit around the sun is about 23.5°. As a result the solar zenith angle (the angle of sun away from vertical) of the sun at the equator at midday is never more than 23.5°. Consequently locations near to the equator see little or no seasonal variations in the climate. Seasonal variation increases with latitude, and this is consistent with the fact that the poles experience 6 months of daylight and 6 months of darkness (although there is a lot of twilight to give partial lighting).

   The daily rotation of the earth and the orbit of the earth around the sun cause the solar zenith angle to be continually changing at any single position on the earth. The solar zenith angle is 0° (or perpendicular to the earth's surface) at sunrise and sunset and about 90° when the noon sun is directly overhead (solar noon). At high solar zenith angles, such as local noon (the maximum solar angle for that location, time, and day), the sunlight passes through relatively little atmosphere, minimizing the scattering of light by the atmosphere and any effects of pollution, haze, or water vapor. At low solar zenith angles for the day (such as after dawn or near dusk), atmospheric scattering is increased, decreasing the amount of shorter wavelength light (blue) in incident sunlight.

   At latitudes where the noon sun is not directly overhead (such as at the poles), the rays are oblique. Oblique rays occur because of the spherical shape of the earth and spread over a larger surface area and pass through a greater depth of atmosphere. The resulting energy from oblique rays to a position on earth is less intense. The further a latitude is from the latitude at which noon sun is directly overhead, the greater the obliquity of the sun's rays and the weaker the intensity.

The maximum solar zenith angle for a location and day depends on the latitude on the earth and the position of the earth in its orbit around the sun. Between latitude 23.5° N (Tropic of Cancer) and 23.5° S (Tropic of Capricorn) the maximum solar zenith angle at midday is 90° at any single location on 2 days (summer solstice and winter solstice) each year. At latitudes greater than 23.5°, the maximum solar zenith angle is always less than 90°.

As the northern midsummer solstice (approximately June 21) approaches, the inclination of the earth's axis ensures that the Northern Hemisphere is tilted increasingly toward the sun. The North Pole is inclined toward the sun 23.5°; therefore, the sun's rays are shifted northward by 23.5° and noon rays are perpendicular at the Tropic of Cancer (23.5° N). During summer solstice, all latitudes north of the Arctic Circle are constant light and latitudes from the Antarctic Circle to the South Pole are constant night. All latitudes except the equator are cut unequally by the circle of illumination. Those in the Northern Hemisphere have larger parts of the circumference toward the sun, so the days are longer than the nights. Photoperiods generally increase with increasing latitudes (i.e., photoperiods for locations lengthen the more poleward from the equator the locations are). Longer days and more perpendicular rays of the sun result in maximum receipt of solar energy (insolation) in the Northern Hemisphere during the summer. The sun's rays strike the Southern Hemisphere more obliquely and therefore less effectively, explaining that winter then occurs in the Southern Hemisphere while the Northern Hemisphere experiences summer conditions. The angle of the sun in the tropics varies little with consistent photoperiod and continuously high year-round temperatures.

The South Pole is tilted toward the sun 23.5° during the northern midwinter solstice (approximately December 21). In this situation, nights are greater than days and the sun's rays are more oblique and must pass through more atmosphere in the Northern Hemisphere and winter occurs. This affects both light levels and temperature. The duration of solar radiation changes seasonally due to changes in day length.

Twice during the year, during vernal equinox and autumnal equinox, the sun's noon rays are perpendicular to the equator and the circle of illumination passes through both poles and cuts all latitudes exactly in half (i.e., day length = night length). This occurs approximately on March 21 (vernal equinox) and September 23 (autumnal equinox).

Atmospheric Transmissivity and Physiography   The atmosphere is only semi-transparent to solar radiation due to the reflectance and absorption by the gases and suspended particles in the atmosphere. Light scattering caused by gas molecules or large particles results in substantial decreases in the amount of radiation reaching the earth. The loss of radiation caused by water in the atmosphere can be especially great, particularly when the water is aggregated into fog or clouds.

The scattering of solar radiation in the atmosphere has other important consequences other than a loss of radiation. Radiation reaching the earth's surface due to

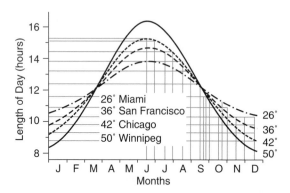

FIGURE 11.8 The general influence of time of year and latitude of representative cities on photoperiod.

atmospheric scattering appears to originate from angles in the sky other than directly from the sun. The sky radiation is the diffuse component of the radiation at the earth's surface. Diffuse radiation as a consequence is often more efficiently intercepted and used by plant canopies than direct radiation.

Atmospheric radiation scattering also influences the quality or color of the radiation reaching the earth's surface. The sky under clear conditions appears blue because the shorter wavelengths (e.g., the blue wavelengths of the visible spectrum) are scattered to a greater degree than longer wavelengths. The sun appears orange or red at lower sun angles because of the loss of shorter wavelengths from the direct radiation. The presence of dust, pollutants, or fog reduces the intensity of and/or changes the spectral distribution of global radiation in specific situations. Also, as the elevation increases, light rays pass through less atmosphere and, as a result, the ultraviolet light is a larger fraction of the incoming radiation.

Photoperiod Due to the tilt of the earth's axis (approximately 23.5°) and its travel around the sun, the length of the light period (also called photoperiod or day length) varies according to the time of the year and latitude (Fig. 11.8). Photoperiod varies from a nearly uniform 12-hour day at the equator (0° latitude) to continuous light or darkness throughout the 24 hours for a part of the year at the poles (180° north and south latitude).

Surface Intensity The amount of light received by a location on earth changes as a consequence of season (day length and solar intensity) and cloud cover. Because atmospheric gases filter some parts of the spectrum more effectively than others, seasonal and even daily differences also occur in the distribution of wavelengths. Seasonal differences in light intensity occur because radiation passes through a thicker layer of atmosphere in winter than in summer.

Solar radiation for December in the northern section of the United States is about 3 to 4 kWh m^{-2} day^{-1}, whereas for June it is 6 to 7 kWh m^{-2} day^{-1} (Fig. 11.9).

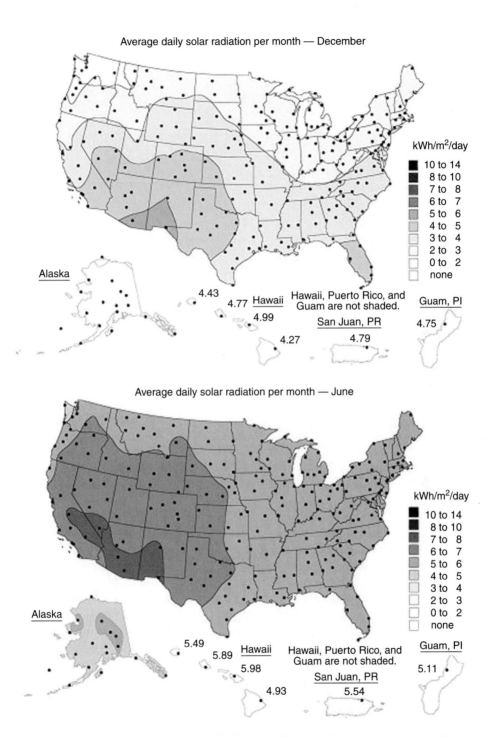

FIGURE 11.9 Maps of average daily global solar radiation during a representative winter month, December (top map), and a representative summer month, June (bottom map), on a south-facing flat surface.

Source: Office of Energy Efficiency and Renewable Energy, U.S. Department of Energy.

FIGURE 11.10   General trends for annual mean percent sunshine in the United States. Source: Office of Satellite Data Processing and Distribution, National Oceanic and Atmospheric Administration.

Atmospheric turbidity (smoke, humidity, and dust) also influences radiation levels. The June solar radiation values in desert climates are about 7 to 8 kWh m^{-2} day^{-1}, whereas the more humid Great Lakes region has values of about 5 to 6 kWh m^{-2} day^{-1}. Also, the dry atmosphere of the southwestern United States has the highest percent sunshine in the United States, with lower levels in the more cloudy northeast and the upper Pacific coast (Fig. 11.10).

Globally, terrestrial radiation gradually decreases over land areas from latitudes of about 30° N and S toward the poles. This decline in radiation results in the seasonality experienced in temperate regions and poleward. From the equator to 30 N and S, radiation is generally high, with limited areas of very high radiation or very low radiation.

Near the poles, because of low solar input during the winter months, annual receipt of solar radiation is only 20% to 25% of that in the tropics. The highest annual input of solar energy is at the Tropic of Cancer and Tropic of Capricorn rather than at the equator. This is because the climate at the equator is generally cloudy and rainy compared to the higher latitudes. Deserts commonly exist at the latitudes of the Tropic of Cancer and Tropic of Capricorn because of the lack of clouds and humidity.

Radiation Measurements and Instrumentation

Accurate and proper measurement of the light environment is important in understanding the role of light in a particular situation. The flux (energy) can be measured by radiometry, spectroradiometry, pyranometry, and photometry, depending on the situation. Photosynthetically active radiation is of particular importance to understanding plant development and can be measured with either a radiometer or a spectroradiometer.

Radiometry, Spectroradiometry, and Pyranometry   Radiometry is the measurement of flux. A radiometer's output is the integration of spectral flux incident on the radiometer across a broad band of wavelengths. Typical units for the radiometer output include W m^{-2} (irradiance), W ster^{-1} m^{-2} (radiance), and W ster^{-1} (intensity). Spectroradiometry measures the radiant flux dependent on the wavelength of the radiation. A spectroradiometer is used to measure the wavelength dependence of radiation and typical units for spectroradiometry are W m^{-2} nm^{-1} or μmol m^{-2} nm^{-1}. A pyranometer is an instrument for measuring solar radiation. It is used for measuring global sun plus sky radiation.

Photometry   Photometry is the measurement of flux typically utilized by the human eye. An ideal photometer has a spectral response with maximum sensitivity to yellow and green (approximately 550 nm). Photometry is used to describe lighting conditions where the eye is the primary sensor, such as illumination of work areas, TV screens, and so on. It is not an appropriate measurement of irradiance for use in plant science. The output is the integration of the spectral flux incident on the photometer with the phototropic spectral response. Typical units for the photometer include lux (illuminance), foot-candle (illuminance), candela m^{-2} (luminance), and candela (luminous intensity).

Photosynthetically Active Radiation (PAR)   Photosynthetically active radiation (PAR) is the measurement of flux utilized by plants for photosynthesis. The waveband across which PAR is measured is between 400 and 700 nm. PAR is a general term that covers both photon and energy measurements. Photosynthetic photon flux (PPF) is the photon flux density of the PAR, and its units are μmol m^{-2} s^{-1} or μE m^{-2} s^{-1}. Photosynthetic irradiance (PI) is the radiant energy flux of PAR, and its unit is Wm^{-2}.

Conversion of Units   The conversion between outputs of the various light measurements is possible but complicated due to the different parameters of wavelengths measured and energy units used. The conversions are often different for each light source, and the spectral emission distribution curve of the radiant output of the source must be taken into consideration.

Effects on Plants

Seed Germination

Germination is the resumption of active plant embryo growth after a dormant period. In general, three conditions must be satisfied for a seed to germinate. First, the seed must be viable (the embryo must be alive and capable of germination). Second, the internal conditions of the seed must be favorable for germination (i.e., any physical, chemical, or physiological barriers to germination must have disappeared or been removed). And finally, the seed must be exposed to appropriate conditions, including water (moisture), temperature, oxygen, and, for some species, light (often, specifically red).

Light can stimulate or inhibit seed germination of some species. Seeds that require light for germination are often small and rich in fat, and usually from nondomesticated species. Most cultivated species have seeds that don't require light for germination, probably a result of breeding programs selecting against a light requirement. Some plants germinate in either light or dark conditions, and some plant species have even been reported to have seeds that respond to photoperiod. For cultivated crops, seed catalogs and seed packets often list germination and cultural information for particular plants, including light requirements.

The light promotion of seed germination is typically a red light/far-red light reversible phenomenon, implying the involvement of phytochrome (discussed in a later section). Typically light-sensitive seed will germinate when exposed to red light, but this stimulation can be negated by subsequent treatment with far-red light (and it is often the final exposure that determines germination).

This may suggest that light-controlled germination is an adaptation of plants that are intolerant of shade, because canopy shade is often high in far-red light as compared to red light (due to selective absorption of red versus far-red wavelengths by the leaf pigments), and light-sensitive seed often will not germinate under leaf canopies. This mechanism to control seed germination may also be used by plants to adjust the timing of germination to take advantage of seasonal "windows" that may occur in the canopy.

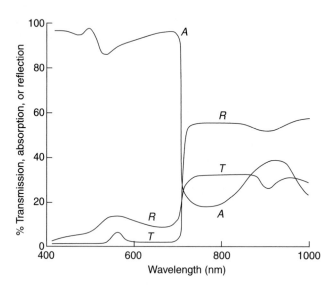

FIGURE 11.11   Generalized spectral characteristics of plant leaves between 400 and 1000 nm. Abbreviations: A, absorption; R, reflection; T, transmission.
Source: Fitter, A. H., and R. K. M. Hay. 1987. *Environmental physiology of plants*, 2nd ed. New York: Academic Press. Used with permission: Elsevier Science.

Uptake of Radiation by Plants

Plants can either reflect, absorb, or transmit solar radiation (Fig. 11.11), and the relative amounts will depend on the wavelength of the radiation, leaf structure, and leaf orientation. The capacity of a plant to reflect visible light depends on the leaf surface. Modifications such as hairs and leaf colorations often increase the reflection of solar energy from the leaf.

Leaves can reflect up to 70% of the infrared, 6% to 12% of the visible, and only 3% of the UV. Green light tends to reflect more strongly (10% to 20%) than the orange or red light (1% to 10%). UV radiation is greatly absorbed by epidermal waxes, cuticles and suberin, and phenolic compounds within the leaf. Infrared radiation (or heat) is readily absorbed by the plant. The chloroplast pigments determine the extent of visible light absorption. Transmission of light through a leaf depends on leaf structure and thickness. Thin leaves will transmit more light than thick leaves.

Ultraviolet Radiation

Ultraviolet radiation is nonionizing electromagnetic radiation of wavelengths just shorter than those normally perceived by the human eye. Electromagnetic radiation from the sun contains about 7% UV radiation at sea level. UV radiation is often categorized in three basic wavebands: UV-A, 315 to 400 nm; UV-B, 280 to 315 nm; and UV-C, shorter than 280 nm. Because of efficient absorption of shorter wavelengths by the ozone layer in the upper atmosphere, most of the sunlight striking the earth's surface is composed of wavelengths longer than 295 nm. Longer UV wavelengths such as UV-A and some of UV-B are attenuated through the atmosphere. The greatest distribution of UV-B at the earth's surface is at low latitudes and high altitudes (Fig. 11.12).

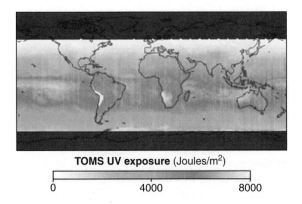

TOMS UV exposure (Joules/m^2)

0   4000   8000

FIGURE 11.12   Estimates of UV-B irradiance at the earth's surface based on the abundance of ozone, as measured by NASA's Total Ozone Mapping Spectrometer (TOMS) instrument during the month of November 2000.
Source: Earth Observatory, NASA, image by Reto Stöckli, based on data from the TOMS.

Because shorter wavelengths have more energy than longer wavelengths, UV radiation is particularly effective in causing photochemical reactions. As a result, UV radiation can be detrimental to biological organisms by increasing loss of DNA activity and accelerating mutation rates. Reduced growth can occur when plants are exposed to UV radiation as a result of inhibition of photosynthesis or reduction in leaf expansion. Plants can reduce damage from UV radiation using screening compounds such as flavonoids and flavones or cuticular waxes.

Plant Uses of Radiant Energy

Plants monitor the environment using specialized pigments that intercept and capture radiant energy and also perceive changes in the quality, quantity, duration, and direction of the light. For example, plants capture radiant energy between 400 and 700 nm using chlorophyll and other accessory pigments during the process of photosynthesis. Light energy is transformed through the process of photosynthesis into chemical energy for production of carbon metabolites.

Plants may adjust their growth through the process of photomorphogenesis to a form/shape or growth rate that might impart an ecological advantage over other plants or organisms. Colors of light important to photomorphogenesis are red (approximately 660 nm), far-red (approximately 730 nm), and blue (400 to 500 nm). Phytochrome is the primary pigment involved in photomorphogenesis, though others (such as cryptochrome) have been suggested or implicated.

Light Quantity   Light quantity or intensity is a major factor governing chloroplast development and the rate of photosynthesis, as well as other plant responses (Fig. 11.13). The quantity or amount of light received by plants in a particular location is affected by the intensity of the incident (incoming) light, length of the photoperiod, elevation, and latitude. The amount of sunlight also varies with season and time of day. Other factors that affect the intensity of sunlight include the presence and amount of clouds, dust, smoke, or fog.

Light intensity effects on chloroplast development and chlorophyll synthesis   The number of chloroplasts in plant cells varies with plant species and certain environmental conditions, predominantly light. In general, the number of chloroplasts per cell increases as the cells grow as long as sufficient light is available. Most of the increases in number of chloroplasts regulated by the amount of light are due to division of existing chloroplasts.

Chloroplasts may also change their location within the cells of the leaves to regulate the amount of light that is absorbed. Chloroplasts typically locate at cell surfaces that are parallel to the plane of the leaf under low light situations. This position maximizes absorption of light. Chloroplasts align along the radial walls of the cell that are parallel to incident light under high light conditions. Chloroplasts in this orientation often become shaded by other chloroplasts, reducing the amount of light absorbed and avoiding excessive absorption of light. The movement of organelles or entire organisms in response to light is referred to as phototaxis, and

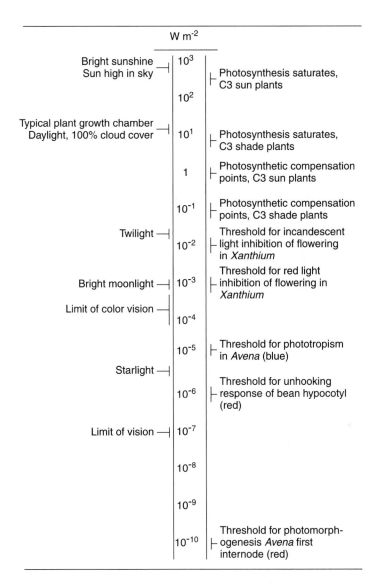

FIGURE 11.13 Variation in irradiance in the natural environment and plant responses to it. Source: Fitter, A. H., and R. K. M. Hay. 1987. *Environmental physiology of plants*, 2nd ed. New York: Academic Press. Used with permission: Elsevier Science.

many phototactic responses (such as orientation of chloroplasts) have action spectra typical of responses to blue light.

Chlorophyll synthesis in most plants is prevented when the plants germinate or grow in the dark or if they are transferred from light to darkness (some conifers can form chlorophyll in darkness). In the absence of chlorophyll, normal dark green plants appear yellow, primarily because many of the carotenoid pigments continue to be formed in the dark while the green chlorophylls do not. Plants grown in the dark exhibit etiolated growth, which is characterized by whitish-yellow, spindly stems with leaves that are not fully extended, internodes that exhibit extreme elongation, and poor root growth. Plastids present in dark-grown seedlings, also called etioplasts, can become chloroplasts upon exposure to light. Chlorophyll development often resumes and plants become green when exposed to light and resume active photosynthesis.

Light intensity effects on rate of photosynthesis Light intensity is an important factor in determining the plant photosynthetic rate (how fast photosynthesis is occurring). The light compensation point is the level of light at which photosynthetic uptake of CO_2 is equal to the amount of CO_2 released by respiration. In general, as the light intensity increases above the light compensation point, photosynthesis rates also increase in a linear relationship with light intensity until a certain maximum or saturating level is reached (Fig. 11.14).

The light saturation point is the point above which further increases in light intensity do not result in an increase in photosynthetic rate. Above the light saturation point, the light-dependent reactions are producing more ATP and NADPH than can be used by the light-independent CO_2 fixation reactions of photosynthesis. The slope of the linear phase of the photosynthetic response curve is a measure of photosynthetic efficiency (how efficiently solar energy is converted into chemical energy).

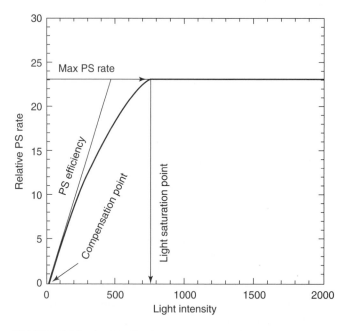

FIGURE 11.14 An idealized representative photosynthetic response curve. As light intensity increases, a light saturation point is reached where the plant is at its maximum photosynthetic rate. The slope of the linear phase of the response curve is a measure of photosynthetic efficiency—how efficiently solar energy is converted into chemical energy.

Photosynthesis and C3 versus C4 Plants   The two broad categories of plants in regard to their net photosynthetic rate are C3 plants and C4 plants. C3 plants fix carbon from ribulose bisphosphate to the 3-carbon acid PGA (phosphoglycerate) using the enzyme RuBP carboxylase-oxygenase (rubisco). C4 plants fix CO_2 to PEP (phosphoenolpyruvate) to yield 4-carbon acids (such as oxaloacetic acids, malic, and aspartic acid) using the enzyme PEP carboxylase (for more in-depth discussion refer to Chapter 6). C4 plants include tropical grasses, maize, sugar cane, and species of *Atriplex* and *Amaranthus* (two genus that include many fast-growing weeds).

C4 plants generally have higher rates of photosynthesis at lower light intensities than C3 plants (Fig. 11.15). C4 plants carry out both C3 and C4 photosynthesis, whereas C3 plants lack the C4 pathway. Plants capable of C4 photosynthesis often exhibit a more efficient form of photosynthesis. The higher light saturation points and lower light compensations points for C4 plants contribute to the ability C4 plants possess to increase the amount of CO_2 available to the Calvin-Benson cycle.

Photorespiration often lowers the apparent efficiency of CO_2 assimilation in C3 plants because the rate of photorespiration increases with temperature faster than gross photosynthesis. As a result, many C3 plants are nonproductive at relatively high temperatures and high light, whereas C4 plants (which don't undergo photorespiration) increase in productivity at higher temperatures and higher light levels. In CO_2-enriched

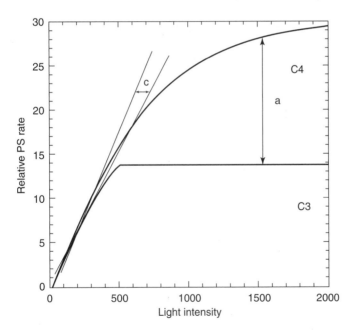

FIGURE 11.15   Idealized representative photosynthetic response curves for C3 versus C4 plants. Plants capable of C4 photosynthesis have a higher light saturation point (*a* illustrates the difference between the C4 and C3 saturation points), a lower light compensation point (not illustrated in this diagram), and a greater rate of photosynthetic efficiency (*c* illustrates the difference between C3 and C4 photosynthetic efficiency rates). These characteristics relate to the ability of C4 plants to increase the amount of CO_2 available to the Calvin-Benson cycle.

atmospheres (e.g., 1% CO_2), C3 photosynthesis becomes quite sensitive to temperature because elevated CO_2 tends to reduce photorespiration.

Effect of season on plant carbohydrates Leaf photosynthesis and carbohydrate production for annuals begins with early seedling growth. Photosynthesis and carbohydrates increase with bud break for most deciduous plants and trees, and with early spring growth for most nondeciduous perennials and evergreens. As the season progresses and the plants or trees grow, carbohydrates from leaf photosynthesis are utilized in the development of new plant parts. Leaf photosynthesis may be insufficient to meet demands of developing flowers and fruits during the reproductive growth phase of biennial and perennial plants and trees, and carbohydrate reserves from roots and shoots may be mobilized to growing points (and other sinks). Often as a result, stored carbohydrate levels as starch decrease from anthesis (flowering) to about 6 to 8 weeks after anthesis. After this, leaf photosynthesis resumes to supply the bulk of the carbohydrates. During fruit development, some carbohydrates may begin to go into starch reserves. After fruit maturity, almost all of the carbohydrates synthesized by fruit trees go into starch. These starch reserves increase until leaves abscise in the fall. The plant uses these reserves throughout the dormant season primarily for respiration.

Effects of light deficits (shade) Because most plants cannot relocate in response to deficient light environments, they are required to adapt to the light intensities in that location (Table 11.2). Light intensities below the compensation point can lead to

TABLE 11.2 *Some Plant Responses to Shade*

Physiological Process	Response to Shade
Extension growth	Accelerated
Internode elongation	Rapidly accelerated
Stem	Thinner and less weight
Petiole elongation	Rapidly accelerated
Leaf development	Changed
Area per leaf	Increased
Leaf thickness	Reduced
Chloroplast development	Retarded
Chloroplast synthesis	Retarded
Chlorophyll a/b ratio	Changed
Apical dominance	Strengthened
Branching	Inhibited
Flowering	Accelerated
Rate of flowering	Increased
Seed set	Reduced
Fruit development	Truncated

plant starvation ("exhaustion") with carbohydrate reserves used as substrates for respiration being depleted. Plant adaptations to shade include increased leaf area (often at the expense of roots) and a decrease in the amount of transmitted and reflected light. Shade leaves are generally thinner but larger in surface area than sun leaves. The increase in potential light absorption area results in an increase in number of chloroplasts/leaf area, increased chlorophyll content of the chloroplasts, and lower concentration of other pigments that may interfere with the light absorption process. Also a change in location and orientation of the chloroplasts within the cells can occur so that chloroplasts locate to cell surfaces parallel to the plane of the leaf with the broad dimension of the chloroplast orientated toward the light. As a result, shade leaves often are more efficient in harvesting sunlight at low light levels (the slope of the line observed under low light conditions is a measure of photosynthetic efficiency), but sun leaves have a higher maximum rate of photosynthesis (Fig. 11.16).

Although some plants do not grow well in low light, numerous others thrive under these conditions and are often referred to as shade plants. The actual amount of shade in a particular location may determine ultimately which plants will grow successfully. As with shade leaves, shade plants are capable of greater photosynthetic rates in lower light intensities. Sun (or high light-preferring) plants have leaves with higher rates of dark respiration (therefore, more light is needed to

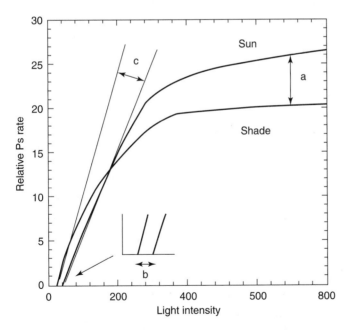

FIGURE 11.16 Idealized representative photosynthetic response curves for sun versus shade leaves. A sun leaf may have a higher light saturation point (*a* illustrates the difference between the sun leaf and shade leaf saturation points), but a lower light compensation point (*b* illustrates the difference between the sun leaf and shade leaf light compensation points). The shade leaf has a greater rate of photosynthetic efficiency (*c* illustrates the difference between the sun and shade leaf photosynthetic efficiency rates) and is often more efficient in harvesting sunlight at low light levels than a sun leaf.

reach the compensation point). Sun plants have light compensation points that range from 10 to 20 mmol m^{-2} s^{-1}, and shade plants have light compensation points that range from 1 to 5 mmol m^{-2} s^{-1}. Light saturation is achieved in shade plants at approximately 5% of maximum daylight, whereas sun plants continue to respond linearly to high light intensities to the light saturation point. Low respiratory rates appear to be a basic adaptation that allows shade plants to survive in light-limited environments.

Landscape plants typically change their degree of shade over time. As trees and shrubs mature, the landscape under or near them receives greater shade. Plants growing in the shade may also need to compete with the shading trees or shrubs for nutrients and water. Shallow-rooted trees such as maples and willows are particularly troublesome to plants growing in the shade. In general, roots competing for limited surface water may cause shaded areas to dry out more quickly than sunny sites during extended dry periods. As an adaptation, some shade-tolerant plants are adapted to low moisture situations, whereas others require moist situations in the shade.

Effects of bright light   High light can cause photoinhibition of photosynthesis and destruction of chlorophyll. Photoinhibition results when excess quanta are absorbed during the light reactions and released as heat or fluorescence. Shade plants have a limited capacity for adjustment to high light intensities and are usually injured ("bleached") or killed. Sun plants gradually increase their capacity for light-saturated photosynthesis.

Plants can develop resistance to high light stress by increasing the amount of light reflected from their leaves. This can be accomplished by the plant through increases in suberin, wax, cuticles, and cell wall materials. Photodestruction of chlorophyll can be reduced by increasing the content of carotenoids or xanthophylls. Heliotropism (or the movement of leaves in response to the moving direction of sunlight during the day) can reduce or increase the amount of light intercepted by a leaf depending on the leaf's orientation. A leaf orientation less than perpendicular to the rays of sun would reduce the amount of light intercepted, whereas a perpendicular leaf orientation would provide maximum light interception.

Leaf Area Index and Crop Growth Rate   The arrangement of leaves can contribute to determining the amount of interception of solar radiation by plants and the overall photosynthetic efficiency of a crop canopy. The fraction of photosynthetically active radiation (the proportion of the sunlight actually absorbed by plants versus the total available energy in those wavelengths of sunlight that a vegetation canopy can absorb) is a measure of the photosynthetic efficiency of plants in a large area (e.g., the United States) that can be estimated using satellite data and analysis (Fig. 11.17). The quantity of leaves of a crop is often expressed as the leaf area index (LAI). LAI provides information on the structure of the plant canopy by determining the amount of land surface area covered by green foliage. The LAI can be measured on a small plot of land as the

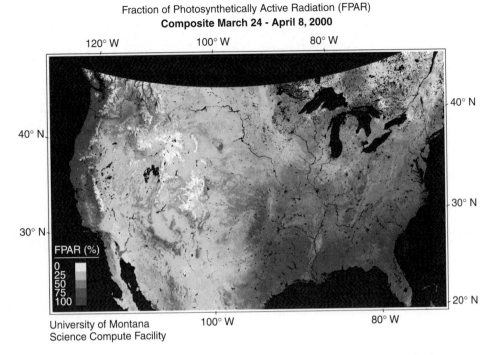

Fraction of Photosynthetically Active Radiation (FPAR)
Composite March 24 - April 8, 2000

University of Montana
Science Compute Facility

FIGURE 11.17 Sunlight absorbed by plants in the United States. This composite image, produced with data acquired by the Moderate-resolution Imaging Spectroradiometer (MODIS) during the period March 24–April 8, 2000, is a map of the photosynthetically available radiation that was absorbed by land vegetation for photosynthesis. In this image, dark red pixels show where land plants are absorbing 100% of the photosynthetically available sunlight, pink pixels show where plants are absorbing anywhere from 25% to 75% of the sunlight, and tan pixels show zero.
Source: Earth Observatory, NASA.

amount of leaf area per ground area at any point in time or it can be estimated over a larger area using satellite data and analysis (Fig. 11.18). Satellite estimated LAI is produced by radiometrically measuring the visible and near infrared energy reflected by vegetation. The high and low satellite data values for locations in the United States for photosynthetic efficiency strongly correlate with high and low values for LAI.

LAI is affected by pruning, population density, development (age), and plant spacing. At low LAI, typically much of solar radiation strikes bare soil and is not used by the plant. At optimum LAI, the highest rate of assimilation per unit area of ground is achieved. At greater than optimum LAI, the assimilation per unit area of ground is lower because some leaves are in deep shade and respire more than they photosynthate. The net assimilation rate (NAR) is the rate of increase in plant weight per amount of leaf area. The crop growth rate (CGR) is defined as the rate of increase in plant weight/unit area of ground or the NAR × LAI.

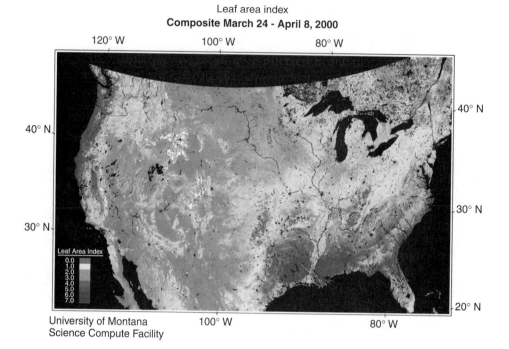

Leaf area index
Composite March 24 - April 8, 2000

University of Montana
Science Compute Facility

FIGURE 11.18   Total U.S. leaf area. This composite image, produced with data acquired by the Moderate-resolution Imaging Spectroradiometer (MODIS) during the period March 24–April 8, 2000, is a map of the density of the plant canopy covering the ground. In this image, dark green pixels indicate areas where more than 80% of the land surface is covered by green vegetation, light green pixels show where leaves cover about 10% to 50% of the land surface, and brown pixels show virtually no leaf coverage.
Source: Earth Observatory, NASA.

For many crops in high-yielding agricultural situations, the LAI is often 4 to 6 (i.e., 4 to 6 m^2 leaf area per m^2 of ground). Most agricultural crops are grown as annuals and do not form a solid canopy until 6 weeks or more after emergence. They also lose lower leaves as the crops approach maturity. In forest and natural ecosystems, the LAI is often much higher. In any situation, LAI values seldom exceed 10 because lower leaves usually senesce in the very low light levels at the bottom of the canopy when new leaves are added above them.

Light Quality   Sunlight is considered as white light and is composed of all the colors of the visible portion of the electromagnetic spectrum. Color refers to the relative distribution of wavelengths from a radiation or reflective source. Each color has its own distinctive wavelength distribution.

Photomorphogenesis   Photomorphogenesis is defined as the ability of light to regulate plant growth and development, independent of photosynthesis. The wavelengths

FIGURE 11.19   The effect of phytochrome regulation on leaf color appearance (chlorophyll and other plant pigments) as suggested by brief exposures to red and far-red light at the end of the daily photoperiod.

involved in generating photosynthesis are generally broader (400 to 700 nm, with peaks in red and blue) and less specific than those involved in photomorphogenesis. Plant processes that appear to be photomorphogenic include internode elongation, chlorophyll development (Fig. 11.19), flowering, abscission, lateral bud outgrowth, seed germination, dormancy, cold hardiness, and root and shoot growth.

Photomorphogenesis differs from photosynthesis in that the plant pigment primarily responsible for photomorphogenic growth responses is phytochrome. Phytochrome is a pigment that is in plants in very small amounts and exists in two photoreversible forms, the red form and the far-red form. The red (600 to 660 nm) and far-red (700 to 740 nm) wavelengths of the electromagnetic spectrum appear to be most important in the light-induced phytochrome-regulated growth of plants (Fig. 11.20). Also a series of well-documented plant responses have been attributed to radiation in the blue portion (400 to 500 nm) of the electromagnetic spectrum. Knowledge of the action or even isolation of the hypothesized blue light pigment ("cryptochrome") is not as advanced as it is for phytochrome. In addition, some of the plant's responsiveness to blue light may be attributed to perception and activation of phytochrome in these wavelengths.

Light quality perception in plants is a sequential process. Light is absorbed by the phytochrome photoreceptor, and the photoreceptor is transformed to the Pr or Pfr form (Fig. 11.21). A ratio of the two forms of phytochrome (Pr and Pfr) is established depending on the spectral characteristics of the light. A message is then perceived by the plant, which influences the balance of endogenous growth regulators and stimulates a plant growth response.

Photomorphogenesis is considered a low energy response, which means that it requires very little light energy to elicit a growth-regulating response, whereas photosynthesis is considered a high energy response. The amount of energy required to

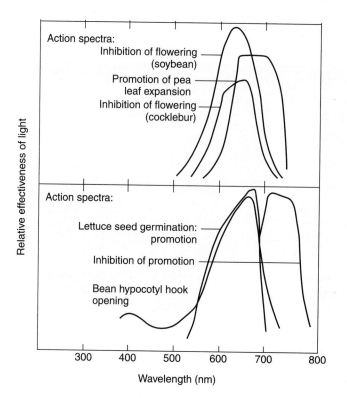

FIGURE 11.20 Action spectra for various physiological processes under phytochrome control.
Source: Modified from Salisbury, F. B., and C. W. Ross. 1978. *Plant physiology*, 2nd ed. Belmont, CA: Wadsworth. Used with permission: Thomson Learning.

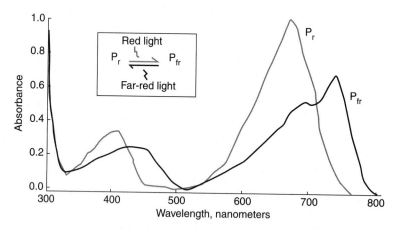

FIGURE 11.21 A generalized absorption spectrum of the two main forms of phytochrome: the red-absorbing form (Pr) and in the far-red-absorbing form (Pfr). The phytochrome pigment is synthesized in the Pr form. If it absorbs red light, it will change to the biologically active Pfr form (illustrated in box insert). The Pfr form can absorb red light and switch back to the Pr form.

FIGURE 11.22   Effect of photoperiod on strawberry in western North Carolina, under normal day (at left) and normal day plus 4, 7, and 14 hours (toward right) and 24 hours artificial light (at right). November 11 to January 31.
Source: National Agricultural Library, USDA.

elicit a photomorphogenic response is often 100x less than the amount of energy for photosynthesis to occur.

Photoperiodism   Photoperiodism refers to the growth responses of a plant to photoperiod (day length) (Fig. 11.22). An important response to day length in some plants is flowering. For example, short-day plants will flower only when the light period is shorter than some critical period. Long-day plants will only flower when the light period is longer than some critical period. Plants that are not affected by day length and flower under any light period are called day-neutral plants. Plants that exhibit photoperiodism actually are responding to the length of the night period (nyctoperiod) and not the length of the light period. This is illustrated in experiments in which the night period is interrupted with a flash of light and plants do not respond to the appropriate light period. Other plant responses that are under photoperiodic control in some species include dormancy of buds and cold hardiness, formation of storage organs, leaf development, stem elongation, and seed germination.

Photoperiodic responses enable plants to time vegetative and floral growth to "match" seasonal changes in the environment. Day length for areas outside of the tropics is often the most reliable indicator for predicting, and hence avoiding or resisting, potentially unfavorable conditions for plant growth. Most temperate zone plants exhibit photoperiodic responses, whereas many tropical plants do not. In general, day-neutral plants mature normally over a wide range of latitudes, whereas those with long-day or short-day adaptations are often restricted to specific latitudes.

Phototropism   Phototropism or phototaxis is the response of plants to directional light rays. For some plants growing in low light intensities, plant movements typically exhibit heliotropism (sun-induced leaf movements) to ensure interception of maximum amount of light (Fig. 11.23). Plants grown in high light may exhibit avoidance reactions and orientate their leaves so light interception is less than maximal.

Phototropism appears to be the result of differential growth due to changes in cell elongation. For example, the illuminated side of a stem has a reduced elongation rate as compared to the shaded side of a stem. This response appears to be controlled by a blue light receptor.

FIGURE 11.23   Light-induced organ movement is exemplified by the orientation of the leaves of these plants in a planting box. The plants oriented their leaves toward the direction where the light intensity was the greatest (the sunlight was directed coming downward from the upper-right-hand corner of the picture).

Distribution of Light in Plant Canopies

Solar radiation is attenuated by the plant habitat. A plant canopy has an overlapped or stacked arrangement of photosynthetic leaves, and the amount of photosynthetically active radiation received by a specific leaf usually depends on its location in the canopy and its leaf angle to incoming sunlight. Typically, the stacked arrangement of leaves effectively absorb most of the incident light, with leaves near the top of a canopy perceiving near maximum sunlight intensity and the lower leaves perceiving sunlight of a lowered intensity and an altered spectral composition. The amount of light absorbed by a leaf varies with its chlorophyll content, but is usually 90% of incident light. Thus, the second layer of leaves receives 10% of full sunlight, and the third layer only about 1%. Also, light of several distinct wavelength ranges is absorbed, reflected, or transmitted by the pigments of leaves, and the resulting wavelengths that strike underlying leaves can affect not only photosynthetic processes but also photomorphogenic processes.

Leaves at the top of the canopy are usually oriented at oblique (acute or obtuse) angles to incident light. At an oblique leaf angle, a given amount of light is distributed over a greater total leaf area of the plant than if all the leaves were at right angles to the direction of light. The light intensity per unit leaf area is therefore reduced, but the overall rate of photosynthesis is not affected, because incoming solar radiation is already superoptimal for the outer leaves. When leaves are at oblique angles to sunlight, more leaves can be accommodated in the uppermost layer of the canopy and more leaf area of the canopy will operate at optimal photosynthetic rates. Although the uppermost leaves are most efficient at utilizing full sunlight when at an oblique angle to the sun's rays, the lower leaves operate best in lower light intensities if the leaf area is at right angles to the light, capturing the greatest sunlight per leaf surface.

Most leaves selectively absorb blue and red wavelengths of light and transmit a relatively large proportion of far-red light. Leaf canopies, therefore, produce a radiative climate varying in both irradiance and spectral distribution (Fig. 11.24). Underlying plants in a canopy or in sunlight filtered through green leaves often use the phytochrome pigment to absorb light and modify plant growth through the process of photomorphogenesis.

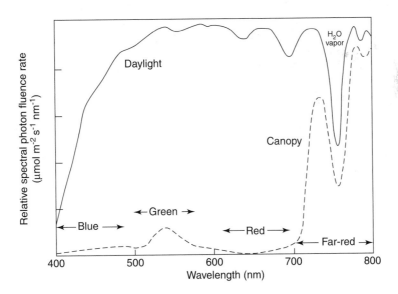

FIGURE 11.24   A generalized spectral photon distribution of sunlight above and within a vegetation canopy.
Source: Smith, H. 1994. Sensing the light environment: The functions of the phytochrome family. In *Photomorphogenesis in plants,* 2nd ed., R. E. Kendrick and G. H. M. Kroneneberg. Netherlands: Kluwer. p. 383. Used with permission: Kluwer.

The quantity and quality of light available to the plant in a planting or in competition with other plants depends on stand density, plant height, and leaf shape. More light will be available to lower leaves in a planting of plants with narrow leaves, such as grasses, as compared to plants with broad leaves.

Agricultural Technologies That Affect Light

Plant Density

The amount of light received by plants is affected by the plant population density (or spacing between neighboring plants). For many field- or nursery-grown crops, the spacing within and between rows of plants determines population density. Crop plants are seldom grown in agricultural situations in such wide plant spacing (or low population densities) that they do not interfere with each other, especially for light. Nursery crops and greenhouse-grown crops are also spaced so that a maximum number of plants are produced with minimal negative effects of competition from neighboring plants. When plants are grown close together, mutual shading may prevent direct light from reaching all but the tops of the plants, resulting in reduced plant growth or quality and altered branching habit. Plants grown in close plant spacings also tend to grow taller with a smaller plant diameter.

Crop spacing determines the resources available to each plant and whether the total resources are fully utilized. As the plant population increases, typically so does the dry matter production per growing land area until a maximum is reached, at which point further increases in population brings about little or no increase in production. Most crops have some optimum population for yield of their economically important part, and selecting the right population to produce the size of the marketable plant part typically required for a specific market is important. Often, the amount of space between planted rows that a grower uses is determined by the dimension and nature of the cultivation equipment and implements used in that system.

Weeds

A weed is any plant growing in the wrong place. Weeds often are detrimental to the growth of the desired crop in a weed–crop interaction in that they may compete and utilize resources, such as light, that could have been used by the crop. Weeds can restrict the planted crop's growth by increasing total (crop + weed) LAI and competition for resources. The extent of competition by weeds in a weed–crop situation is affected by crop species, crop cultivar, crop density, weed species, weed density, the relative time of emergence of the crop and weed, duration of the weed presence, the efficacy of weed control, and other environmental factors. Planted crops are often slower growing and more susceptible to competition than weed species. This may occur because the crop plants were selected for high-yielding capacity and not for their fast-growing ability to survive competition. For many agricultural crops, crop yields decrease as weed specie density increases, often irrespective of soil fertility. The use of weed control, either by mechanical means or through the use of herbicides (a chemical that kills unwanted plants), can reduce the weed populations in a planting. In some crop–weed combinations, the crop may exhibit some allelopathy on the weed specie, resulting in reduced presence or influence of the weed.

Row Orientation

In some situations, the orientation of the planted crop rows in a field can affect the amount of light the plants receive. In these situations, often the crops in east–west rows utilize light more efficiently than those planted in north–south rows due to the travel of the sun in the sky from east to west. This occurs because the plants in the north–south rows shade each other more than plants in the east–west rows. In many field situations though, the optimal direction of the row is usually governed by the prevailing slope of the land or by convenience.

Supplemental Lighting

Supplemental lighting has the potential of increasing yield or modifying plant growth. Supplemental lighting can be used to increase intensity, affect day length, and affect light quality, but it is often costly, requires access to electricity, and not all light sources are suitable to use to successfully grow plants. As a result, there has been limited work on supplemental lighting for field-grown or nursery crops, whereas supplemental lighting of greenhouses and controlled-environment situations is common.

Type of Lighting for Plant Growth   The three basic types of lights for plant growth include thermal radiators, fluorescent light, and multi-vapor lights (such as high-intensity discharge lamps).

Thermal radiators   A thermal radiator light source has a continuous spectral distribution of energy with respect to the absolute temperature of the source. Incandescent and halogen lamps are types of thermal radiators. An electric current is passed through a tungsten filament placed in a vacuum or in an inert gas such as nitrogen or argon causing the tungsten to glow. Tungsten acts as a black body of 3200° K. A considerable amount of infrared is emitted in addition to visible light.

The sun is a natural thermal radiator. The sun obtains its energy from nuclear fusion in which matter is transformed into energy that is emitted into space as electromagnetic radiation. The sun is the source of all visible light energy on the earth with a wavelength maximum energy per unit wavelengths at about 470 nm with a rapid decline on both sides of this peak. As the sun rays enter the atmosphere, much of the UV (300 to 400) is absorbed by the ozone layer and the infrared is absorbed by H_2O molecules. As the sun moves to the horizon, the solar energy travels through more atmosphere and, as a result, its total energy is lower, often with changes in light quality.

Fluorescent lamps   Ultraviolet radiation produced by a low pressure discharge of gases in fluorescent lamps (or tubes) is absorbed by a phosphor coating the inside of the tubes. It is then reemitted as fluorescents at longer wavelengths resulting in the emitting of light.

Discharge sources (often called HID)   Electric current flows between two electrodes through metal vapors or an ionized gas. Electrons emitted from a cathode strike the metal or gas atoms and excite an electron to a higher orbital level. After excitation, the electrons spontaneously return to their ground state in the form of fluorescence. Discharge sources includes arc lamps and discharges in metallic vapors. Radiant energy is composed of lines characteristic of the emission spectra of the elements in the discharges, superimposed on a thermal background from the hot gases and electrodes. The first discharge lamps were street lights. In theory fluorescent lamps are luminous discharge lamps.

Filtering Light

The emitted radiation from a light source (including the sun) often is not optimal for plant growth and development. This emitted radiation can be modified with the use of filters or shading materials. Filters can shade or decrease the energy uniformly over a specified wavelength range (neutral density) or can selectively suppress specific wavelengths (color filters).

Neutral Density Filters   Neutral density filters are often nets or meshes, which transmit only a certain percentage of incident radiation proportional to the percentage of open areas. The disadvantages of nets and meshes are that the light may not be uniformly transmitted and the degree of absorption is not predictable. Nets or meshes cannot usually be stacked to obtain higher absorption.

Color Filters (Wavelength Selective Filters)   Color filters transmit only certain wavelength bands. Before modern glass filters were available, solutions or suspensions of minerals such as chromium or copper salts or organic substances in water or gelatin were used. Copper sulfate solutions are often used to filter the infrared components of radiation.

Wavelength selective filters can be divided into three groups. Band filters absorb and transmit certain broad wavelength bands. Cut-off filters block short wavelength radiation below or above a certain wavelength. Interference filters transmit wavelength bands only a few nanometers wide.

Practical Uses of Shade

Shading is important in many plant production operations to reduce light and temperature and to lower moisture requirements. In permanent slat houses in nurseries, shading is determined by the spacing between slats and their thickness

FIGURE 11.25   Slat house in a nursery providing shade to plants.
Courtesy of Dr. Dave Beattie, Penn State University.

(Fig. 11.25). Various meshes also can be placed on a structure to reduce light intensity to the crop below it. Also day length may be shortened by shading with black, opaque cloth. This is a regular practice in the production of many floricultural crops in greenhouses to create short days. Shading is also used to inhibit pigment formation in which lack of color or blanching is a preference by the consumer. Asparagus or celery is blanched by mounding the base of the crop to produce white stalks.

Planting Considerations in the Shade Environment

Some plants grow well in the shade and appear to be adapted to these relatively low light environments. The following are some generalizations on the response of various groupings of plants when they are grown in the shade.

Perennials   Many herbaceous perennials bloom reliably in light shade, but some will blossom in fairly dense shade. Most of these are woodland plants that usually blossom very early in the season, though there are some exceptions. Most woodland flowers are subtlety colored.

Unlike the annuals, which tend to bloom throughout the growing season, most perennials only flower for a few weeks. When not in bloom, though, their foliage is still important in the shade garden by adding variety in form and texture as well as in shades of green. Flowers often are followed by seed pods or bright berries. Some perennials, such as hosta lilies, usually are planted for their attractive leaves rather than for their flowers, which in most species are not particularly colorful or showy.

Bulbs   Spring-flowering bulbs can be planted in deep shade if they are grown as annuals (planting new bulbs each fall and then digging them up and discarding them once they've bloomed) or as perennials in light shade. Fresh bulbs have miniature flowers inside the bulb. A cold winter in the ground is required for root growth and for those flower buds to emerge in spring. To repeat the performance the following year, though, the leaves must receive full sunlight for most of the day until they die back naturally. This builds up food reserves for the next blooming cycle.

Some spring bulbs such as crocus, scillas, snowdrops, and tulips bloom and produce leaves early in the season, before surrounding trees leaf out, so that they receive adequate amounts of sun to blossom annually in a lightly shaded area.

Annuals   Almost all crops grow best in sunny locations. Most do best in bright sunlight from early morning to nightfall, but a few can be grown in partial shade. These include plants that are grown for greens rather than for fruits or roots.

Annual crops such as leaf lettuce, spinach, Swiss chard, kale, mustard greens, and beet greens will be thinner leaved and less robust when grown in light shade rather than full sunlight, but this does not appear to affect their taste. Cool-season salad vegetables such as lettuce, spinach, and radishes often benefit from light shading through the heat of the summer. Beans, beets, broccoli, cabbage, kohlrabi, peas, potatoes, rhubarb and turnips will grow in light shade but not be as productive as plants growing in full sun. Currants and gooseberries tolerate medium shade and still produce a crop. Bramble fruits such as blackberries and raspberries grow in light shade, but yields will be reduced.

Turfgrass   Many home lawns have heavily shaded areas where high-quality turfgrass is not easy to establish or maintain. The competition for light, nutrients, and water can stress and weaken turfgrass. Because shaded areas also attract people during sunny, hot weather, this can also contribute to increased soil compaction and wear on the turf in the area.

Trees that develop a dense but shallow, fibrous root system, such as silver maple, are extremely competitive with turfgrass for moisture. Trees with heavy shade patterns such as some oaks, maples, and lindens, create a poor environment for turfgrass. Trees such as honeylocust and Golden Raintree that have an open canopy are more conducive to turfgrass growth and more light reaches the grass.

The shade canopy tends to moderate temperature fluctuations by lowering daytime temperature and keeping it warmer at night by reducing reradiative heat loss. This moderating effect on temperature means humidity remains high both day and night. The relatively constant temperature in the presence of high humidity and lower light encourages disease (probably due to free liquid on surfaces). Turfgrass growing in this environment can be susceptible to heat, drought, and disease. In addition, the reduced sunlight promotes turfgrass that is less able to recuperate from foot traffic, mower damage, or plant pests.

Turfgrass growing in shade generally requires less total nitrogen than grass in full sunlight. Heavy applications of nitrogen on shaded grasses can cause stress and disease. Late fall fertilization of cool-season grasses is extremely beneficial in shaded environments. This is the only time of the year when the grass plants under the trees can efficiently utilize the applied nitrogen without competition from trees for moisture, nutrients, and light.

Pruning and Training

Pruning, the judicious removal of plant parts, is a means of reducing competition between the parts of a single plant or among neighboring plants. Thus, the optimum population or canopy shape for production of some crops can be maintained by pruning. Proper pruning techniques also can control the balance between vegetative and reproductive growth of a plant.

Training is orientating or directing plant growth in a distinct way in space. This can be done to help maximize plant growth by maximizing light penetration and the overall photosynthetic rate of the plant. Training is particularly important in optimizing fruit load in fruit trees and in the desired configuration of espaliers and topiaries (pruning and training are covered more in depth in Chapter 15).

Reflective Plastic Mulches

Plastic (polyethylene) mulches are commonly used in the production of several vegetable and fruit crops. Reported benefits with the use of plastic mulches include earlier yields, better fruit quality, and greater total yields. In general, these responses have been attributed to enhanced soil warming, more efficient use of water and fertilizers, and better weed control.

A new generation of selectively permeable and wavelength selective reflective mulches has been recently developed. Selectively permeable mulches allow certain wavelengths of light to pass through the mulch to warm the soil without encouraging weed growth. Wavelength selective or colored mulches reflect wavelengths of light into the plant canopy to influence plant growth and production or insect pressures (Fig. 11.26).

FIGURE 11.26   Reflective colored mulch being evaluated for its ability to enhance growth and increase yields.

Summary

Irradiance is electromagnetic energy from a portion of the spectrum (primarily the visible and neighboring wavelengths) that has characteristics of wavelengths and packets of energy (quanta). The ozone layer and atmospheric gases filter sunlight as it passes through the atmosphere, and as a result only 47% of radiation emanating from the sun reaches the earth's surface. The light energy received at a particular location on the earth is further affected by its geographical location (latitude and physiography) and the season of the year (day length and intensity). Accurate and proper measurement of the light environment can be accomplished by radiometry, spectroradiometry, pyranometry, and photometry. Photosynthetically active radiation is the measurement of flux utilized by plants for photosynthesis (generally energy of wavelengths between 400 and 700 nm).

Seed germination, depending on species, can be affected by light. The seeds of some plant species require light for germination and the light promotion of germination is often under phytochrome regulation (with the wavelengths of red and far-red exhibiting strong influence). Light can be reflected, absorbed, or transmitted by leaves. Canopy shade is often characteristically high in far-red light compared to red light. Plants may adjust their germination rate and growth when in the shade of a canopy through the process of photomorphogenesis to a rate or form that might impart an advantage over other organisms or in an attempt to survive the shaded environment. Light intensity exerts an effect on plant chloroplast development and rate of photosynthesis. In general, there are two broad categories of plants according to their net photosynthetic rates: C3 plants and C4 plants. C4 plants (which fix carbon during photosynthesis into 4-carbon intermediate) generally have higher rates of net photosynthesis and are more productive at higher light levels than C3 plants (which fix carbon during photosynthesis into a 3-carbon intermediate).

Plant light environments below the compensation point can lead to depletion of carbohydrate reserves in the plant ("starvation") and plant death. Plant adaptations to shade include increased leaf area at the expense of roots, decreased amount of transmitted and reflected light, thinner leaves, and increased number of chloroplasts per leaf area and chlorophyll content of the chloroplasts. Shade plants have adaptations to survive in low light environments, including having higher photosynthetic rates with lower light intensities and lower rates of dark respiration. High light can cause photoinhibition of photosynthesis and destruction of chlorophyll. Sun plants tend to do well in high light environments and gradually increase their capacity for light-saturated photosynthesis.

Photomorphogenesis is the ability of light to regulate plant development, independent of photosynthesis. Plant processes that appear to be photomorphogenic include elongation, chlorophyll development, flowering, abscission, lateral bud growth, seed germination, dormancy, cold hardiness, and root and shoot growth. Phytochrome and cryptochrome appear to be the primary pigments involved in photomorphogenesis. Photoperiodism is the growth response of a plant to photoperiod. Plant responses that are photoperiodic include flowering, cold hardiness, formation of storage organs, stem elongation, and seed germination. Phototropism is the response of plants to directional light rays. Heliotropism (or the movement of leaves to follow the direction of sunlight) is a phototropic response.

Crop spacing determines the resources available to each plant and whether the total resources are fully used. Most crops have some optimum plant population density for maximum yield. Mutual shading from closely planted crops may prevent light from reaching all plants to an optimum level and growth and production may be reduced. Weeds can also reduce plant production by increasing competition for resources. For many agricultural crops, crop yields decrease as weed specie density increases. In some situations, the orientation of the planted row can affect the amount of light the plant receives, with crops in east–west rows utilizing light more efficiently than those planted in north–south rows. Supplemental lighting can be used in some situations to increase intensity, affect day length, and add light quality, but it is usually restricted to greenhouses and controlled environment situations. If the emitted radiation from a light source is not optimal for plant growth and development, filters can decrease energy either uniformly over a specified wavelength range (neutral density filters) or selectively suppress specific wavelengths (colored filters). Pruning is a means of reducing competition between parts of a single plant or among neighboring plants, and training is orientating plant growth in a specific way, potentially to maximize light penetration through the canopy. Mulches have been developed to enhance soil warming and influence light reflection back into the canopy to influence plant growth and production or insect pressures.

Review Questions

1. What is the approximate surface temperature of the sun?
2. The visible spectrum of light falls between what range?
   a. 200 to 1200 nm
   b. 380 to 770 nm
   c. All radiation is visible light
   d. 700 to 1100 nm
3. Define the following as they pertain to the propagation and description of radiation:
   a. Wavelength
   b. Frequency
   c. Wave number
4. Explain why the light meter on an old camera might not be an appropriate instrument for measuring light in plant research.
5. Why is UV light generally considered dangerous to biological organisms?
6. Define phototropism.
7. Which is the best equation that describes photosynthetic efficiency?
   a. Gross photosynthesis—respiration
   b. Light phase/dark phase $\times$ 100
   c. Photosynthetic rate/temperature
   d. Photosynthetic rate/(light respiration—dark respiration)
8. What are some disadvantages of the following light sources for growing plants?
   a. Incandescent bulb
   b. Fluorescent tube
   c. High intensity discharge lamp (such as a metal halide)

9. List three differences between C3 and C4 plants.
10. True (T) or False (F)
    a. _____ Solar energy is the ultimate source of free energy for virtually all life on this planet.
    b. _____ Photoperiodism is how plants respond to the length of the dark period.
    c. _____ Landscape plants typically change their degree of shade over time.
    d. _____ Row orientation in a field can affect the amount of light that plants receive in these rows.
11. Define the following as they pertain to photosynthesis:
    a. Light saturation point
    b. Light compensation point
12. What are some characteristics of shade plants?
13. What are some mechanisms that plants can develop to resist damage from high light stress?
14. Define photomorphogenesis. How does it differ from photosynthesis?
15. How is the optimum plant spacing for a crop grown in the field a balance between the plant's biological needs and economic reality of farming?
16. What are some effects of weeds in a planting on crop growth and production?
17. How does a neutral density light filter differ from a color filter?
18. What effect does latitude and altitude have on UV light?
19. What effect does the solar zenith angle have on the light available to plants at a location?
20. Where in the United States do we find the greatest mean percent sunshine (either region or states would be fine) and what weather factor about this general location may be responsible for this occurrence?
21. Why would plants evolve a mechanism to change the location of their chloroplasts?
22. Define heliotropism and discuss how it is used by plants to optimize their survival.
23. What is LAI and how can we use that information to grow plants optimally?
24. How can pruning affect light levels with the canopy of trees?
25. Which row orientation in general most efficiently utilizes light and why?
26. What are some general effects of the following colors of light on plant development?
    a. Far-red
    b. Red
    c. Blue

Selected References

Bickford, E. D., and S. Dunn. 1973. *Lighting for plant growth.* Kent, OH: Kent State University Press.

Bleasdale, J. 1973. *Plant physiology in relation to horticulture.* Westport, CT: AVI Publishing.

Cavigelli, M. A., S. R. Deming, L. K. Probyn, and R. R. Harwood, eds. 1998. *Michigan field crop ecology: Managing biological processes for productivity and environmental quality.* Michigan State University Extension Bulletin E-2646.

Fitter, A. H., and R. K. M. Hay. 1987. *Environmental physiology of plants,* 2nd ed. New York: Academic Press.

Galston, A. W., P. J. Davies, and R. L. Satter. 1980. *The life of the green plant,* 3rd ed. Englewood Cliffs, NJ: Prentice Hall.

Greving, A., D. Steinegger, D. Janssen, R. Gaussoin, and S. Rodie. 1997. *Landscapes for shade.* University of Nebraska Coop. Ext. Serv. Publ. G97–1341-A.

Hale, M. G., and D. M. Orcutt. 1987. *The Physiology of Plants under Stress.* New York: John Wiley.

Janick, J. 1979. *Horticultural science,* 3rd ed. San Francisco: W. H. Freeman.

LI-COR (anon.). 1982. *Radiation measurements and instrumentation.* Publication No. 8208-LM.

Smith, H. 1994. Sensing the light environment: The functions of the phytochrome family. In *Photomorphogenesis in plants, 2nd ed.,* R. E. Kendrick and G. H. M. Kroneneberg, pp. 377–416. Netherlands: Kluwer.

Nobel, P. S. 1991. *Physicochemical and environmental plant physiology.* New York: Academic Press.

Salisbury, F. B., and C. W. Ross. 1978. *Plant physiology,* 2nd ed. Belmont, CA: Wadsworth.

Schrock, D. S. 1998. *Gardening in the shade.* University of Missouri Coop. Ext. Serv. Ag. Publ. G06911.

Taiz, L., and E. Zeigler. 1998. *Plant physiology,* 2nd ed. Sunderland, MA: Sinauer.

Selected Internet Sites

www.cie.co.at/cie International Commission on Illumination (excellent source of procedures and publications on measuring light).

www.epa.gov/ Environmental Protection Agency (EPA).

www.epa.gov/oar/oarhome.html EPA Office of Air and Radiation.

www.nrcs.usda.gov/ Natural Resources Conservation Service.

www.wcc.nrcs.usda.gov/wcc.html National Water and Climate Center, USDA.

www.nws.noaa.gov/ National Weather Service, National Oceanic and Atmospheric Administration.

http://virtual.clemson.edu/groups/hort/sctop/photomor/photo.htm Clemson University Photomorphogenesis Research Program.

12

Temperature

Matter is composed of molecules in motion. The motion of these molecules contributes to the energy of the molecule and is referred to as kinetic energy. Temperature is the measure of the average kinetic energy of molecules. A molecule may also possess potential energy (the possibility of that molecule acquiring kinetic energy).

This chapter begins with a discussion of how air temperature is affected by changes in altitude, season and latitude, and time of day. This is followed by information on frost and freeze conditions. The response of plants to temperature is discussed with emphasis on temperature coefficients of reactions, chilling and freeze responses, and high-temperature responses. The chapter concludes with an overview of production practices that can be used by crop producers to minimize or avert cold-temperature stresses and high-temperature stresses.

Background Information

Radiation and Heat

Radiation, energy in the form of either electromagnetic waves or discrete packets (quanta) (see Characteristics of Sunlight in Chapter 11), travels at the speed of light ($3 \times 10^8 \text{ m s}^{-1}$) in a vacuum and slightly slower in the atmosphere. As a result, the transport of radiation throughout the surface atmosphere is effectively instantaneous. The amount of radiant energy emitted by an object is a strong function of temperature. In addition, radiation absorbed by matter can result in an increase in temperature of that material.

Sensible heat is the energy gain or loss from molecules by transfer of kinetic energy to adjacent molecules. Conduction is the transfer of kinetic energy (heat) in a fixed media. If the medium is in fluid motion, the energy transfer is referred to as convection.

Temperature Changes with Altitude, Seasons and Latitude, and Time of Day

Altitude   Air temperature generally decreases with increasing altitude or height above sea level (Fig. 12.1). As a result, mountains and other high-elevation areas generally have lower surface temperatures than lower-elevation areas of the same latitude. There is an approximate reduction of about 10°C in average dry-air temperature for each 1000 m rise in elevation. The temperature reduction in

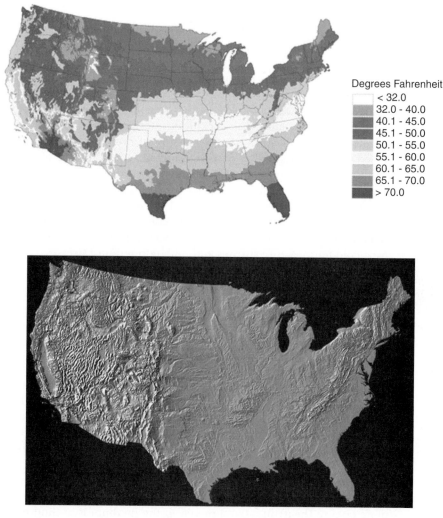

Degrees Fahrenheit
- < 32.0
- 32.0 - 40.0
- 40.1 - 45.0
- 45.1 - 50.0
- 50.1 - 55.0
- 55.1 - 60.0
- 60.1 - 65.0
- 65.1 - 70.0
- > 70.0

FIGURE 12.1   Annual mean daily average temperature (top) and a relief map of the United States (bottom). Conversion: °C = 5/9 (°F − 32).
Source: Office of Satellite Data Processing and Distribution, National Oceanic and Atmospheric Administration (top) and U.S. Geological Survey (bottom).

"moist" air is less, about 6°C for each 1000 m. The decrease in temperature with increasing altitude occurs because the surface of the earth is the source of heat, the result of absorbing radiation from the sun and atmosphere. As an object moves away from the surface, the distance from the source of the heat is increased and the temperature decreases. Solar energy intensity generally increases with elevation. This is due to the depth of the atmosphere decreasing with elevation, which also results in a decreasing amount of radiation absorbing gases.

Seasons and Latitude   The path, or orbit, that the earth follows as it moves around the sun is an ellipse. In addition to orbiting around the sun, the earth also spins (rotates) on its own axis, making one complete rotation in a 24-hour period. The earth's axis is tilted with respect to the plane in which it orbits. These planetary processes control the daily and seasonal variation in the reception of solar energy at locations on the earth's surface (Fig. 12.2), and exert a major influence on the relative lengths of day and night from location to location and from time to time.

   In most areas of the world (especially in the temperate climates), the calendar year is divided into the seasons of spring, summer, autumn (fall), and winter. The differentiation of the seasons in most areas is based on changes in temperature and light (for more information see Solar Zenith Angle and Latitude in Chapter 11). Summer in the United States typically has the warmest temperatures (and longest photoperiods), whereas winter has the coolest temperatures (and shortest

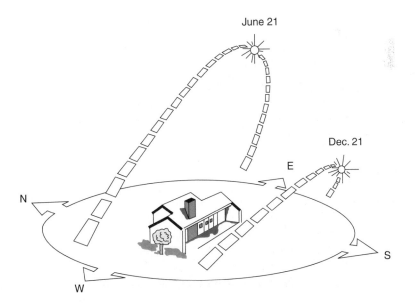

FIGURE 12.2   The sun's path across the horizon changes with the seasons as does day length. During summer the sun is higher on the horizon than during winter months. Source: NC State Cooperative Extension Service Horticultural Information Leaflet HIL #8554, Indoor Plant Selection and Care, by Douglas Bailey, 1999.

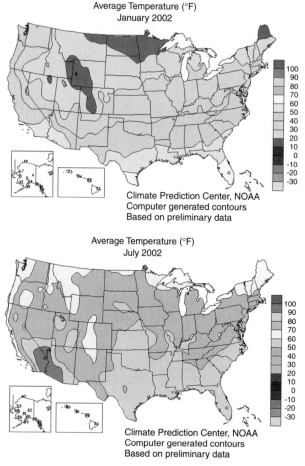

FIGURE 12.3   Average U.S. 2002 monthly temperatures for a representative winter month (January) and summer month (July). Conversion: °C = 5/9 (°F − 32).
Source: Climate Prediction Center, NOAA.

photoperiods) (Fig. 12.3). Spring and autumn have intermediate temperatures and photoperiods, with spring temperatures and photoperiods increasing during the year and autumn's decreasing. Seasonal effects on temperature and light are influenced by latitude. When one moves poleward from the equator (increases in latitude), temperatures decline, total solar radiation decreases (Fig. 12.4), and summer photoperiod increases. The further from the equator toward the poles the greater the seasonal variation.

The rate at which photoperiod changes varies during the year. Near the times of summer solstice and winter solstice, when photoperiods are longest and shortest, there is little change from day to day. During spring and autumn the rate of change is much more rapid as the photoperiods become longer during the spring and shorter during the autumn.

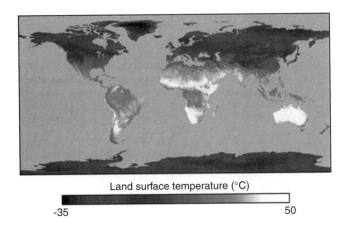

Land surface temperature (°C)

-35   50

FIGURE 12.4 Land surface temperature by location on the earth. The map shows average land surface temperature for December 2001 through February 2002. Dark areas represent the coldest temperatures; grays are progressively warmer. Light gray and white colors indicate the hottest areas on the surface of the earth.
Source: Earth Observatory, NASA (Image by Robert Simmon).

Time of Day Maximum day temperatures usually occur in the early afternoon, after maximum incoming radiation has occurred. During daylight, solar radiation is absorbed by surfaces on earth. Much of the solar radiation is converted to heat, which in turn increases surface temperatures above air temperatures. By conduction, this heat is transmitted to the air and the heated air rises and convection currents are formed as cooler air moves downward. This cooler air is heated by the surface and moves up resulting in a continuous cycle.

After sunset, incoming direct solar radiation does not occur, only outgoing radiation, and surfaces cool. The air at ground level becomes cooler than the air above it because cool air is dense and there are often no convection currents produced as there were during the day. If it is windy, then the wind will mix the air and create more homogenous temperatures in the lower atmosphere. If there is no wind, temperature inversions could occur. A temperature inversion is the inversion of the normal situation during the day (i.e., during inversions, temperature increases with increasing elevation). Conditions that promote maximum energy loss promote inversions: long nights; clear skies; cool, dry air; calm conditions (especially in winter).

Effect of Temperatures on Physical and Chemical Processes

Temperature controls many physical and chemical processes. Temperature determines the diffusion rates of gases and liquids. As temperature decreases, the viscosity of water and gases increases. Also the solubility of various substances is temperature dependent. When temperatures rise, oxygen solubility decreases resulting in fish kills in lakes and rivers and reduced availability of oxygen in hydroponic systems. The inverse is true of many solids. For example, solid inorganic fertilizers are more soluble in warm water than in cold water.

Frost Types

There are two types of frost: white or hoar frost and black frost. White or hoar frost is the type most commonly observed in humid climates and results from a thin coating of ice crystals deposited directly on surfaces. In the formation of white frost, the liquid stage between vapor and solid is very quick and seemingly nonexistent. The surface temperature where frost forms is always 0°C or less than below the dew point temperature (a nonfrozen water deposit on a surface is called dew).

Black frost is usually confined to dry climates. When the dew point is several degrees below 0°C, the water or cell sap in the living vegetation may freeze before sublimation takes place on the surfaces. Then when the temperature increases, the vegetation blackens, thus the term *black frost.*

Weather Conditions Causing Frosts

Depending on the climate and location, frost may be observed during all seasons. In some areas, the probability of frost in the spring and autumn greatly influences crop production practices, such as planting dates and the use of season-extending methods. The frost-free season is that portion of the year with temperatures continually above 0°C.

Frosts in the early fall and late spring are due to excessive radiant heat loss from the soil and crop surfaces under clear skies and subnormal air temperatures. Radiation frosts and freezes require cool, dry air; clear conditions; and little to no wind during the night and early morning hours. Frosts in the early spring and late fall are caused by advancing cold air from the northern climates that is several degrees below 0°C or high radiational loss.

Frost versus Freeze

An advective or wind-borne freeze occurs when a cold air mass moves into an area resulting in freezing temperatures. Wind speeds are usually above 8 km/h and clouds may be present. The thickness of the cold layer ranges from 30 m to more than 300 m above the surface. Options to protect crops by modifying the environment are limited under these conditions.

A radiation frost occurs when a clear sky and calm winds (less than 8 km/h) allow an inversion to develop and temperatures near the earth's surface drop below freezing. The thickness of the inversion layer varies from 10 to 60 m.

The Freezing Process

Water undergoes a phase transition (liquid to solid or solid to liquid) at 0°C. Dilute solutions that are found in plants undergo melting or freezing at temperatures usually below 0°C. Because the solutions in the cells are more concentrated than the water outside of the cells, a phase separation occurs during exposure to freezing temperatures. As the phase separation occurs, both liquid and solid coexist.

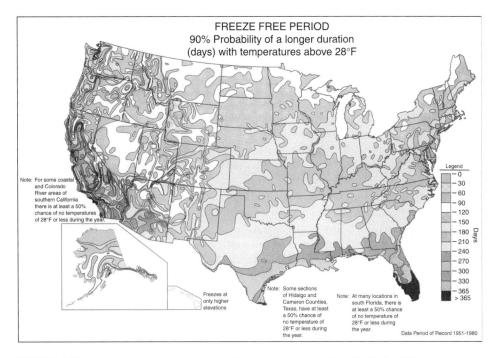

FIGURE 12.5   Average length of the frost (freeze)-free period in the United States.
Conversion: °C = 5/9 (°F − 32).
Source: Office of Satellite Data Processing and Distribution, National Oceanic and
Atmospheric Administration.

Length of Growing Season

The average growing season for a location is measured as the number of days between the average date of last spring frost and the first fall frost. Generally, the length of the growing season decreases as you move from equator to poles. The length of the growing season in the continental United States varies from 330 days in subtropical areas to 70 to 80 days in the northern section of the United States (Fig. 12.5). Hawaii has a 365-day growing season.

Topographic Factors That Affect Frosts and Freezes

Lake or Large Body of Water Proximity   Water is a good temperature buffer compared to land (i.e., water changes temperature much slower than land). Large bodies of water serve as heat reservoirs (hold heat longer than surrounding air and land areas) in fall and heat sinks (take longer to heat than surrounding air and land areas) in spring. As a result, a large lake's warming influence on nearby land in the fall can delay the first freeze by weeks. A lake effect can also delay the advancement of spring. This tempering effect of the lake reduces the risk of both early fall frosts and unusually late spring frosts.

Terrain   During the daylight hours, higher elevations generally have lower surface air temperatures than nearby lower elevations. Also, when the sun begins to set in the evening and the solar zenith angle (angle formed by the direction of the sun and the horizon) approaches 0°, soil, plant, and other exposed surfaces often experience reduced temperatures due to lower solar energy absorption. Their temperatures may drop below surrounding air temperature, and the air directly above these surfaces can become cooler, resulting in a temperature inversion. Because cool air is denser than warm, it flows to low-lying areas or valleys. The air settling in the low-lying areas can be up to 8°C cooler than surrounding high ground, and temperatures at the soil or crop surface can be as much as 3°C to 5°C cooler than those at 1.5 m above the ground. Very low lands in a valley that often experiences reduced temperatures are often referred to as frost pockets. Such areas may be too risky for certain crops without frost protective measures. Surface inversions usually end after sunrise. Also slopes in the Northern Hemisphere exposed to the north are normally the coldest, whereas slopes exposed to the south or southwest are the warmest.

Temperatures in urban areas are often warmer than in surrounding country-side. This appears to be at least partially due to an increase in the number of hard

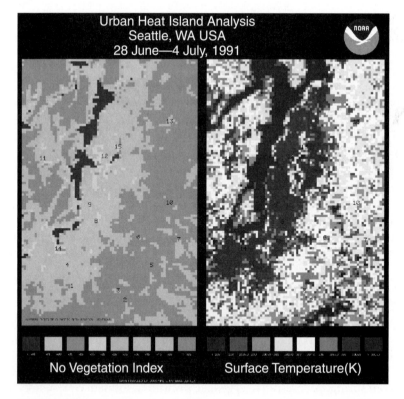

FIGURE 12.6   Urban heat island analysis of Seattle, WA. The relative amount of vegetation cover [expressed as the normalized difference (ND) vegetation index] affects the radiant surface temperature. Data derived from a one-week (28 June–4 July 1991) composite of satellite data. Conversion: °C = 1.00 (°K − 273).
Source: National Climate Data Center, NOAA.

surfaces, such as roofs and pavement common in many urban situations, and a reduction in the amount of vegetation in the area (Fig. 12.6). These hard surfaces absorb solar energy and reradiate it to the surrounding environment as heat. These higher temperature urban areas are often referred to as urban heat islands (see Green Roof Technology section later in chapter).

Soil Type   The soil type or texture affects its maximum and minimum daily temperatures. These characteristics are largely determined by the properties that define a soil's ability to hold heat, such as water and organic content, texture, and color. A wet soil holds more heat than a dry soil. Clays and loams, which have a considerable amount of tightly bound water, tend to warm slowly in the spring and cool slowly in the fall. Sandy soils, which have low water-holding capacity and little permanently bound water, tend to warm quickly in the spring and cool rapidly in the fall.

Heat conductivity is largely determined by the soil's parent material. Heat flow is relatively fast through soils containing large particles (and large pores). As a consequence, large-particle sandy and stony soils normally heat and cool each day to greater depths than other soils. Darker colored soils also tend to absorb a greater amount of heat than lighter colored soils of similar particle size and texture distribution.

Mucks and peats are soil types that are the most subject to frost and freezing temperatures. Sands and loams are also highly susceptible, especially if very dry and/or at low elevations. Clays and clay loams normally have smaller daily temperature changes and thus are less susceptible. However, elevation, slope direction, exposure to wind, soil color and dryness, and nature of the vegetative cover all can override the effects of soil type in determining the occurrence or extent of frosts and freezing at any particular site.

Wind Chill

Wind chill refers to the cooling effect of moving air on a warm body and is expressed in terms of the amount of heat lost per unit area per unit time. This was developed to estimate heat-loss rate from humans and other warm-blooded organisms. Even though wind chill does not apply to plants, wind can remove heat rapidly from an area where plants are being grown, such as a field or a nursery.

Effects on Plants

Effects of Temperatures on Biochemical Processes

Temperature drastically affects the stability of enzymes and biochemical processes. At optimum temperatures, enzyme systems function well and remain stable for long periods. At lower temperatures, the enzymes remain stable but are nonfunctioning, whereas at high temperatures they completely break down.

Within certain temperature ranges, increases in temperature generally result in an increase in chemical reaction rates, whereas decreases in temperature result in reduced enzymatic activity and reduced reaction rates. The term used to describe

TABLE 12.1   *Temperature coefficients (Q_{10}) for selected plant processes at varying intervals within the range 0°C to 30°C*

Process	Q_{10}
Diffusion of small molecules in water	1.2–1.5
Water flow through the seed coat of *Arachis hypogaea*	1.3–1.6
Water movements into germinating seeds, various species	1.5–1.8
Hydrolysis reactions catalyzed by enzymes	1.5–2.3
Respiration	2.1–2.6
Photosynthesis (light reactions)	approx. 1
(CO_2 fixation reactions)	2.0–3.0
Phosphate uptake into maize seedlings	2.0–5.0

Source: A. H. Fitter and R. K. M. Hay. *Environmental physiology of plants*, 2nd ed. (New York: Academic Press, 1987). Used with permission: Academic Press.

the effect of temperature on a chemical reaction process is the temperature coefficient or Q_{10}.

$$Q_{10} = \text{(rate at temperature T + 10°C)}/\text{(rate at temperature T)},$$

where T = temperature in °C.

A Q_{10} of 2 indicates that the reaction rate doubles with a 10°C increase in temperature. Nonenzymatic reactions have a Q_{10} of about 1.2, but enzymatic reactions generally have a Q_{10} of 2 or more (Table 12.1). Between 0°C and 30°C, respiration usually has a Q_{10} of 2 to 3. Above 30°C, respiration declines because of heat inactivation of enzymes.

Cardinal Temperatures

Every physical and chemical process in plants is influenced by temperature. Three general cardinal temperatures for plants are minimum, optimum, and maximum. Cardinal temperatures for plants vary with species (e.g., temperate versus tropical, and cool season versus warm season), stage of development (young tissue is usually more temperature sensitive than older tissue), tissue (flower bud is more sensitive than vegetative tissue), process of concern (e.g., germination, photosynthesis, respiration) (Table 12.2), and environmental factors (e.g., lower relative humidity increases tolerance to higher temperatures because the heat is dissipated as long wavelengths more efficiently at lower relative humidity).

Minimum Temperature   Minimum temperature is the temperature at which the plant response will still occur but very slowly. If the temperature goes much lower, the response will not occur. For temperate plants, the minimum temperature for growth is generally around 5°C, whereas the minimum temperature for tropical plants is around 10°C.

TABLE 12.2 *Cardinal temperatures for germinating seeds of plant species*

Group	Temperature (°C)		
	Minimum	Optimum	Maximum
Grasses			
Meadow grasses	3–4	25	30
Temperate zone grain	2–5	20–25	30–37
Rice	10–12	30–37	40–47
C4 grasses of tropics	10–20	32–40	45–50
Herbaceous dicotyledons			
Plants of tundra	5–10	20–30	
Meadow herbs	2–5	20–30	35–45
Cultivated plants in the temperate zone	1–3	15–25	30–40
Desert plants			
Summer germinating		20–30	
Winter germinating		10–20	
Cacti		15–30	
Temperate zone trees			
Conifers	4–10	15–25	35–40
Broad-leaved trees	< 10	20–30	

Source: K. J. Boote and F. P. Gardner. Temperature. In *Principles of ecology in plant production*, ed. T. R. Sinclair and F. P. Gardner. (New York: CABI Publishing, 1998). Used with permission: CABI Publishing.

Optimum Temperature Optimum temperature is defined as the temperature at which the specific plant response (often overall growth or flowering) will occur most rapidly. Plants often do not have one optimum temperature, but instead have a range of temperatures. For temperate plants, the optimum temperature for growth is generally between 25°C and 30°C, whereas the optimum temperature for tropical plants is generally between 30°C and 35°C. Crop species are often placed in two general categories depending on their optimum temperatures, cool-season crops and warm-season crops. The optimum temperatures for the two groups differ by about 10°C (15°C to 25°C optimum temperatures for most cool-season crops and 25°C to 35°C for many warm-season crops). Also, vegetative and reproductive processes within a plant often differ in their optimal temperatures. For many reproductive processes the optimal temperature is at least 5°C lower than for the vegetative processes.

Maximum Temperature Maximum temperature is defined as the temperature above which the plant response will not occur. For most temperate plants, the maximum temperature for growth is generally between 35°C and 40°C, whereas the maximum temperature for tropical plants is around 45°C. If temperatures increase above maximum, the response may be inhibited.

Potential Differences between Air and Plant Surface Temperatures

Plants are poikilotherms (i.e., they tend to assume the temperature of their environment), but differences between air and plant surface temperatures often exist. Although air temperature is relatively easy to measure, leaf or plant temperature is often harder to measure and may differ greatly from air temperature because of microclimate variables.

Leaf temperature is the product of the energy budgets of the leaf. The energy budget of a representative leaf (EB_{leaf}) can be expressed as

$$EB_{leaf} = [S \pm L - rs - R] \pm H \pm E + A,$$

where S = solar input
L = longwave radiation
rs = reflected solar energy
R = reradiated solar energy
H = sensible heat transfer (conduction and convection to cool air)
E = latent heat transfer (evaporative cooling from water loss)
A = assimilation
$[S \pm L - rs - R]$ = net radiation

When leaf temperature during the daytime is in equilibrium with air temperature (a rare occurrence), EB_{leaf} is equal to 0. When the environment is warm and windy with low humidity, the leaf temperatures tend to be lower than the air temperature. Cool and sunny conditions with high humidity and little air movement often result in leaf temperatures higher than air temperatures. This occurs because the sun warms the leaf and there is little air movement for transpiration to dissipate the heat. At night, there is little to no transpiration and leaf temperatures are more often in equilibrium with air temperatures.

Temperature Effects on Photosynthesis and Respiration

Light and carbon dioxide primarily control the rate of photosynthesis, but temperature also has an effect. The effect of temperature on photosynthesis is roughly equivalent to the effect of temperature on enzymatic reactions and depends on species, the environmental conditions under which the plant was grown, and the environmental conditions during measurement. In general, as temperature increases over a specific range, the rate of photosynthesis increases if other factors are not limiting. The increase is linear at lower temperatures, but the rate starts to decrease at higher temperatures. The overall plotted effect of temperature on photosynthesis generally resembles a bell curve (Fig. 12.7).

The temperature at which the highest photosynthesis rate is observed represents the optimum temperature response. The optimum temperature is the point at which the capacities of the various stages of photosynthesis are optimally balanced, with some steps becoming limiting as temperatures decrease or increase. Desert species generally have higher optimum temperatures than arctic or alpine species, and annual desert plants that develop during the hot summer months (mostly C4 plants)

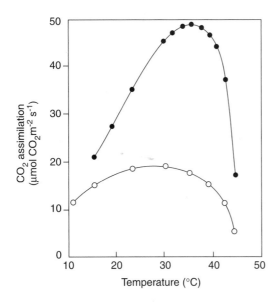

FIGURE 12.7 Idealized representation of the influence of temperature on leaf photosynthetic rates. The bottom line (open dots) is a representative response for a C3 plant, and the top line (solid dots) is a representative response for a C4 plant.

have higher optimum temperatures than those that grow only during the winter or spring (mostly C3 species). Plants acclimated (or adjusted) to growing at low temperatures maintain higher photosynthetic rates at low temperatures than plants grown at high temperatures. In general, the optimum temperatures for photosynthesis are similar to the daytime temperatures at which the plants normally grow.

In general, C4 plants tend to have higher optimum temperatures for photosynthesis than C3 plants, and this difference is primarily the result of the low rates of photorespiration often present in C4 plants. Increases in temperature usually stimulate photosynthetic rates until stomata close or enzymes denature. However, respiratory CO_2 loss also increases with temperature, and this is especially pronounced for the chemical reactions for photorespiration (primarily because a temperature rise raises the ratio of dissolved O_2 compounds compared to CO_2). For C3 plants the stimulating effect of a temperature rise is nearly balanced by increased respiration and photorespiration over much of the temperature range at which C3 plants normally grow. A rather flat and broad temperature response curve between 15°C and 30°C often occurs. Because photorespiration is of less importance in C4 plants, they often exhibit optima in the 30°C to 40°C range.

Respiration is the process of the controlled release and utilization of stored energy by the plant. All life requires a source of energy and respiration provides that source of energy. One of the primary factors that controls respiration is temperature. As temperatures increase the rate of respiration also increases, until a maximum temperature is reached and the rate then decreases. The Q_{10} for respiration rate for most plants is between 2 and 2.6 across the temperature range in which respiration is potentially affected by temperature (Fig. 12.8).

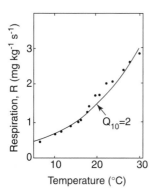

FIGURE 12.8   Idealized representation of the effect of temperature on respiration.

Photosynthesis and respiration are generally considered to be competing processes, but photosynthesis rates are generally 6- to 20-fold higher than respiration rates. The net effect is that plants fix more CO_2 through photosynthesis than they lose through respiration. Apparent photosynthesis or net photosynthesis is the difference between true photosynthesis and respiration. Apparent photosynthesis is important because the plant grows by using the products of apparent photosynthesis as substrate. Enhanced respiration rates relative to photosynthesis at high temperatures are more detrimental in C3 plants than in C4 or CAM plants because the rates of both dark respiration and photorespiration are increased in C3 plants.

The temperature compensation point is the temperature (often a high temperature) at which photosynthesis just balances respiration (net CO_2 exchange is zero). When the plant is above the temperature compensation point, more CO_2 is being respired than is being produced by photosynthesis, resulting in a net loss of stored carbohydrates. This imbalance between photosynthesis and respiration is one of the main causes of the deleterious effects of high temperatures on plant growth. On the same plant, the temperature compensation point is usually lower for leaves that developed in the shade than for leaves that developed exposed to full sun (and heat).

Besides effects on rate of plant growth, temperature has a role in controlling the pattern and timing of plant development (e.g., vernalization and seed dormancy). Temperatures can also affect the morphology and dimensions of plant parts as well as partitioning of dry matter within the plant.

Membranes

Membranes are important components of plant cells, and their integrity is necessary for proper cellular metabolism. During freezing and chilling injury, there appears to be a critical temperature below which injury to cellular membranes occurs. The membrane hypothesis for chilling injury suggests that cellular membranes in sensitive plants undergo a physical phase transition from normal flexible liquid-crystalline to a solid-gel structure at the temperature critical for chilling injury. This change in phase to solid-gel results in cracks or channels that could lead to increased membrane permeability, with resulting ion imbalances and disruption of enzyme systems. The ratio of saturated to unsaturated fatty acids in the cell membrane generally determines membrane fluidity. The membrane phase transition appears to be somewhat reversible, as return to temperatures above chilling (within a certain time period of chilling exposure) can result in the membranes returning to the liquid-crystalline state, and the cell will recover if metabolism has not been too disrupted during the chilling period.

Chilling Injury

Some plants are injured at temperatures at or slightly above freezing. This is referred to as chilling injury. The primary cause of most chilling injuries is the loss of membrane properties resulting from changes in membrane fluidity. Direct chilling injury to plants includes necrosis, discoloration, tissue breakdown, browning, reduced growth, or failure to germinate seeds. Indirect chilling injury includes reduced yields, delayed harvest, and reduced net photosynthesis.

Susceptibility to cold damage varies with different species and among varieties of the same species. The susceptibility to cold damage varies to some degree with stage of plant development. Plants tend to be more sensitive to cold temperatures shortly before flowering through a few weeks later.

Tropical plants are injured when the temperature drops to some point above freezing but low enough to cause damage. For many sensitive plants this happens after exposure to about 10°C.

Some plants can be "hardened" to withstand temperatures that would otherwise cause chilling injury. This is accomplished by a gradual exposure of plants to low but not injurious temperatures.

Freezing Injury

Freezing injury occurs when ice forms within the cells and organs of the plant at temperatures slightly below 0°C and disrupts the structure of the protoplasm. Freezing temperatures can also cause tissue injury due to the dehydration of the protoplasm as the water evaporates to the extracellular ice. As water leaves the cell the concentration of solutes increases to the point where some of them may become toxic. Factors that are involved in determining the type of injury that occurs include the duration of the freeze, the degree of supercooling, the number of cycles of freezing and thawing, and a number of conditions of the plant (such as stage of growth and development, anatomy of tissue, moisture content of the tissue, and the growth habit of the plant). The extent of damage also depends on species of plants, age of tissue, and degree of acclimation before hardening. Freezing injury is the result of damage to membranes through loss of fluidity and rupture. Plant tissues that

FIGURE 12.9   Freeze injury (bleaching of tissue) on broccoli.

freeze generally appear dark green and water soaked at first, later becoming black-ened and necrotic or bleached in appearance (Fig. 12.9).

Plants with little to no freezing tolerance, such as corn, cucurbits, and beans, are subject to frost damage. As a result of frost damage, the foliage becomes flaccid and water soaked. Membranes become leaky and the compartmentalization of cells is de-stroyed. The external temperatures often need to be no lower than −1°C to −3°C for this to occur. Slight supercooling may occur, but when nucleation is initiated freezing of susceptible tissue of these types of plants occurs.

Plants with limited freezing tolerance such as potatoes and cabbage can with-stand some ice formation. Plants with the growing points below the surface of the soil or snow cover are also somewhat protected. These plants seldom supercool more than −1°C to −2°C. Death as a result of cold in these plants is a result of de-hydration as water moves to the intercellular ice and freezes.

Woody plants are tolerant to freezing and are often capable of supercooling. Most deciduous forest trees and fruit tree cultivars have some tissue that may super-cool to about −40°C. Seasonal changes occur within the plant allowing them to with-stand cold temperatures.

Winter Desiccation   Winter desiccation injury of plants occurs when the absorption of water by the roots cannot keep up with the amount of moisture lost through the foliage by transpiration. This frequently occurs on sunny days, especially when it is windy and when the soil water is frozen, or if liquid water is otherwise in short supply. Injury appears as brown leaf margins or needle tips at the onset of the first period of warm weather. In severe cases, all of the leaves and buds are killed. More commonly, though, the leaves are injured or killed, but the buds survive.

Root Injury   Roots of trees and shrubs are more sensitive to cold injury than stem tissues. In the landscape or orchard, the roots are not commonly injured because the soil and snow cover protect them from exposure to freezing air temperatures. Containerized plants in nurseries are very susceptible to root freezing, because they are more exposed. Cold hardiness of roots varies with species and rootstocks. Cold injury to roots appears to be greater in sandy soils than in clay, because cold temperatures penetrate more deeply into soils with lots of air spaces (and less water). For the same reason, injury is more likely in dry soils than in moist.

Frost heaving is another type of winter root injury, and it is caused by the repeated freezing and thawing of the water in the soil physically (expansion/contraction) displacing plants, especially the smaller ones (strawberries, shrubs, young trees), often upward in the soil. This displacement of the plant can result in damage to the fine feeder roots. If sufficient roots are broken and/or exposed and subsequently dehydrate, injury or death of the plant may follow.

Vernalization

Vernalization is the induced or accelerated (premature) flowering that occurs in certain plants due to exposure to low temperatures. Premature flowering (also referred to as bolting) can cause substantial yield losses in certain crops. This is particularly true for crops that require little cold exposure to induce vernalization, such as heat tolerant Chinese cabbage.

The biennials initiate flower formation after extended (several weeks or months) exposure to low temperature at the end of the first growing season. The required length of low temperature exposure for vernalization varies with species. In some species, seedlings and young plants in the juvenile stage are insensitive to conditions that promote flowering in older plants. Many biennial species in which flower formation is induced by vernalization need long days for further development of seed stalk. Therefore, there is an important interaction between vernalization and photoperiod.

In some species, seeds can be vernalized. The seeds of these species must have sufficient water to allow the vernalization process to occur. Certain tubers, corms, and bulbs require low temperatures following moderately high temperatures before growth occurs.

Crop-Specific Responses to Cold Conditions

Annuals   Annual crops differ in their hardiness to low temperatures, depending on their genetics and origin. Warm-season crops, such as tomatoes, snap beans, and the cucurbits, originated in tropical areas and can be severely injured by even a light frost. Cool-season crops, such as broccoli, cabbage, peas, and onions, originated in temperate areas and can tolerate frost and light freezes of short durations with little damage.

FIGURE 12.10   Winter sunscald and frost cracking on a maple tree.
Courtesy Dr. Rick Bates, Penn State University.

Some crops, such as cabbage, broccoli, cauliflower, and onion sets respond to cold weather by producing a seed stalk during bolting. This occurs when young plants are exposed to low temperatures for several days and vernalization occurs (vegetative buds convert to reproductive within the growing points). When warmer weather returns, the buds develop into flower and seed stalks. This greatly reduces the quality and marketability of the affected crop.

Bolting occurs on plants that may have been placed in the field too early in the spring. These plants may have put on enough growth to advance past the juvenile stage (pencil thick stems for cabbage, half-inch wide bulbs for onion), but at this point, the plant is sensitive to a cold spell. Bolted plants are often discarded as cutting the flower stalks off will not prevent the deterioration in flavor and/or quality of the plant.

FIGURE 12.11   Winter burn (browning of leaves) and winter drying (reddening, browning, and, in some cases, drooping of foliage) of arbor vitae observed in late winter and early spring.
Courtesy Dr. Rick Bates, Penn State University.

Woody Plants and Perennials   Frost cracking is a problem of trees with thin bark, such as peach or silver maple (Fig. 12.10). This occurs when the bark and underlying cambium, usually on the south or southwest side of the tree, heat up on cold, bright days. When the sun sets or is blocked by a cloud, the bark and cambium quickly return to air temperature, which can cause physical and physiological damage. The bark and the wood underneath contract at different rates as they cool, causing mechanical stress. Eventually the wood splits, sometimes violently enough to produce a rifle-like noise. Wood-decaying organisms and insects can enter the damaged areas. The cracks may heal over the following season, but are likely to split again the following winter.

Sunscald is a type of physiological damage to the tree trunk caused by extreme temperature fluctuations. The elevated temperature of the trunk in dormant plants causes the cambium to become active and thus lose its hardiness. The drop in temperature kills the nonhardy cambial tissues. The scalded bark may split, forming an entry point for decay-causing organisms.

Winter burn and winter drying are problems often associated with evergreens, such as pine (Fig. 12.11) and rhododendron. Containerized nursery stock, whose small, above-ground root balls freeze easily, and newly planted bare-root or balled-and-burlapped plants with their reduced root systems, are also vulnerable. Winter burn is the browning of needles caused by a rapid temperature change in winter, particularly on the south side of trees where there is more exposure to the sun. Rapid temperature changes often occur at sunset and sunrise or when sunlight is suddenly blocked by other trees, hills, or buildings.

Winter drying is caused by the desiccation of foliage and twigs by warm dry winds, when water conduction is restricted by frozen plant tissues or frozen ground. Reddening, browning, and, in some cases, drooping of foliage become apparent in late winter and early spring. Often a combination of winter burn and winter drying will occur, occasionally complicated by drought. If severe enough, the entire tree may die.

Fruit Crops   Cold injury is a common cause of economic loss in fruit crops. Winter injury occurs when the trees are dormant; and spring frost injury occurs when the trees are no longer dormant, but in various stages of flower, fruit, and/or leaf development. Both types of injury occur when temperatures drop below certain threshold levels. The injury threshold temperature is lower for dormant than nondormant tissues, and varies for different species, varieties, and stages of development.

Spring frosts and freezes are an annual threat to the buds of many fruit crops. As the weather warms, the buds begin to come out of (break) dormancy, thereby losing their cold hardiness. The further developed the buds are the more susceptible they are to injury if the temperature should drop. Also, the critical temperature at which injury can be expected depends on the stage of bud development, as well as the length of time the temperature stays at or below the critical temperature.

Not all the buds in an orchard, or even on the same plant, develop at the same rate. The stage of development of the buds depends on species, cultivar, location on the shoot, orchard site, and management practices. Therefore, it's rare that all of the buds in a field are at the same level of hardiness. If a spring frost occurs, the most advanced buds may be injured, while the less developed ones may survive. However, if critical low temperatures occur after the 100% bloom stage, then all fruit and flowers are essentially equally susceptible to damage. If buds are injured sufficiently during the prebloom or bloom stages, they will desiccate and eventually abscise.

A certain number of days of low temperatures are needed by some fruit trees to grow properly. For example, most cultivars of peaches require 700 to 1,000 hours below 7°C and above 0°C before they break their rest period and begin flowering and growth. If this cold requirement is not met, then small, misshapen leaves and fruit will result (fruit may not develop at all).

High Temperature

Heat Stress   The stability of cellular membranes is important during high-temperature stress, just as it is during cold-temperature stress (chilling and freezing).

Excessive fluidity of membrane lipids at high temperatures is correlated with loss of physiological function. Elevated temperatures can result in increased plant respiration rates and decreased photosynthetic rates. Plant tissues may die because of the absence of supporting energy and/or altered activity of enzymes or membranes. Enzymes also have a specific range of temperatures at which they are active. Denaturation of proteins due to high temperatures can occur if temperatures are excessively high. Dehydration caused by high temperatures can cause a more rapid water loss than can be replaced.

High temperatures can damage trunks of trees, especially those with thin bark. Scald can also be a problem on fruit such as apples and tomatoes. Sunscald damage on plant tissue (such as fruits) is a discoloration on the surface with collapsed or damaged cells under the surface (Fig. 12.12). When temperatures rise too high (45°C to 50°C), cell death results as the protoplasts in the plant cells are destroyed. Tomato fruits exposed on vines to high temperatures and high solar radiation can reach 50°C to 52°C. If fruits are exposed to these temperatures for an hour or more, they become sunburned.

Heat stress at flowering or fruit set can result in increased flower or fruit abortion. Some crops such as onions and radishes become more pungent and asparagus and beans are more fibrous at high temperatures. Symptoms of heat injury are the

FIGURE 12.12 Sunscald of tomato.

FIGURE 12.13 High temperature and drying winds cause rapid loss of water, especially in maple leaves, resulting in leaf margins turning yellow or brown (leaf scorch) and leaves falling prematurely (heat defoliation).
Courtesy Dr. Rick Bates, Penn State University.

appearance of necrotic (dead) areas on young leaves, chlorotic mottling of leaves, and death (Fig. 12.13). Heat injury occurs over a wide range of crops depending on the species or tissue.

Tolerance to High Temperatures In general, heat tolerance of most plants varies directly with the temperature in their natural environment. Plants that tolerate high temperatures have the ability to maintain high rates of photosynthesis. Stability of proteins also is imperative in these plants. In addition, some plants avoid high temperatures by adjusting the angle and arrangement of leaves toward incident radiation (often in an attempt to reduce the amount of solar radiation intercepted). Coloration and reflectance of leaves is also important. For example, desert species may reflect as much as 70% of incident radiation. The presence of leaf hairs and thick cuticles alter the reflectance properties of the leaf surface and also improves cooling. Also, transpiration from the leaf stomata helps cool leaves.

As is the case with freezing tolerance for plant species in natural ecosystems, a relationship between environmental temperature and heat tolerance is suggested by the seasonal cycles of the plant's tolerance to high temperatures.

Temperature Acclimation or Hardening

Temperature acclimation or hardening is the development of tolerance to injury from either low or high temperatures. This is also referred to as acquisition of hardiness. Hardening refers to the processes that increase the ability of a plant to survive the impact of a too high or too low temperature stress.

Cold hardiness refers to the ability of plants to withstand cold injury. Several factors affect cold hardiness, including solar energy (intensity and photoperiod), temperature and temperature duration, tissue water content, and nutrition (N fertilization decreases hardiness by increasing tenderness and succulence).

One of the most common techniques to condition plants for cold weather is to withhold water and fertilizer, especially nitrogen, toward the end of the growing season. Nitrogen applications enhance young succulent growth, which is very susceptible to cold injury. Transplants are hardened by any treatment that slows new growth. This can be accomplished by gradually exposing plants to cold, by withholding moisture, or by a combination of these two treatments. In general, about 10 days are required to harden plants.

Plants with high levels of reserve carbohydrates are better able to withstand cold weather than those in poor condition. Promoting plant vigor during the growing season increases the supply of carbohydrates. Respiration continues to some extent throughout the winter, and there must be a supply of carbohydrates to maintain life processes during this period. As a general rule, resistance to cold injury increases with age until senescence.

As with cold resistance, the plant cells can become gradually acclimated to heat to a certain extent by slowly raising the temperature for a period of time during the day and lengthening the daily exposure to the raised temperature. An important component to high-temperature acclimation is a reduction in water content and transpiration rates of the tissue. This is especially true for seeds, as dry seeds of some species are able to survive temperatures as high as 120°C. High-temperature hardening also generally increases thermotolerance of leaves and thermostability of enzymes.

Agricultural Technologies That Affect Temperature

A number of factors can affect the risk of frost and freeze injury to crops. The regional and local climates are important, as are the area's topography and the microenvironment of the specific site. The actual hardiness of the plant also determines the risk of injury.

Frost Protection

Four general strategies can be used to protect plants from frost. These include escape by selectively choosing planting dates and growing location; reduction of heat loss with the use of row covers, high tunnels, cold frames (season extenders); addition of heat with the use various heaters and heating systems; and the application of chemicals.

Escape   Escape strategies for frost protection include choosing appropriate planting dates and location of growing sites.

Planting date   The date in which a seed or transplant is placed in the field (planting date) can determine to a considerable extent the success or failure of the crop. Planting dates are chosen for a crop to minimize potential injury from low temperatures or frost. The time of planting is usually determined with reference to the soil and weather conditions, the type of the crop, and the time of desired harvest for the crop. Frost defines the crop growing season for many annual crops. Many crops can be grouped into three classes with respect to cold resistance: (1) hardy, or those that will withstand hard frosts, (2) half-hardy, or those that will withstand light frosts and the seeds will germinate at low temperatures, and (3) tender, or those unable to withstand any frost and the seeds of which will not germinate in cold soil.

Hardiness zones   Low temperature is one of the most critical environmental limitations for plants. Some plants, such as the annuals, simply avoid the cold of winter by dying at the end of summer. Perennial plants (including trees, shrubs, and herbaceous perennials) must be able to survive the lowest temperature each winter in order to live into the next growing season. Such plants have a threshold temperature below which they will die. These threshold temperatures determine how far north (in the Northern Hemisphere) plants will survive.

Hardiness zones ratings, developed by the U.S. Department of Agriculture, help in determining appropriate plant material to grow in an area (Fig. 12.14). The zones are based solely on the average annual minimum temperatures for each zone. There are 11 different zones, ranging from Zone 1, which is the coldest (below $-45.6°C$), to Zone 11, which has the highest average temperature of all 11 zones (above $40°C$).

Many seed and nursery plant catalogs and plant references utilize plant hardiness zone maps to indicate whether a plant is likely to survive in a location. For example, an Eastern White Pine is rated hardy to Zone 3 and would be expected to do well in many parts of the northeastern United States (3 and higher numbered zones). A Japanese maple, on the other hand, is rated hardy to Zone 5 or 6 (depending on the type) and might not survive in Zone 3.

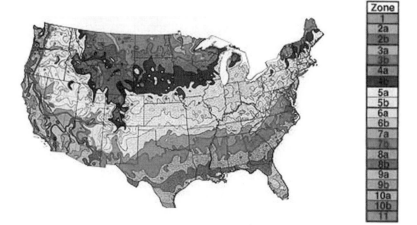

FIGURE 12.14   Plant hardiness zones for the continental United States.
Source: USDA.

Although hardiness zones are often used as a guide to determine which plant materials might grow in an area (Table 12.3), additional factors (such as soil pH, average relative humidity, saline conditions, winter sun intensity) that must be considered are the occasional deviations from the average minimum temperature or other weather occurrences not considered in the zonal delineations that could be detrimental to plant growth and survival. For example, some years the temperature in an area may fall below the average minimum for a particular zone, and it is such extreme low temperatures, not averages, that injure sensitive plants. Also, within a zone, temperatures, including winter minimums, vary with elevation, proximity to water, and microclimatic features. There is also a difference between a plant surviving and a plant thriving. A plant exposed to the lowest temperatures it can tolerate may survive the winter but may lose so much vigor that it does not grow and function well for its purpose, such as in a landscape. As a result, landscape plants are often selected for an area that is rated a zone warmer than indicated by hardiness alone.

Site exposure   Southern or southeastern exposures on gentle slopes are preferred for early spring and fall crops in temperate climates. In general, a sunny southern or southeastern exposure slope dries and warms earlier in the spring than a northern exposure.

Slope   A sloped site generally has good air drainage thereby reducing the likelihood of early autumn frosts. It is this microclimate factor that makes slopes good sites for fruit plantings. Low areas collect cold air and become "frost pockets." Sufficient air and water drainage also will reduce disease problems.

TABLE 12.3 *Cold hardiness ratings for selected woody plants*

Zone	Botanical (and Common) Names
1	*Betula glandulosa* (Dwarf birch)
	Empetrum nigrum (Crowberry)
	Populus tremuloides (Quaking aspen)
	Potentilla pennsylvanica (Pennsylvania cinquefoil)
	Rhododendron lapponicum (Lapland rhododendron)
	Salix reticulata (Netleaf willow)
2	*Betula papyrifera* (Paper birch)
	Cornus canadensis (Bunchberry dogwood)
	Elaeagnus commutata (Silverberry)
	Larix laricina (Eastern larch)
	Potentilla fruticosa (Bush cinquefoil)
	Viburnum trilobum (American cranberry bush)
3	*Berberis thunbergii* (Japanese bayberry)
	Elaeagnus angustifolia (Russian olive)
	Juniperus communis (Common juniper)
	Lonicera tatarica (Tartarian honeysuckle)
	Malus baccata (Siberian crabapple)
	Thuja occidentalis (American arborvitae)
4	*Acer saccharum* (Sugar maple)
	Hydrangea paniculata (Panicle hydrangea)
	Juniperus chinensis (Chinese juniper)
	Ligustrum amurense (Amur River privet)
	Parthenocissus quinquefolia (Virginia creeper)
	Spiraea x vanhouttei (Vanhoutte spirea)
5	*Cornus florida* (Flowering dogwood)
	Deutzia gracilis (Slender deutzia)
	Ligustrum vulgare (Common privet)
	Parthenocissus tricuspidata (Boston ivy)
	Rosa multiflora (Japanese rose)
	Taxus cuspidata (Japanese yew)
6	*Acer palmatum* (Japanese maple)
	Buxus sempervirens (Common boxwood)
	Euonymus follunei (Winter creeper)
	Hedera helix (English ivy)
	Ilex opaca (American holly)
	Ligustrum ovalifolium (California privet)
7	*Acer macrophyllum* (Bigleaf maple)
	Rhododendron Kurume hybrids (Kurume azalea)
	Cedrus atlantica (Atlas cedar)

Zone	Botanical (and Common) Names
	Cotoneaster microphylla (Small-leaf cotoneaster)
	Ilex aquifolium (English holly)
	Taxus baccata (English yew)
8	*Arbutus unedo* (Strawberry tree)
	Choisya temata (Mexican orange)
	Olearia haastii (New Zealand daisybush)
	Pittosporum tobira (Japanese pittosporum)
	Prunus laurocerasus (Cherry laurel)
	Viburnum tinus (Laurustinus)
9	*Asparagus setaceus* (Asparagus fern)
	Eucalyptus globulus (Tasmanian blue gum)
	Syzygium paniculatum (Australian bush cherry)
	Fuchsia hybrids (Fuchsia)
	Grevillea robusta (Silk-oak)
	Schinus molle (California pepper tree)
10	*Bougainvillea spectabilis* (Bougainvillea)
	Cassia fistula (Golden shower)
	Eucalyptus citriodora (Lemon eucalyptus)
	Ficus elastica (Rubber plant)
	Ensete ventricosum (Ensete)
	Roystonea regia (Royal palm)

Source: USDA, ARS.
Representative plants listed under the coldest zones in which they normally succeed.

Large bodies of water   Frost protection afforded by large bodies of water is due to the high specific heat of water compared to that of land. Large bodies of water become heat reservoirs in the fall and cold reservoirs in the spring, exhibiting a moderating effect on air temperatures.

Reduction of heat loss   Strategies used to reduce heat loss include using season extenders.

Season extenders   Growers often attempt to extend the growing season for selected crops to obtain a marketing advantage. For example, crops that produce early in the spring or early summer often command a greater price on the market. Also, producing a crop when large quantities of the crop are not available (typically referred to as off season) can command greater prices and increased demand.

Extending the growing season can be accomplished by planting the field early in the spring (before the recommended planting date for the region) with the use of protective structures such as row covers (low tunnels) or high tunnels. Earlier production can also be accomplished with the use of plastic mulches, which can hasten crop maturity. Later production in the fall is possible with the use of protective structures, such as row covers, that are placed on top of the crop when lower outdoor temperatures prevail. Greenhouse production potentially allows year-round crop production.

Important factors in deciding whether to try to extend the season are the increased costs of using season extender production systems, potential increase in sale prices of the crop if produced either earlier or later, and suitability of the crop to the season extender production system.

Mulches   Mulching is the practice of covering the soil around plants to improve crop growth and development. Mulch materials may be organic (leaves, straw, grass) or synthetic (plastic). Using a mulch can modify soil surface temperatures and reduce evaporation. Other specific types of mulches are used to repel insects and regulate plant growth by modifying the plant light environment.

Synthetic mulches   Plastic mulches have been used commercially on crops since the early 1960s and have substantially increased earliness and total marketable yields of some selected (primarily warm-season) crops (Fig. 12.15).

The plastic provides a greenhouse effect during the day, warming the soil considerably relative to unmulched soil. The plastic also modifies fluctuations in soil moisture (partially by eliminating evaporation), minimizes crusting from hard rains, and reduces the leaching of nutrients. Synthetic mulches have the disadvantage that they can be expensive and must be removed and disposed of at the end of the growing season. Also recent evidence shows that they increase runoff of water and pesticides in the field.

FIGURE 12.15   Plastic mulch being placed in the field.

- *Black plastic mulch*—A dark-colored mulch, such as black, can trap heat from the sun's rays and warm the soil. This can result in a warmer root zone sooner in the spring than in bare ground, which encourages earlier plant growth. The increased value of the earlier production of some crops with mulches can easily offset the costs of the plastic mulch.
- *Clear mulch*—A clear (transparent) plastic mulch warms the soil the most of any of the plastic mulches, but allows weed growth. A herbicide is often needed with the clear plastic. In addition, clear plastic could warm the soil to such an extreme as to adversely affect plant growth. In some areas of the country, such as the southern United States, clear plastic is also used as a means of solarizing or sterilizing soil in the field.
- *White mulch*—White mulch reflects more light back to the plant than black mulch and has little to no effect on soil temperature. White mulch may repel some insects, although it is not as effective in this regard as reflective mulches.
- *Degradable mulches*—Degradable plastic mulches have many of the properties and provide the usual benefits of nondegradable polyethylene mulches except that they degrade after the film has received a critical amount of sunlight. When the film has received sufficient light, it becomes brittle and develops cracks and holes (Fig. 12.16). Small sections of the film, as a result, may tear off and be blown by the wind. The film eventually breaks down into small flakes and disappears into the soil. The edges of the mulch covered by soil retain their

FIGURE 12.16   Example of a slow degrading photodegradable mulch (left) and a relatively rapidly degrading photodegradable mulch (right). Photodegradable mulches are formulated to begin degrading upon receiving a certain minimum amount of accumulated UV radiation.

FIGURE 12.17   Bark mulch used for a planted tree in landscape.

strength and break down more slowly. Thus it can be more difficult to remove from the field than conventional mulches. Rate of breakdown is also hard to predict in difficult climates.

- *Selectively permeable or reflective (colored) mulch*—Other colors of plastic mulch are available, including a new generation of selectively permeable and wavelength selective reflective mulches. Selectively permeable mulches allow certain wavelengths of light to pass through the mulch to warm the soil without encouraging weed growth. Wavelength selective or colored mulches reflect wavelengths of light into the plant canopy to influence plant growth or insect populations. These mulches are more expensive than conventional mulches and effects can be variable depending on location, season, and crop.

Organic mulches   Grasses, leaves, and straw are some of the materials that can be used as organic mulches (Fig. 12.17). Organic mulches suppress weeds, reduce crusting and preserve soil moisture. The gradual decomposition of organic mulches can also add organic matter and mineral nutrients to the soil.

Living mulches   Grasses, legumes, and other crops that are seeded into an established crop, usually mid- to late-season, are considered living mulches. These living mulches provide erosion control and some weed control. Crops usually chosen to be used as living mulches have relatively slow growth rates and vigor compared to the established crop so as to not compete aggressively with the crop for water and nutrients.

Row covers   Row covers are flexible, transparent coverings that are installed over single or multiple rows of crops to enhance plant growth by warming the air around the plants in the field. Other terms used for plastic row covers include cloches, tunnels, and low tunnels. Row covers work by trapping radiant heat during the day

FIGURE 12.18   Example of row covers used for crop production in the field: clear row cover supported by wire hoops (top) and fabric row cover laying directly on top of plants (bottom).

and retarding its loss at night. They also block wind, which can accelerate cold damage. Row covers can warm the soil and protect the plant from hail and wind injury.

Row covers can be supported above the plant using hoops of wire. A clear plastic is often stretched over the hoops and the sides are secured by placing in soil (Fig. 12.18). The plastic can be slit or perforated prior to purchase or can be mechanically slit for venting during the growing season as the temperatures increase within the row cover environment. Row covers can also be made from fabric material. These row covers are often used as floating row covers, meaning that the material is placed directly on the top of the developing plants without the need for wire hoops. This fabric material is usually spunbonded or extruded and colored white with about 80% light transmission. Floating row covers can be narrow and cover a single row of crops, be medium wide and cover a crop bed with three to four closely spaced rows, or wide and cover multiple rows in widths of 3 to 15 m.

The advantage of floating row covers is their relative ease of application and the reduction in labor as compared with using plastic row covers and the hoops associated with their use. A disadvantage with the use of floating row covers is the potential abrasion of the leaf surface from the cover material during windy weather.

In newly established annual crops, row covers can protect plants from temperatures 1°C to 3°C below freezing for short durations. They reduce damage but do not offer complete protection when temperatures drop below −4°C. Slit row covers protect crops from −1°C to −2°C frosts only. The floating row cover protects against frosts of −2°C to −3°C.

High tunnels High tunnels are used by growers, particularly in the northern United States, who want earliness and improved quality and yields of a wide variety of high-value crops (Fig. 12.19). High tunnels generally have no or very low sophisticated heating systems.

High tunnels differ from low tunnels (row covers over single rows) in that they are constructed using hooped metal bows connected to metal posts, which are driven into the ground and evenly spaced. The tunnel often has a height of about

FIGURE 12.19 High tunnel used for season extension.
Courtesy of Dr. Bill Lamont, Penn State University.

2 m and a width of 4.3 m. The end walls use minimal framing to allow easy removal to accommodate free movement of power tillage equipment. The unit is covered with greenhouse-grade polyethylene, which is left on year-round or can be removed during the winter. Ventilation is often accomplished by rolling up the sides of the tunnel as far up as desired. Trickle irrigation tubing beneath a black plastic mulch cover is placed on the tilled soil within the high tunnel. The high tunnel is then planted with the desired crops.

Greenhouses   Greenhouses are used to provide the greatest control over many of the environmental conditions surrounding the plant. The greenhouse structure modifies the environmental conditions and allows crop production in regions and at times when outdoor production would not be optimum. Greenhouse production also involves the growing of specialty or niche marketed crops (for example, greenhouse-grown tomatoes or cucumbers).

Thermal blankets   Thermal blankets or fabrics (often polyethylene) are often used for overwintering containerized perennials (Fig. 12.20). The potted perennials

FIGURE 12.20   Thermal blankets used to reduce desiccation and provide some freeze protection for nursery crops.
Courtesy of Dr. Rick Bates, Penn State University.

FIGURE 12.21 Example of cold frames.

that are ready for storage are placed adjacent to one another in an upright position and covered with a material having insulating qualities. The insulating material can be the blanket or straw (often with a polyethylene sheet placed on top). Tall plants may be leaned over on to one another and then covered. The thermal blanket sheets are usually oriented in a north–south direction. Thermal blankets also trap and contain moisture sufficiently to eliminate the need for supplemental irrigation during the winter.

Cold frames and hot beds   A cold frame is a protected ground bed, usually sunken, with a removable glass or plastic roof that has no artificial heat source (Fig. 12.21). Heat is provided through the trapping of solar energy, the result of the greenhouse effect. Heat is stored in the soil and plants can be protected (often during the night) even though outside air temperatures may drop below freezing. A mat or blanket may be placed over the frame on cold nights to conserve heat. A cold frame is used to provide shelter for tender perennials, to harden off seedling plants, or to start cold-tolerant plants earlier than they can be started in open soil. They may also be used to overwinter summer-rooted cuttings of woody plants.

A hot bed is a heated cold frame and resembles in many ways a miniature greenhouse, providing the same benefits with limited space at minimal expense. A hot bed is a means for extending the growing season and is most often used to give an early start to warm-season vegetables. It may also be used to root cuttings of some woody plants.

Addition of Heat   Additional strategies for frost protection include heaters, wind machines, and sprinkler irrigation.

FIGURE 12.22   Wind machine for frost protection in a vineyard.

Heaters   Heaters have been used for cold protection of various crops for centuries. Most heaters are designed to burn oil and can be placed as free-standing units or connected by a pipeline network throughout the crop area. Propane, liquid petroleum, and natural gas systems have been used as energy sources for the heaters. Heaters are commonly used in areas where tree fruits are exposed to occasional frosts.

The hot gases emitted from the top of the burners initiate convective mixing in the crop area, tapping the important warm air source above the inversion. About 75% of a heater's energy is released in this form. The remaining 25% of the total energy is released by infrared radiation from the hot metal stack. This heat is not affected by wind and will reach any solid object not blocked by another solid object.

In the past, tree fruit growers have also burned old rubber tires, petroleum blocks and other smudges for frost protection. Some heat is added to the crop area by these fires, but there has been a misconception that the smoke acts like a cloud. Smoke does not provide the greenhouse effect of water vapor, because smoke particles are too large to block longwave radiation loss. In fact, smoke not only has no effect on outgoing radiation, but actually impedes warming in the morning, because smoke particles block incoming solar radiation.

Wind machines   Wind machines capitalize on the inversion development in radiation frost (Fig. 12.22). Their purpose is to circulate the warmer air down to crop level. They are not effective in an advective freeze. A single wind machine can protect approximately 4 ha, if that area is relatively flat and round. A typical wind machine is a large fan about 5 m in diameter mounted on a 10 m steel tower. The fan is powered by an industrial engine. Wind machines are sometimes used in conjunction with heaters. This combination is more efficient than heaters alone. Helicopters have been used as wind machines. They hover in one spot until the temperature has been increased sufficiently and then they move to the next area. Repeated visits to the same location are usually required.

FIGURE 12.23 Sprinkler irrigation for frost protection of apple trees. Courtesy of Kathy Demchak, Penn State University.

Sprinkler irrigation The freezing of water releases 80 calories of heat (latent heat of fusion). This phenomenon is exploited in the use of sprinkler irrigation as a method of frost protection (Fig. 12.23). The ice forming on the plant (Fig. 12.24) releases heat into tissue (and all other directions).

The water film on plants from sprinkle irrigation has to be maintained continuously as long as temperatures are low enough to freeze water, or until the ice starts to melt rapidly. Poor irrigation distribution can cause an excessive buildup of ice and under long periods of freezing, this excessive weight can cause limb breakage.

Sprinklers begin when the temperature reaches 0°C or 1°C and the dew point temperature is high. Some growers begin sprinkling at higher temperatures when the dew point is extremely low, because dry air cools faster than moist air. Sprinklers are typically used in fruit crops to protect flowers.

Chemicals Numerous materials have been evaluated since the mid-1950s for their effect on reducing frost and freeze injury. These fall into several categories but, in general, they have been materials that allegedly: (1) change the freezing point of the plant tissue, (2) reduce the ice-nucleating bacteria on the crop and thereby inhibit ice/frost formation, (3) affect growth, or (4) work by some "undetermined mode of action." Growth regulator applications that delay flowering will delay development of the flower and reduce frost and freeze injury.

Certain bacteria (such as *Pseudomonas syringae* and *Erwinia herbicola*) that naturally inhabit leaf surfaces can act as ice nucleators and accelerate the formation of

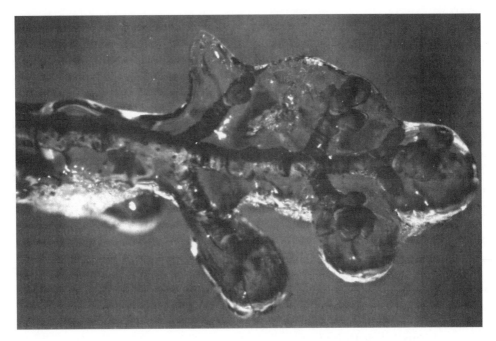

FIGURE 12.24   Plant limb protected from freeze injury by sprinkler protection and ice encasement.
Courtesy of Dr. Rick Bates, Penn State University.

ice crystals on the leaf surfaces. The surface ice quickly spreads to the intercellular spaces within the leaf leading to cellular dehydration. Genetically modified bacterial strains with reduced ice-nucleating characteristics can be sprayed on the foliage of valuable frost-sensitive crops such as strawberries to compete with native bacterial strains, minimize the number of ice-nucleating points, and provide a measure of frost protection to the plants.

Methods of Alleviating Excessive Heat

Shade is used to alleviate excessive heat for high-value crops. This can be done with plastic fabrics with varying percentage shade or with lath (wood or aluminum strips) house. Shading also decreases the photosynthetic rate.

Shading is practiced in greenhouse production during the warmer seasons of the year. Overhead irrigation of field-grown crops can alleviate high-temperature stress based on the principle that as water evaporates, it absorbs heat (580 cal/g at 20°C) and cools the plant surface. The process of cooling an air mixture with the addition of water is called evaporative cooling. A disadvantage of overhead irrigation for cooling is that it can also encourage diseases. Overhead irrigation is commonly used on wine grapes in California.

Influence of Vegetation Cover on Surface Temperatures

Energy available to a soil surface or covering is either used to move and store heat into the soil, move heat into the atmosphere, or evaporate water. The partitioning of energy into these processes depends on the properties of the surface. One of the most critical properties is the amount of green vegetation.

During High Temperatures   During warm weather, surfaces with vegetation utilize much of the available energy for evapotranspiration. This results in lower temperatures, as less energy is available to heat the air and foliage.

During Low Temperatures   Depending on leaf area index (LAI), vegetation reflects more solar radiation than most soils during the day. Transpiration by the vegetation lowers its temperature. This reduces the heat that it stores and that which is stored in the soil below it. The frost and freeze protective disadvantages of ground cover management must be weighed against the benefits such as erosion control and dust reduction.

Use of Plants for Energy Conservation around Homes and Buildings

Proper use of trees, shrubs, vines, and structures can modify the climate around a home dwelling to reduce heat gains in the summer and heat losses in winter (Fig. 12.25). Plants can protect a house from winter winds and provide shade from summer sun. Winter heating bills may be reduced by as much as 25% and summer cooling bills by 50% or more.

Summer Shade   Summer shade is provided by strategically locating plants along the sunny borders of a home. Vegetation that shades walls that would otherwise be exposed to sun intercepts solar energy that could otherwise overheat a home. In addition, the shade maintains an environment several degrees cooler than the temperature in sunlight. Tree arrangements that provide shade in the summer may be detrimental in the winter if they block solar heating.

Wind Protection   Up to 25% energy savings for heating of homes is possible if landscaping is used as windbreaks. An unprotected home loses more heat on a cold, windy day than on an equally cold, still day. Up to one-third of the heat loss may escape through the walls and roof by conduction. Wind also increases the convective air currents along outside walls and roof, thus increasing heat loss.

Infiltration or air leakage can account for as much as one-third of heating losses in some buildings. Cold outside air flows in through cracks around windows and doors, and even through pores in walls. Both windbreaks and foundation plantings can cut down the penetrative power of the wind. For example, an evergreen, properly placed, can divert cold winds from the home. The amount of money saved by a

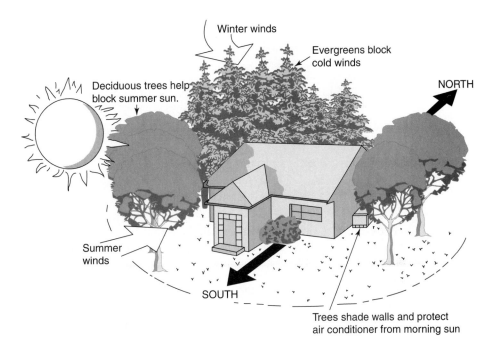

FIGURE 12.25   Effective placement of trees around a building can conserve energy and lower the cost of utility bills. In this example, three well-placed deciduous (leaf-losing) trees on the east, south, and west sides of a home will shade it from summer sun and potentially lower cooling costs. Trees can also produce savings in cold weather. Staggered rows of evergreen trees on the northwest side of the house (or the side with prevailing winter winds) will block harsh winds and potentially lower heating costs.
Used with permission from American Forests.

windbreak will vary depending on the climate of the area, the location of the home, and the construction material and quality of the home.

Green Roof Technology   Green roof technology (sometimes called green roofs, eco-roofs, nature roofs, or roof greening systems) originated in Germany over 30 years ago, and green roofs have become quite popular throughout Europe, mainly due to their positive environmental impacts. Green roof development involves the creation of contained green space on top of human-made structures (Fig. 12.26). A green roof is an extension of the existing roof, which involves a special root repelling membrane, a drainage system, lightweight growing medium, and plants.

Green roofs are living, vegetative roofing alternatives. Their greatest potential lies in their capacity to cover impervious surfaces with permeable plant material. Through the daily dew and evaporation cycle, plants on vertical and horizontal surfaces are able to reduce the urban heat island effect in the summer. The urban heat island effect is the difference in temperature between a city and the surrounding

FIGURE 12.26   Green roof on top of a building.
Courtesy of Dr. Dave Beattie, Penn State University.

countryside (Fig. 12.27). It is mainly due to the expanse in urban areas of hard and reflective surfaces, such as roofs and pavement, which absorb solar radiation and reradiate it as heat. Reduction of the urban heat island effect will also reduce the distribution of dust and particulate matter throughout the city.

Harvesting According to Accumulated Heat Units

For some crops, the time required to reach a harvestable stage may be expressed in terms of temperature-time values called heat units. Heat units are calculated in relation to time above a certain minimum. For example, if the minimum temperature for growth of a particular crop were 10°C, then a 24-hour period with a mean temperature of 15°C would provide 5 degree-days of heat units (as measured in °C). The harvest date is determined by an accounting of accumulated heat units. Harvesting according to accumulated heat units works well for some crops (such as corn and peas), but not so well for others.

Commodity Cooling to Extend Postharvest Life

Temperature is perhaps the single most important factor in prolonging the postharvest life of crops, and recommended storage temperatures and cooling methods will

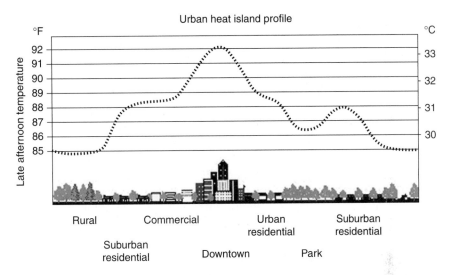

FIGURE 12.27   Diagram of a representative urban heat island temperature profile. On hot summer days, urban air temperatures can be 1°C to 4°C hotter than the surrounding countryside.
Source: EPA.

vary by the crop. The higher the temperature, the faster the metabolic rate and depletion of reserves in the harvested crop. Some crops are maintained at 0.5°C to 2°C immediately after harvest. Auxiliary cooling is required to reach and maintain recommended storage temperatures. Auxiliary cooling is often not available at the field. Proper handling procedures in the field, however, can reduce high-temperature stress of many crops.

Harvest Procedures to Reduce Temperature Stress

Many growers harvest their crops during cool morning or evening hours or when it is cloudy. This results in less heat in the crops (also called field heat) when the crops are harvested than if they were harvested during the middle of the day when exposure to sunlight and heat may be maximal and potentially intense.

Harvested crops or commodities are placed out of direct sunlight before and after loading onto trucks. Perishable commodities are generally shipped in refrigerated trucks. If the trucks are not refrigerated, they are usually vented to prevent buildup of heat and detrimental gases, such as CO_2 and ethylene. Although venting is important to prevent increases in temperature, excessive airflow may cause desiccation and reduce the quality of some crops. A loose-fitting or air-permeable, heat-reflective tarpaulin is alternatively used when transporting crops in open trucks during sunny conditions.

Cooling Procedures after Harvest

Several types of cooling procedures can be used on harvested crops and commodities. These differ in their rate and effectiveness of cooling, compatibility with specific commodities, and cost.

Room Cooling   Many crops are typically packed in bins and transported to a refrigerated room. Rapid cooling is dependent on the flow of cooled air around the bins. Proper stacking of the bins is important for effective cooling. Room cooling can be used with most commodities but may be too slow for some that require quick cooling. Room cooling is generally effective for storing precooled crops but in some cases it cannot remove field heat rapidly enough. For bins of product, room cooling is often inefficient because the density of the packed crops allows for little surface contact with the cooled air. Dehydration can also be a problem because of the constant airflow over the surface of the plant material.

Forced-Air Cooling   Forced-air cooling is similar to room cooling except the bins are vented and stacked. Cooled air is forced by a pressure gradient through the bins, allowing better contact of cool air with warm product. Pressure gradients are created by fans that draw air around and through the bins. Forced-air cooling is faster than room cooling. Although the cooling rate depends on air temperature and relative humidity and the rate of airflow through the packages, forced-air cooling is usually 75% to 90% faster than room cooling. Once the product is cooled, the forced air can be stopped and the product can be kept under refrigeration with little air movement and water loss. Forced-air cooling can be very energy efficient and is an effective way to increase the heat removal rate of a cooling room.

Hydrocooling   Hydrocooling is achieved by flowing chilled water over the product, rapidly removing the heat. At typical flow rates and temperature differences, water removes heat about 15 times faster than air. Hydrocooling can be used on most commodities that are not sensitive to wetting. Excessive free moisture can also encourage diseases.

Vacuum Cooling   Vacuum cooling is the best method for cooling crops that have a high ratio of surface area to volume, such as the leafy vegetables. Air is pumped out of a chamber containing the product. The partial vacuum causes water to evaporate from the surface of the produce. As water leaves the product due to reduced air pressure, it carries heat with it. Cooling is very rapid. Vacuum coolers can be energy efficient but are expensive to purchase and operate. Some vacuum coolers are portable and can be used in more than one location. This is especially cost effective when the harvest moves from one location to another with the change of season.

For each 6°C of vacuum cooling, about 1% of water is lost from the product. Some vacuum coolers add a fine spray of water to reduce water loss. Water loss from vacuum cooling of lettuce is only 0.5% to 1.5% and is usually not considered a problem for the product. After cooling, the commodity is moved to a cold room or refrigerated truck.

Evaporative Cooling   Evaporative cooling is accomplished by misting or wetting the commodity in the presence of a stream of dry air. It works best when the relative humidity of the air is below 65%. At best, however, it reduces the temperature of the product only 5°C to 8°C and does not provide consistent and thorough cooling. It is an inexpensive way of cooling produce.

Top or Liquid Icing   In top icing, crushed ice is added over the top of selected crops (commonly sweet corn and broccoli) by hand or machine. For liquid icing, a slurry of water and ice is injected into the produce packages through vents or hand holds without removing the package tops. Icing is very effective on dense packages that cannot be cooled with forced air. Two kg of ice will cool about 7 kg of produce from 30°C to 5°C.

Summary

Temperature is the measure of the average kinetic energy of molecules, whereas sensible heat is the energy gain or loss from molecules by the transfer of kinetic energy. Temperature in the field is affected by changes in altitude, season and latitude, and time of day. In general, air temperatures generally decrease with increasing elevation in an area. As latitudes increase, total solar radiation and temperatures decline and a greater seasonal variation in temperature is experienced. The maximum daily temperatures in a location often occur in the early afternoon.

Cold temperatures in a location can result in either frost or freeze conditions. White frost results from a thin coating of ice crystals deposited directly on surfaces at 0°C. Black frost occurs when the dew point is several degrees below 0°C and the water or cell sap in the vegetation freezes before sublimation on the surface occurs. When temperatures warm, the vegetation often blackens. Freezes occur in an area when cold air masses move into an area resulting in freezing temperatures.

Topographic factors that affect frosts and freezes include proximity to lakes and large bodies of water, terrain, and soil type. Large bodies of water serve as heat reservoirs in the fall and heat sinks in the spring, often resulting in reducing the risk of early fall frosts and unusually late spring frosts. Although higher elevations generally have lower surface air temperatures during the daylight hours, after sunset the cooler, denser air will settle in the lower elevations, resulting in temperature inversions and the potential formation of frost pockets. Clay or loam soils tend to warm more slowly in the spring and cool more slowly in the fall, whereas sandy soils tend to warm quickly in the spring and cool rapidly in the fall. Mucks and peat are soil types most subject to frost and freezes, particularly when soil surfaces are dry.

The stability of enzymes and biological processes are dramatically affected by temperature. Within certain temperature ranges, increases in temperature generally result in an increase in chemical reaction rates. The temperature coefficient or Q_{10} is used to describe the effect of temperature on a chemical process. Nonenzymatic reactions have a Q_{10} of about 1.2, whereas enzymatic reactions have a Q_{10} of 2 or more. The three cardinal temperatures for plants (minimum, optimum, and maximum) vary with plant species, stage of development, tissue, process of concern, and environmental factors.

Chilling injury to plants occurs at temperatures at or slightly above 0°C, whereas freezing injury occurs when ice forms inside the plant at temperatures slightly below 0°C. Tolerance to chilling and freezing injury varies with plant species and varieties within species. Other cold-temperature effects on plants include winter desiccation, root injury, and vernalization. High temperatures can increase plant respiration rates and decrease photosynthetic rates, alter activity of enzymes, and denature proteins. Other plant responses to high temperature include dehydration, sunscald, and reduced quality of marketable product.

A number of factors can affect the risk of frost and freeze injury to field-grown crops including regional and local climates, the area's topography, the microenvironment of the site, and the hardiness of the plants in the area. Crop producers can minimize or avert cold-temperature stress to growing crops by using season extenders (such as row covers, high tunnels, and cold frames), utilizing heaters or heating systems, or applying chemicals to the crop that help reduce frost and freeze injury. Methods to alleviate excessive heat also include use of shade and overhead irrigation.

Review Questions

1. What is the difference between white frost and black frost?
2. Match the following.

_____ Calorie

_____ Temperature

_____ Heat

_____ Specific heat

_____ Heat of fusion

_____ Heat of evaporation

a. The amount of heat required to change 1 g of substance from solid to liquid

b. The movement of kinetic energy from warmer to cooler regions

c. The number of calories needed to change the temperature of 1 g of a substance by 1°C

d. The amount of heat required to raise 1 g of water 1°C

e. The amount of heat required to change 1 g of substance from liquid to gas at its boiling point

f. The velocity or kinetic energy of molecules

3. Where would you normally find frost pockets in a field?
4. What does a $Q_{10} = 2$ mean?
5. What conditions could cause frost cracking of trees?

6. Match the following.

____ Plants with little to no freezing tolerance

____ Plants with limited freezing tolerance

____ Plants capable of deep supercooling

a. Deciduous forest and fruit trees
b. Corn, cucumbers, beans
c. Potatoes, cabbage

7. List the three cardinal temperatures.
8. How does chilling injury differ from freezing injury?
9. Match the following.

____ Cold period needed to induce flowering in some plants

____ Point where photosynthesis equals respiration

____ Process by which light is converted into chemical energy

____ Process of the controlled release and utilization of stored energy

____ Period of rest

a. Photosynthesis
b. Respiration
c. Compensation point
d. Dormancy
e. Vernalization

10. Why are some fruit trees planted on the sides of hills?
11. List three season extenders for plants.
12. How do large bodies of water affect the temperature in a region?
13. Define heat unit and how it may be used.
14. How is kinetic energy different from potential energy?
15. How is air temperature generally affected by latitude, altitude, slope, season, and time of day?
16. What is a temperature inversion and what are some potential concerns when an inversion occurs in a region?

Selected References

Boote, K. J., and F. P. Gardner. 1998. Temperature. In *Principles of ecology in plant production,* ed. T. R. Sinclair and F. P. Gardner, 208. New York: CABI Publishing.

Caplan, L. A. 1988. *Effects of cold weather on horticultural plants in Indiana.* Purdue University Cooperative Extension Service Publication HO-203.

Fitter, A. H., and R. K. M. Hay. 1987. *Environmental physiology of plants.* 2nd ed. New York: Academic Press.

Hale, M. G., and D. M. Orcutt. 1987. *The physiology of plants under stress.* New York: John Wiley.

Janick, J., R. W. Schery, F. W. Woods, and V. W. Ruttan. 1974. *Plant science.* 2nd ed. San Francisco: W. H. Freeman.

Lambers, H. F., S. Stuart III, and T. L. Pons. 1998. *Plant physiological ecology.* New York: Springer-Verlag.

Lehninger, A. L. 1975. *Biochemistry.* 2nd ed. New York: Worth.

Levitt, J. 1980. *Responses of plants to environmental stresses. Vol. 1: Chilling, freezing, and high temperature stress.* 2nd ed. New York: Academic Press.

Parsons, L. R. 1995. *Cold protection by irrigation: Dew point and humidity terminology.* University of Florida Cooperative Extension Service Fact Sheet H.S.-76.

Perry, K. B. 1997. *Frost/freeze protection for horticultural crops.* North Carolina Cooperative Extension Service Publication HIL-705.

Salisbury F. B., and C. W. Ross. 1992. *Plant physiology.* 4th ed. Belmont, CA: Wadsworth.

Stern, K. R. 1991. *Introductory plant biology.* 5th ed. Dubuque, IA: Wm. C. Brown.

Taiz, L., and E. Zeigler. 1998. *Plant physiology.* 2nd ed. Sunderland, MA: Sinauer.

Talbot, M. T., and D. Baird. 1998. *Psychometrics and postharvest operations.* University of Florida Cooperative Extension Service Circular 1097.

Selected Internet Sites

http://www.epa.gov/   Home site of the Environmental Protection Agency (EPA).

http://www.epa.gov/oar/   Government site providing information on air pollution, clean air, and air quality. Good links for acid rain, ozone depletion, and climate change. Office of Air and Radiation, EPA.

http://www.eia.doe.gov/   Official energy statistics of the U.S. government. Energy Information Administration, DOE.

http://www.nrcs.usda.gov/technical/airquality.html   Air quality information and report of Agricultural Air Quality Task Force. Natural Resources Conservation Service, USDA.

http://www.wcc.nrcs.usda.gov/   Government weather site containing observations, forecasts, maps and models, weather safety, and education. National Water and Climate Center, USDA.

http://www.nws.noaa.gov/   Government weather site containing observations, forecasts, maps and models, weather safety, and education. National Weather Service, National Oceanic and Atmospheric Administration.

13

Atmospheric Gases

Ninety-five percent of the earth's atmosphere and all of its life are concentrated in a layer only 11 to 20 km thick surrounding the earth's crust. Atmospheric air contains many gases as well as water vapor (Fig. 13.1). Dry air is a mixture of nitrogen (78% by volume), oxygen (21%), argon (0.93%), carbon dioxide (0.03%), and other minor constituents (0.04%). Moist air is a mixture of these gases and water vapor. The relative percentage of water vapor in the air can vary from zero (dry air) to 100% (saturation), and the amount of water vapor in atmospheric air depends on temperature and pressure.

This chapter begins by describing the common atmospheric gases with an overview of their role in plant growth and development. The role of greenhouse gases in global warming is also presented. A discussion on atmospheric gas effects on plants primarily centers on their effects on stomata and photosynthesis, transpiration, wilting, and incidence of diseases and insects. The chapter concludes with a presentation on selected agricultural practices that can affect crop growth and production by influencing the gases around the plant, including plant spacing, carbon dioxide enrichment in the greenhouse, the use of antitranspirants to reduce water loss, and modified atmosphere storage of fruits and vegetables.

Background Information

Evolution of the Earth's Atmosphere

The present gaseous composition of the earth's atmosphere is the result of various processes over the past 4.6 billion years. Most of the gases in the atmosphere probably were created in the first few 100 million years of the earth's life. It is believed that the earth's early atmosphere did not contain any free oxygen and that all water probably existed as water vapor. After about 500 million years, the earth is suspected to have cooled

FIGURE 13.1   Water vapor image of the earth from space (image from September 5, 1995). Source: Earth Observatory, NASA.

sufficiently so that some of the water vapor condensed to a liquid. The oceans were subsequently formed and large amounts of CO_2 dissolved into it, substantially lowering the atmospheric CO_2 concentration.

The relatively simpler forms of life coupled with changes in environmental factors of the ecosphere altered the atmosphere to the present condition. It is speculated that living organisms appeared on the earth about 3.5 million years ago. The photosynthetic blue-green algae were probably the first organisms to exist, and they removed additional CO_2 from the atmosphere and released O_2. As a consequence of increased atmospheric O_2, the ozone layer was formed. The current atmospheric composition probably was attained before the appearance of highly complex life forms.

Some Basic Properties of Gases

A gas is defined as that form of matter that expands without limit to fill the available space of a container. The pressure of a gas is uniform on all walls of any container regardless of its shape or size. According to Boyle's law, the product of the volume (V) and the pressure (P) is constant (which is proportional to the number of moles of the gas) for a particular mass (or number of moles) of a gas at a single temperature, or

$$P \times V = \text{a constant, at constant temperature,}$$

where P = pressure,
       V = volume

Because the pressure on the walls of a container is the result of collisions of the gas molecules with the walls of the container, an increase in volume of the container would result in a decrease in the number of collisions of gas molecules per unit area, proportional to the volume increase.

Because $P \times V$ equals a constant that is proportional to the number of moles of a gas and because both P and V are linear functions of T, the ideal gas law states that at a constant pressure and temperature, the volume of the gas is proportional to the number of moles of the gas, or

$$PV = nRT,$$

where P = pressure,
  V = volume,
  n = number of moles of the gas,
  R = ideal gas constant $(8.314 \, \text{J K}^{-1} \, \text{mol}^{-1})$,
  T = temperature

In a mixture of gases, the total pressure of the mixture is the sum of the pressures each gas would exert if it were the only gas present. Dalton's law of partial pressures states that the total pressure of a mixture of gases is equal to the sum of the partial pressures of its components, or

$$P = P_1 + P_2 + ... + P_n,$$

where P = total pressure of the gas mixture,
  P_n = partial pressures of each gas present

According to the ideal gas law, Dalton's law of partial pressures then becomes

$$P = n_1RT/V + n_2RT/V + ... + n_nRT/V = \Sigma n_1RT/V.$$

The most familiar measure of a concentration of gas is the mole fraction and not partial pressure. The mole fraction of a gas in a mixture is the number of moles of the gas of interest in a mole of the mixture. Other expressions of mole fraction include partial volume of a gas divided by the volume of the mixture, and the partial pressure of a gas divided by the total pressure.

Common Atmospheric Gases

Nitrogen Gas (N_2)   Gaseous nitrogen (N_2) comprises about 78% of the volume of the earth's atmosphere. Other nitrogen gases (NH_3, N_2O, NO, NO_2) are released into the atmosphere as a consequence of increasing consumption of fossil fuels. These gases (especially N_2O) absorb infrared radiation and influence global warming, and they are often classified as greenhouse gases.

Certain kinds of nitrogen-fixing bacteria and some blue-green algae are the only organisms that can convert (reduce) or fix N_2 directly into useful organic forms. Most plants absorb nitrogen in the form of nitrate (NO_3^-) and ammonium (NH_4^+) salts dissolved in the soil solution. Nitrogen is a component of many plant compounds including chlorophyll, amino acids, proteins, nucleic acids, and organic acids. Because N is contained in the chlorophyll molecule and proteins, a deficiency of N typically results in a chlorosis (or yellowing) of the plant.

The transformation of nitrogen-containing compounds to its various forms is referred to as the nitrogen cycle (see Important Nutrient Cycles in Chapter 18). Nitrogen is removed from the atmosphere and deposited at the earth's surface mainly by nitrogen-fixing organisms and by way of lightning through precipitation. The addition of this nitrogen to the earth's surface soils and various water bodies supplies the needed nutrition for plant growth. Nitrogen returns to the atmosphere primarily through biomass combustion and denitrification.

Nitrification occurs as a result of NH_4^+ in the presence of *Nitrococcus* or *Nitrosomas* bacteria being transformed to NO_2^- (nitrite ion). This is then converted to NO_3^- (nitrate ion). Denitrification is accomplished by certain bacteria that convert nitrates back to atmospheric nitrogen, often in waterlogged soils (hypoxic or anoxic conditions).

Oxygen Gas (O_2)   Oxygen (O) is derived from H_2O during the Hill reaction of photosynthesis and is utilized during respiration. It is also a requirement of oxidative-reductive chemical reactions and is a component of organic compounds, such as simple sugars. Oxygen is necessary for all oxygen-requiring reactions in plants, including nutrient uptake by roots. Two important oxygen reactions are photosynthesis ($12H_2O + 6CO_2$ + light energy and chlorophyll $\rightarrow C_6H_{12}O_6 + 6O_2 + 6H_2O$) and cellular respiration ($C_6H_{12}O_6 + 6O_2 \rightarrow 6CO_2 + 6H_2O$ + energy).

Carbon Dioxide (CO_2)   Although the atmospheric CO_2 concentration is relatively low, the carbon cycle (see Important Nutrient Cycles in Chapter 18) is very important to plants and animals. The movement of CO_2 between the atmosphere and carbon sinks on land and in the oceans is primarily by photosynthesis and respiration. These processes remove some but not all of the anthropogenic CO_2 emissions produced each year.

Atmospheric CO_2 also contributes to the radiation balance and temperature of the earth as a greenhouse gas, contributing to the greenhouse effect and global warming. Larger amounts of CO_2 are dissolved in the water that covers approximately two-thirds of the earth's surface. CO_2 is a component of organic compounds such as sugars, proteins, and organic acids. These compounds are used in structural components, enzymatic reactions, and nucleic acids, among others.

CO_2 is an essential plant nutrient for photosynthesis and is a bioregulator of stomatal conductance and leaf gas exchange. CO_2 enters the leaf via the stomata and interacts with the photoreaction pigments of photosynthesis in the chloroplast. To reach the chloroplast, CO_2 dissolves and diffuses in the cellular water. As a result of the subsequent carbon assimilation reactions of photosynthesis, plants are ultimately composed of about 40% to 45% carbon on a dry-weight basis.

Water Vapor (H_2O)   The water cycle (see The Water Cycle and the Groundwater System in Chapter 16) illustrates water's interrelationships with soil, plants, and the atmosphere. Evaporation of water occurs when molecules of liquid water acquire sufficient kinetic energy so that they enter the atmosphere as water vapor. The amount of water vapor that can possibly be present in the air increases with temperature. Condensation of water occurs when water vapor molecules with insufficient energy become liquid. The atmosphere also may have small particles (often called condensation nuclei or "seed" particles) that facilitate condensation at humidities at or below 100%.

The physical and thermodynamic properties of moist air can be expressed as humidity, relative humidity, dry bulb and wet bulb temperatures, and dew point temperature. Humidity is a generic term for the amount of water vapor in air. The most commonly used terms for describing moist air are relative humidity and dew point temperature. Relative humidity is the amount of water present in the air as a percentage of what could be held at saturation (100% relative humidity) at the same temperature and pressure.

Although commonly used, relative humidity may be misleading because it depends on the air temperature. Warm air holds more water vapor than cold air. For example, the relative humidity could be 70% at 4°C or 70% at 32°C, but the amount of water vapor in the air would be less at the cooler temperature even though the relative humidity values were identical.

Dew point temperature is the temperature at which air reaches water vapor saturation, condensation begins, and dew forms. This is the often preferred term to quantify the atmospheric water content. When air temperature is below the dew point temperature (but above 0°C), water vapor condenses as a liquid. In general, the more water vapor in the air, the higher the dew point temperature.

Dew forms during a clear night because the infrared (heat) loss from a surface will result in the surface cooling faster than the air above it. As the surface cools due to infrared loss, the air in direct contact with the surface will also cool. Eventually the surface cools sufficiently so that the dew point temperature of the air in contact with the surface is reached (or is lower). At this point, the air next to the surface is water vapor saturated, and water droplets as dew will condense on the surface. When the temperature is lower than 0°C, the water vapor cannot change into a liquid and instead changes directly into the ice crystals of frost. When water changes directly between the vapor and the solid states, skipping the liquid state, the process is called sublimation. When the dew point temperature is below 0°C, it is often called the frost point temperature. The dew point is important on freeze nights because water vapor in the air can slow the rate of temperature fall and it sets the lower limit to which air temperature can drop.

A psychrometer is used to determine atmospheric humidity by comparing the reading of a dry bulb thermometer and a wet bulb thermometer. The wet bulb thermometer is maintained moist with the use of a wetted sleeve and requires air to move over it for an accurate reading. With a sling psychrometer, air flow is created by rotating the thermometers through the air by hand. With a fan-ventilated psychrometer, a fan blows air across the two thermometers. Sling psychrometers work well at temperatures above freezing, but are more difficult to operate at temperatures below freezing and in confined spaces. The dry bulb temperature is the temperature of the ambient air. The wet bulb temperature is defined as the lowest temperature to which air can be cooled solely by the addition of water. When the air is saturated with water vapor (i.e., the relative humidity is 100%), the dew point temperature, wet bulb thermometer, and dry bulb thermometer are equal. The cooler the reading of the wet bulb thermometer is than the dry bulb thermometer, the lower the air humidity.

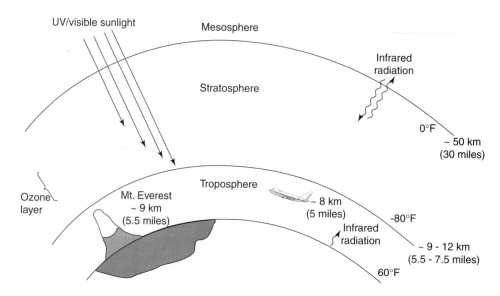

FIGURE 13.2   The earth's atmosphere has layers, which are actually characterized by how the temperature of the atmosphere changes with altitude. The troposphere begins at the earth's surface, which acts as a source of heat resulting from absorption of visible sunlight. The temperature decreases with height in the troposphere, and so the air is well mixed in this region. Weather phenomena such as thunderstorms and clouds occur in this layer. The stratosphere begins between 9 and 12 km (or 16 and 17 km in the tropics). The stratosphere is heated from above (absorption of solar ultraviolet radiation by oxygen and ozone) and temperature increases with altitude. This region has much slower mixing. The ozone layer resides in the stratosphere. At about 50 km (30 mi), temperature begins to decrease with altitude again and the mesosphere begins.
Source: The Aeronomy Laboratory, Office of Oceanic and Atmospheric Research (OAR), National Oceanic and Atmospheric Administration (NOAA).

The Vertical Structure of the Atmosphere

As altitude above the earth's surface increases, atmospheric pressure decreases and the number of molecules of any gas decreases per unit volume of atmosphere. Many humans, animals, and plants can become adapted to the low oxygen partial pressures of high mountains if they are slowly acclimated to lower atmospheric pressures. CO_2 can also become limiting at high altitudes, although its relative concentration remains at about 0.03%.

The earth's atmosphere can be divided into several vertical components, each with its own characteristic influence on temperature (Fig. 13.2). The atmospheric components are the troposphere, stratosphere, and mesosphere, as well as the ozone layer. The troposphere is the only component that directly influences plant growth and development, though disturbances or changes in other portions of the atmosphere can affect climate (including solar energy), which can influence plant growth.

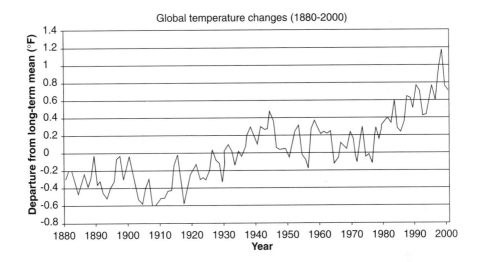

FIGURE 13.3   Global temperature changes from 1880 to 2000.
Source: U.S. National Climate Center, Environmental Protection Agency.

Troposphere   The troposphere is the lower layer (~ 9 to 12 km) of the earth's atmosphere, where temperatures decrease with height. The surface of the earth, which is heated by radiation, is the source of heat for the lower atmosphere. The troposphere is the area where weather occurs.

Stratosphere   The stratosphere is located above the troposphere and extends to about 50 km above the earth's surface. The temperature in this region initially stops decreasing with height and soon begins to increase. Large amounts of ozone in this region form the ozone layer, which absorbs a large amount of often harmful UV radiation from the sun. Very little water vapor occurs in this region.

Mesosphere   The mesosphere is located above the stratosphere from 50 to 80 km. In this area temperature falls greatly with increasing height above the earth.

The Greenhouse Effect

A slight but gradual increase in global warming has occurred since about 1880 (Fig. 13.3) in both the Northern and Southern Hemispheres. This warming has been attributed to an increase of selected atmospheric gases, commonly referred to as greenhouses gases (Fig. 13.4). This is often illustrated by the correlation of the global temperature with increasing global concentrations of CO_2 (Fig. 13.5).

Carbon dioxide and certain other trace gases (CH_4, N_2O, O_3, and chlorofluorocarbons) absorb (and emit) thermal infrared radiation in various wavebands across the 3 to 100 μm (3,000 to 300,000 nm) wavelengths. Thus, as energy is emitted from the earth's surface, some of it can be absorbed by CO_2 and other thermal

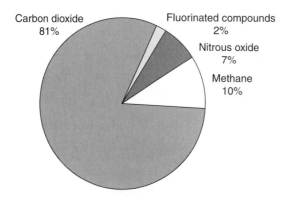

**1998 Greenhouse gas emissions
in the United States**

Carbon dioxide
81%

Fluorinated compounds
2%

Nitrous oxide
7%

Methane
10%

FIGURE 13.4   Greenhouse gases and their relative percent emission in 1998.
Source: U.S. Environmental Protection Agency.

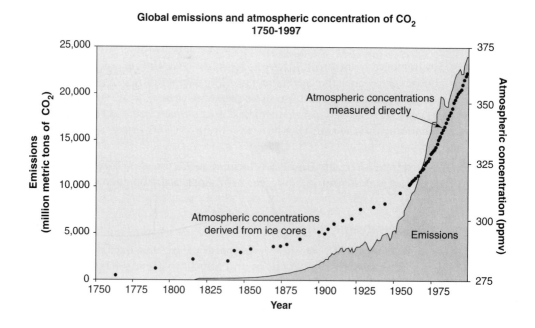

**Global emissions and atmospheric concentration of CO_2
1750-1997**

Atmospheric concentrations
measured directly

Atmospheric concentrations
derived from ice cores

Emissions

FIGURE 13.5   Global emissions and atmospheric concentration of CO_2 from 1750 to 1997.
Source: Carbon Dioxide Information Center, U.S. Environmental Protection Agency.

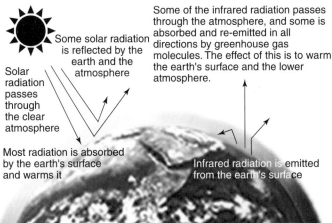

Greenhouse effect

Some of the infrared radiation passes through the atmosphere, and some is absorbed and re-emitted in all directions by greenhouse gas molecules. The effect of this is to warm the earth's surface and the lower atmosphere.

Some solar radiation is reflected by the earth and the atmosphere

Solar radiation passes through the clear atmosphere

Most radiation is absorbed by the earth's surface and warms it

Infrared radiation is emitted from the earth's surface

FIGURE 13.6   Diagram of the greenhouse effect.
Source: U.S. Environmental Protection Agency.

radiation absorbing gases. After this energy has been absorbed, it has an equal probability of being reradiated downward and upward. This process of absorption and reradiation can occur repeatedly so that a substantial portion of the thermal radiation is reradiated back to the earth's surface. Consequently, the earth's surface is prevented from being cooled to the extent that would occur if there were no atmosphere with greenhouse gases. This process is often called the greenhouse effect (Fig. 13.6). The earth system exists in thermal equilibrium over the long term. That is, incoming solar radiation energy at the outer limits of the atmosphere must be balanced by the same amount of total outgoing radiation energy at the top of the atmosphere.

The concentrations of the greenhouse gases in the atmosphere provide the earth with an average surface temperature of about 15°C. Without the greenhouse gases in the atmosphere, the average temperature at the earth's surface would be about -18°C. Therefore, some concentration of greenhouse gases is essential for life on earth. However, a rise in these gas concentrations would be expected to cause additional global warming. A temperature increase of 1.5°C to 5.5°C is frequently projected as a result of a doubling of ambient CO_2 concentration. A global warming of 10°C would seriously affect climates, ocean levels, glaciers, as well as crop production and food supplies.

Effects on Plants

Stomata

Stomata consist of two kidney-shaped guard cells in the epidermal layers. As the guard cells absorb water and swell, the stomata open. Because the cuticles of the leaf greatly restrict diffusion or water loss from leaves, water vapor and other gases must

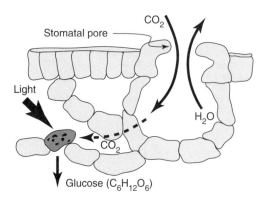

FIGURE 13.7   For photosynthesis to occur, CO_2 diffuses through the stomata, then through the intercellular air spaces, and ultimately into cells and chloroplasts.
Source: Earth Observatory, NASA.

pass through the opening between the guard cells. For photosynthesis to occur, CO_2 diffuses through the stomata, then through the intercellular air spaces, and ultimately into cells and chloroplasts (Fig. 13.7). Stomata are often found in top and bottom leaf surfaces (epidermis) but vary in shape and frequency with plant species.

Environmental Effects on Stomata   For most plants that undergo C3 and C4 photosynthesis, stomata open at sunrise and close in darkness, allowing for entry of CO_2 needed for photosynthesis during the daytime. Bright light often induces a wide opening of the stomata. In general, high temperatures cause stomata to close probably due to water stress.

The aperture of stomata is also affected by the CO_2 concentration in the air and air movement. The boundary layer around a leaf (the thin film of still air that hugs the surface of the leaf) is relatively thick in still air, and this constitutes a major resistance to the flux of H_2O as well as CO_2. A slight increase in the air turbulence or wind speed around the leaf will greatly reduce the thickness of the boundary layer, increase stomatal opening, and increase transpiration water loss from the plant. However, further increases in wind speed may reduce transpiration, especially for sunlit leaves, because the wind speed will cool the leaf, reducing the vapor pressure gradient and stomata may close or partially close. In some situations, stomata sometimes partially close when the leaf is exposed to gentle breezes, probably because of the additional CO_2 provided by the breeze.

Photosynthesis

CO_2 is a substrate for photosynthesis and an increase in CO_2 results in an increase in photosynthetic rate. This occurs because ambient CO_2 levels are below concentrations that saturate photosynthesis. In general the photosynthetic rate will increase with increases in CO_2 until a maximum rate is reached.

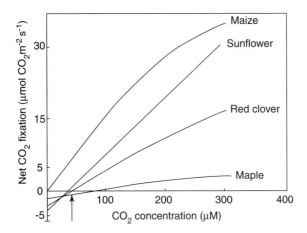

FIGURE 13.8   Representative photosynthesis response curves of C3 plants (sunflower, red clover, and maple) and a C4 plant (maize) to CO_2. C4 plants have a lower CO_2 compensation point (where net photosynthesis is zero) than C3 plants. C4 plants have a higher photosynthetic rate (at least in high light) than C3 plants with CO_2 concentrations at or below current ambient levels (which is now about 360 ppm).

Plant response to CO_2 differs according to whether they are C3 plants, which utilize the 3-carbon pathway for converting CO_2 to carbohydrates, or C4 plants, which utilize the 4-carbon pathway (C3 and C4 plants are discussed in depth in Chapter 6). In general, C3 plants (such as sunflowers, red clover, and maple trees) respond more to CO_2 enrichment than C4 plants (such as maize) (Fig. 13.8).

For C3 plants, increasing the CO_2 above the CO_2 compensation point (the point where CO_2 fixed by photosynthesis and CO_2 produced by respiration is equal) stimulates photosynthesis over a wide concentration range. At low to intermediate CO_2 concentration, photosynthesis is limited by rubisco (D-ribulose-1,5-bisphosphate carboxylase/oxygenase) capacity. At high CO_2 concentration, photosynthesis is limited by the ability of the Calvin cycle enzymes to regenerate acceptor molecules.

In C4 plants and plants that undergo crassulacean acid metabolism (CAM plants), photosynthesis rates saturate at lower CO_2 concentrations (approximately current atmospheric levels) than C3 plants. This reflects effective CO_2 concentrating mechanisms in C4 and CAM plants. In addition, plants with C4 metabolism have a CO_2 compensation point of nearly zero, reflecting low levels of photorespiration. The differences between C3 and C4 plants are not apparent at low CO_2 concentrations, a condition that shifts the activity of the chloroplast enzyme rubisco toward carboxylation.

Plants with C4 metabolism require less rubisco to achieve a given rate of photosynthesis, therefore they often require less nitrogen to grow. They also require lower rates of stomatal conductance for a given rate of photosynthesis and tend to use water more efficiently. Plants with CAM metabolism open their stomata at night and close them during the day. The CO_2 assimilated during the night is fixed into malate. The plant's limited capacity to store malate further restricts CO_2 uptake.

Transpiration

Transpiration is the evaporative loss of water (as vapor) from plants, mostly through their stomata. Stomata in most non-CAM plants open at dawn in response to increasing light intensity. As leaf temperatures continue to increase with increases in light intensity, stomata may eventually close.

The driving force for transpiration is the differences in vapor pressure of water within the leaf and in the atmosphere. The vapor pressure gradient is influenced by humidity and temperature. The vapor pressure doubles for every 10°C increase in temperature. For example, air at 100% relative humidity at 21°C will have 50% relative humidity at about 33°C. Therefore, warm air can hold more water than cold air.

Another important positive factor of an increase in CO_2 is an overall increase in plant water-use efficiency. This increase generally occurs as increased CO_2 leads to the partial closure of stomata. Thus, transpiration (water evaporation from foliage) may be reduced even while the rate of photosynthesis rises. This reduction in transpiration with increased CO_2 levels could also contribute to future global warming as land areas with intense vegetation warm to a level above what is predicted due to the greenhouse effect alone (Fig. 13.9).

Wilting

Wilting is the result of a reduction in cell turgor that reduces the rigidity (turgidity) of the plant. This occurs as a result of water stress or other stresses to the plants. Incipient wilting occurs when plants wilt under weather conditions favoring rapid transpiration, even when soil moisture is at field capacity. A plant can recover from

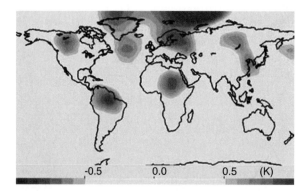

FIGURE 13.9 Computer model calculation of the effect of carbon dioxide on plant physiology and global climate. As carbon dioxide increases, vegetation may evaporate less water and cause the land to heat up. This map shows additional heating (over and above the conventional carbon dioxide greenhouse effect) over the continents due to this phenomenon, for doubled current carbon dioxide concentrations (700 ppm). Source: Earth Observatory, NASA.

incipient wilting if transpiration is stopped or reduced. Permanent wilting point occurs when water uptake ceases and plant tissue and cells are damaged. Plants may recover from permanent wilting (to some degree) if water is provided to the plant roots.

Diseases and Insects

An increase in relative humidity around the plant often increases the incidence of diseases and insects. Higher humidity environments are generally preferred conditions for the reproduction and growth of many insects and diseases. Higher humidities increase the likelihood that dew point will be reached and more free liquid precipitates on leaves.

Agricultural Technologies That Affect Gases

Spacing and Crowding

CO_2 and water in the presence of light is utilized by plants during photosynthesis to form carbohydrates. Plants usually receive a sufficient amount of CO_2 for photosynthesis as long as there is adequate ventilation and freely circulating air. CO_2 can become limiting in the field when a great amount of plant material is crowded into a small volume or space (such as plants in close spacings) and in high light and optimum temperatures. Slight winds are often sufficient to replenish the CO_2 to ambient levels, but if winds are not present CO_2 levels can be reduced to levels that could reduce photosynthetic rates. Irregardless, CO_2 enrichment in the field does not appear to be feasible and is not generally practiced.

CO_2 Enrichment in the Greenhouse

CO_2 can become limiting in greenhouses when temperatures and light levels are optimum for photosynthesis but the greenhouse has inadequate ventilation to replenish the CO_2 used by the plants (often the result of the house being closed to reduce heat loss and CO_2 from respiration is not enough). This occurs primarily in the winter and results in reduced photosynthesis rates. The greenhouse environment can be enriched with CO_2 during the daylight with CO_2 generated by the burning of propane or natural gas or by using liquid-solid CO_2 injectors. The goal of carbon dioxide enrichment is to raise its concentration to about 1,000 to 1,500 ppm. This high level of CO_2 must be provided along with bright light for it to be beneficial to plants.

Antitranspirants

Antitranspirants are materials that are applied to plants to reduce water loss from transpiration by covering the surfaces of the foliage with a barrier impermeable to water. Because the materials are impermeable to water, their use may also close stomata (reducing transpiration), increase leaf heating, and reduce photosynthesis. Another chemical approach to reduce transpiration would be to use a substance that is more permeable to CO_2 than water. Antitranspirant use is currently limited to such crops as Christmas trees and nursery crops (to reduce transplant shock).

Modified Atmosphere Storage of Fruits and Vegetables

Modified atmosphere (MA) or controlled atmosphere (CA) refers to alteration of gases in a storage chamber. Alteration of the gases is typically done in airtight rooms in conjunction with normal cooling procedures for the commodities (such as apples and pears). The primary function of modifying gases is to reduce respiration and plant ethylene (C_2H_4) production. This is usually accomplished by increasing CO_2 levels and reducing O_2 levels as compared to normal atmospheric concentrations. Ordinarily, air is 78% N_2, 21% O_2, and 0.03% CO_2. In MA or CA for some leafy vegetables, the O_2 is reduced to 1% to 3%, and the CO_2 is increased to 15% to 18%, with N_2 making up the difference. Carbon monoxide (CO) may also be injected in an MA/CA facility at $\approx 5\%$ to inhibit some pathogens. Modified atmospheres can be very hazardous to human health and extreme precautions are taken by workers in MA/CA facilities.

Air Pollution

Air pollution, depending on the location, day, and weather conditions, can be a constituent of the atmosphere. When air pollutants are present in relatively high concentrations and for a sufficient period of time, plants may exhibit visual injury and, as a consequence, have reduced yields or marketability. Air pollutants at concentrations less than that required to result in visual injury can still reduce yields due to air pollutants' effects on photosynthesis or other metabolic processes. Air pollution is covered more in-depth in Chapter 14.

Summary

The earth's atmosphere contains many elements and compounds needed by plants for growth and development. The forms and pathways of nitrogen, oxygen, carbon, and water as they cycle through the ecosystem is illustrated in the nitrogen cycle, oxygen cycle, carbon dioxide cycle, and water cycle, respectively. Nitrogen is a component of chlorophylls, amino acids, proteins, nucleic acids, and organic acids. Oxygen is required for respiration and carbon dioxide is essential for photosynthesis. Water is needed by plants for transpiration and nutrient uptake, as well as for photosynthesis. The greenhouse effect is a product of the effects of carbon dioxide and other trace gases on the energy balance within the earth atmosphere. Increasing concentrations of the greenhouse gases has contributed to the slight global warming that has been observed since 1880.

Stomata, often found on the top and bottom leaf surfaces, regulate CO_2 uptake and water vapor loss by opening and closing as regulated by certain environmental stimuli. Stomata generally open at sunrise in response to increasing intensities of light and close in darkness (except for CAM plants). Temperatures and winds also influence stomatal aperture, with high temperatures and great wind speeds independently generally causing stomata to close. Plant responses to CO_2 differ according to whether they are C3 or C4 plants, with C3 plants responding more to CO_2 increases than C4 plants. C4 and CAM plants saturate their photosynthetic rates at lower concentrations of CO_2 than C3 plants, reflecting CO_2 concentration mechanisms in these plants.

Transpiration and wilting involve the loss of water from the plant. Differences in vapor pressure of water within the leaf and the atmosphere serve as the driving force for transpiration. If transpiration is occurring rapidly, plants may lose cell turgor pressure and wilt. As the humidity increases around the plants, an increase in the incidence of injury from insects and diseases often occur.

Although not a common occurrence, CO_2 could become limiting in the field if a great amount of plant material is crowded into a small amount of space and conditions are optimal for photosynthesis. A slight breeze is often sufficient to replenish CO_2 in this situation. CO_2 enrichment is practiced under certain greenhouse conditions to increase photosynthetic rates and plant production.

Antitranspirants are applied to some crops to reduce transpiration and water loss, but they may also reduce overall plant growth. Increased storage life of some fruits and vegetables can be accomplished through modifying the atmosphere surrounding the fruit in storage. In general, increasing CO_2 levels and reducing O_2 levels (as compared to ambient levels) reduces transpiration loss of water from the fruits and reduces the effect of ethylene in promoting senescence during storage.

Review Questions

1. Match the following gases with the appropriate percent composition of air.

   a. Nitrogen   _____ 0.04%

   b. Carbon dioxide   _____ 21%

   c. Argon   _____ 78%

   d. Oxygen   _____ 0.03%

   e. Other gases   _____ 0.93%

2. What is the difference between C3 and C4 plants?
3. Define the following terms.
   a. Relative humidity
   b. Dew point
   c. Incipient wilting
   d. Permanent wilting point
4. Although carbon dioxide is not normally limiting for field-grown crops, certain environmental conditions may occur and cause carbon dioxide to become limiting. What are these environmental conditions?

5. How can modified atmospheres increase the time of storage of some fruits and vegetables?
6. How are the terms *relative humidity* and *dew point temperature* similar and how are they different?
7. What conditions led to the current atmospheric gas composition of the earth's atmosphere?
8. What are some properties of a gas?

Selected References

Allen, L. H. 1998. Carbon dioxide and other atmospheric gases. In *Principles of ecology in plant production,* ed. T. R. Sinclair and F. P. Gardner. Eds. New York: CABI Publishing.

Dole, J. M., and H. F. Wilkins. 1999. *Floriculture: Principles and species.* Upper Saddle River, NJ: Prentice Hall.

Freifelder, D. M. 1982. *Physical chemistry for students of biology and chemistry.* Boston: Science Books International.

Parsons, L. R. 1995. *Cold protection by irrigation: Dew point and humidity terminology.* University of Florida Cooperative Extension Service Fact Sheet H.S.-76.

Taiz, L., and E. Zeigler. 1998. *Plant physiology.* 2nd ed. Sunderland, MA: Sinauer.

Talbot, M. T., and D. Baird. 1998. *Psychometrics and postharvest operations.* University of Florida Cooperative Extension Service Circular 1097.

Selected Internet Sites

http://www.epa.gov/ Home site of the Environmental Protection Agency (EPA).

http://www.epa.gov/oar/ Government site providing information on air pollution, clean air, and air quality information. Good links for acid rain, ozone depletion, and climate change. Office of Air and Radiation, EPA.

http://www.eia.doe.gov/ Official energy statistics of the U.S. government. Energy Information Administration, DOE.

http://www.nrcs.usda.gov/technical/airquality.html Air quality information and report of Agricultural Air Quality Task Force. Natural Resources Conservation Service, USDA.

http://www.wcc.nrcs.usda.gov/ Government weather site containing observations, forecasts, maps and models, weather safety, and education. National Water and Climate Center, USDA.

http://www.nws.noaa.gov/ Government weather site containing observations, forecasts, maps and models, weather safety, and education. National Weather Service, National Oceanic and Atmospheric Administration.

14

Air Pollutants

Air pollution is defined as gaseous or particulate substances released into the atmosphere in sufficient quantities or concentrations to cause injury to plants, animals, or humans. Air pollutants are typically emitted into the atmosphere and transported from the source to the affected organism.

This chapter begins by describing the sources, history, and formation of common air pollutants. A description of the variables that affect the occurrence and dispersion of air pollutants follows. Information on visual injury symptoms of the major air pollutants on plants is presented. The chapter concludes with an overview of the effect of air pollutants on plant growth and yield and the use of plant bioindicators for monitoring the presence of air pollutants.

Background Information

Sources of Air Pollutants

Air pollution comes from both natural and anthropogenic (human-generated) sources. Natural sources of air pollution evolved with the development of the ecosphere and include volcanoes, forest fires, dust storms, decaying marshes, salt spray from oceans, and hydrocarbon emission from some trees. These sources are usually intermittent and are normally cleaned or minimized by the weather and chemical cycles.

Anthropogenic sources of air pollution probably did not cause problems until the late 1200s. In response to increasing occurrences of "bad" air, King Edward I enacted air pollution laws in London in 1273, which prohibited the burning of a certain type of coal. In 1300, King Richard III placed a heavy tax on coal to discourage its use. In 1911, pollution from burning coal killed 1,150 people in London. Physicians reporting on the episode introduced the term *smog* to describe the mixture of smoke and fog that made up the air in the city of London during the disaster. In America, the industrial revolution also increased emissions of air pollutants, with the first known air pollution disaster in the United States occurring in Donora, Pennsylvania, in 1948 killing several people.

TABLE 14.1   *The National Ambient Air Quality Standards as set by the U.S. Environmental Protection Agency*

Pollutant	Standard value*	Standard type
Carbon monoxide (CO)		
8-hour average	9 ppm (10 mg/m3)	Primary
1-hour average	35 ppm (40 mg/m3)	Primary
Nitrogen dioxide (NO$_2$)		
Annual arithmetic mean	0.053 ppm (100 µg/m3)	Primary & secondary
Ozone (O$_3$)		
1-hour average	0.12 ppm (235 µg/m3)	Primary & secondary
8-hour average **	0.08 ppm (157 µg/m3)	Primary & secondary
Lead (Pb)		
Quarterly average	1.5 µg/m3	Primary & secondary
Particulate (PM$_{10}$)	*Particles with diameters of 10 micrometers or less*	
Annual arithmetic mean	50 µg/m3	Primary & secondary
24-hour average	150 µg/m3	Primary & secondary
Particulate (PM$_{2.5}$)	*Particles with diameters of 2.5 micrometers or less*	
Annual arithmetic mean	15 µg/m3	Primary & secondary
24-hour average	65 µg/m3	Primary & secondary
Sulfur Dioxide (SO$_2$)		
Annual arithmetic mean	0.03 ppm (80 µg/m3)	Primary
24-hour average	0.14 ppm (365 µg/m3)	Primary
3-hour average	0.50 ppm (1300 µg/m3)	Secondary

Parenthetical values are approximately equivalent concentrations. *Primary standards* set limits to protect public health, including the health of "sensitive" populations such as asthmatics, children, and the elderly. *Secondary standards* set limits to protect public welfare, including protection against decreased visibility and damage to animals, crops, vegetation, and buildings. Units of measure for the standards are parts per million (ppm) by volume, milligrams per cubic meter of air (mg/m3), and micrograms per cubic meter of air (µg/m3).

Source: Environmental Protection Agency, http://www.epa.gov/air/criteria.html.

The Clean Air Act requires the U.S. Environmental Protection Agency (EPA) to set National Ambient Air Quality Standards (NAAQS) for six common air pollutants: ozone, lead, carbon monoxide, sulfur dioxide, nitrogen dioxide, and respirable particulate matter (Table 14.1). The NAAQS were established to protect the public from exposure to harmful amounts of pollutants. When the pollutant levels in an area are in violation of a particular standard, the area is classified as "nonattainment" for that pollutant. The EPA then imposes federal regulations on pollutant emissions and designates a time period after which the pollutant levels must fall below these standards.

Common Air Pollutants

More than one air pollutant is often present in the air at the same time, and plants, as a result, are often exposed to a mixture of pollutants. The effect of these mixtures on plants may be less than, equal to, or greater than the effects of any one pollutant. Some of the more common air pollutants include photochemical oxidants, sulfur dioxide, oxides of carbon, and fluorides.

Photochemical Oxidants   Photochemical oxidant levels are most severe during inversions and stagnant weather, and maximum levels often occur in mid to late summer. The oxidant levels also typically exhibit a diurnal pattern—low levels during the night, beginning to rise in morning, peaking early afternoon, and then dropping as evening approaches. It is not unusual for oxidants formed over large urban areas to cause plant injury many kilometers away. The two most important photochemical oxidants are ozone (O_3) and peroxyacetyl nitrate (PAN).

Ozone (O_3)   There are two general categories of ozone in the environment: the natural ozone present in the upper atmosphere, and the human-induced air pollutant ozone, which we breathe. The natural source of ozone is formed in the upper atmosphere by the action of sunlight on oxygen. It is also formed by electrical discharges during thunderstorms, though this source of ozone does not contribute significantly to an air pollution problem.

Ozone resulting from human activities is the principal cause of plant injury. Ozone is formed in the air by the action of sunlight on hydrocarbons and oxides of nitrogen. Hydrocarbons and nitrogen oxides usually arise during incomplete combustion of natural gas, gasoline, oil, and other fuels (Figure 14.1). Vehicles typically

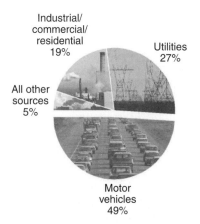

FIGURE 14.1   Nitrogen oxides form when fuel is burned at high temperatures, as in a combustion process. The primary sources of NO_x are motor vehicles, electric utilities, and other industrial, commercial, and residential sources that burn fuels.
Source: Environmental Protection Agency.

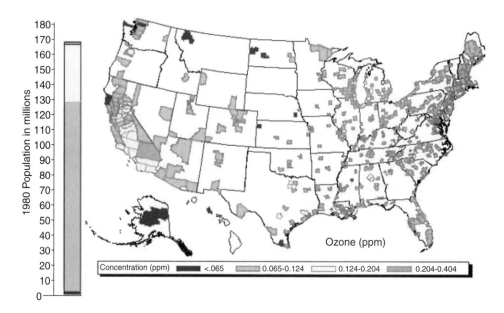

FIGURE 14.2   Variations in NO_2 concentrations (top) and the highest second daily maximum 1-hour ozone concentration (bottom) across the United States in 1996. Higher ambient NO_2 levels are typically observed in urban areas, and in 1996 approximately 39 million people lived in 52 counties where the second daily maximum 1-hour concentration was above the quality standard for ozone.
Source: Environmental Protection Agency.

account for about one-half of the total emissions of hydrocarbons and nitrogen oxides and highly populated areas tend to have ozone concentrations (Figure 14.2).

Normal background levels of ozone are 0.02 to 0.04 ppm. Ozone peaks of 0.15 to 0.25 ppm are common in many urban areas during inversions, and these levels are capable of severely injuring sensitive plants.

Peroxyacetyl nitrate (PAN)   Like ozone, PAN is produced when sunlight reacts with various gases. PAN is formed in the presence of light when nitrogen oxides react with unsaturated hydrocarbons. Peroxypropionyl nitrate and peroxybutyryl nitrate may also be present in urban air and may produce symptoms similar to those caused by peroxyacetyl nitrate.

Sulfur Dioxide (SO_2)   Sulfur dioxide was one of the first human-induced pollutants shown to cause economic damage to plants. Combustion of fossil fuels is the primary source for SO_2 (Figure 14.3), with SO_2 emitted when ores containing sulfur go through the smelting process, when oils are refined, and in the manufacturing of sulfuric acid. It is suspected that over 67% of SO_2 released to the air originates from electric utilities, especially those that burn coal. Other sources of SO_2 are industrial

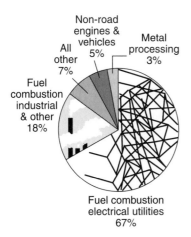

FIGURE 14.3   Sources of SO_2.
Source: Environmental Protection Agency.

facilities that burn coal or oil to produce process heat. Also, trains, large ships, and some off-road diesel equipment burn high sulfur-containing fuel, which releases SO_2.

Industrial manufacturing plants are required to have tall stacks (200 to 400 m tall) to disperse some of the gases over a wider area to minimize adverse effects of released SO_2 to the immediate area and to facilitate dilution of the pollutant. The SO_2 in the atmosphere may be subsequently oxidized to SO_3. SO_3 rapidly combines with H_2O to form sulfuric acid (H_2SO_4), which is a component of acid rain.

Oxides of Carbon   Oxides of carbon as air pollutants are not often injurious to plants but can have profound effects on other aspects of the environment. The two most prevalent oxides of carbon are carbon monoxide and carbon dioxide. Carbon monoxide is the most widely distributed and most commonly occurring air pollutant, and increased concentrations can have deleterious effects on human and animal health (with little or no toxic effects on plants until high concentrations are reached and sustained). Although carbon dioxide is a normal component of air, increases in ambient concentration contribute to global warming.

Carbon monoxide (CO)   Carbon monoxide is a component of motor vehicle exhaust, which contributes about 56% of all CO emissions nationwide (Figure 14.4). Other non-road engines and vehicles (such as construction equipment and boats) contribute about 22% of the CO emissions nationwide. Higher levels of CO generally occur in areas with heavy traffic congestion. In cities, 85% to 95% of all CO emissions may arise from motor vehicle exhaust. Other sources of CO emissions include industrial processes (such as metals processing and chemical manufacturing), residential wood burning, and natural sources such as forest fires. Wood stoves, gas stoves, cigarette smoke, and unvented gas and kerosene space heaters are sources of

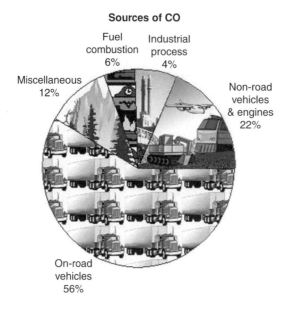

Sources of CO

FIGURE 14.4   Sources of CO.
Source: Environmental Protection Agency.

CO indoors. The highest levels of CO in the outside air typically occur when inversion conditions are more frequent. The CO becomes trapped near the ground beneath a layer of warm air. Natural sources, especially marshlands, also account for CO in the atmosphere.

Carbon monoxide is not often injurious to plants but can be harmful to humans and animals. Carbon monoxide reacts with the hemoglobin in the blood and causes oxygen deprivation. Cigarette smoke containing 200 to 400 ppm CO is estimated at tying up 5% to 15% of the smoker's hemoglobin during smoking.

Carbon dioxide (CO_2)   The average global concentration of CO_2 over the last century has risen slowly. Measurements of CO_2 concentrations in ice cores indicate a preindustrial value of about 0.028%. Since the beginning of the industrial revolution in the 18th century, the atmospheric CO_2 concentration has increased to the current level of 0.035%. Considerable amounts of CO_2 have also been released into the atmosphere as a result of defoliation of forests, plowing of prairies, and other land use changes. This increase in CO_2 contributes to the greenhouse effect and global warming.

Fluorides (Fl)   Fluorides are compounds containing the element fluorine (Fl). Fluorides are produced by glass, aluminum, pottery, brick, and ceramic industries and by refineries, metal ore smelters, and phosphate fertilizer factories. Volcanoes are natural sources of gaseous and particulate fluorides.

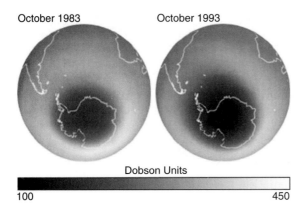

October 1983   October 1993

Dobson Units

100   450

FIGURE 14.5   Ozone levels measured by the total ozone mapping spectrometer (TOMS) indicating a thinning of the protective ozone layer (identified by smaller measured Dobson units).
Source: Earth Observatory, NASA.

The extent and magnitude of atmospheric concentrations of fluorides over an area depends on the rate of emissions from that area. The pattern of dispersion in time and space is determined by the combination of source, topography, and meteorology. Occasional high concentrations of fluorides may occur due to failure of emission controls and accidental spills.

Trace Gases

The concentrations of some trace gases (such as chlorofluorocarbons, CO_2, and CH_4) are increasing, presumably as a result of human activities, and could potentially alter the radiation balance of the earth (see The Greenhouse Effect in Chapter 13), despite their very low concentrations.

Chlorofluorocarbons (CFCs) have been used for refrigeration and cooling, as well as for cleaning electronic circuits. CFCs are not reactive in the lower atmosphere, but they induce a cyclical set of reactions that destroy the ozone layer (Figure 14.5), which in turn increases the amount of UV radiation reaching the earth's surface. Because life evolved under the protective ozone layer, increases in UV radiation are injurious to most plants and especially humans. Increases in incidences of skin cancer have already been observed in many regions.

Variables That Can Affect the Various Air Pollutants

Air pollutants from the time they are emitted from a source until they reach the affected organisms are influenced by many factors. Pollutants can be emitted from a point source such as a power plant, a line source such as a highway, an area source such as a city or large industrial complex, or a nonpoint source such as agricultural pollutants (e.g., ammonia from intensive animal operations). In addition, the emission point may be close to the ground (e.g., tail pipe of a car) or over a thousand meters in height (e.g., high stacks of a power plant). Elevation has a tremendous influence on how rapidly the pollutant will be dispersed and diluted before it reaches the affected organism. Time of emission is also important because weather conditions vary throughout the day. During the day, sunlight exerts an important influence in the transforming of chemical species of pollutants.

Sources of Air Pollution

The sources of air pollutants can be categorized into four groups: industrial, urban, mobile, and rural.

Industrial Sources   Industrial sources emit many different types of pollutants, depending on the type of industry. In many industrial areas, several industries may emit the same type of air pollutant.

Urban Sources   Urban sources include commercial incinerators and small furnaces used for domestic heating purposes. Even though these sources emit smaller amounts of pollutants than industrial sources, they can be important if there are a larger number in a given area.

Mobile Sources   The automobile is the primary mobile source. This category is responsible for several pollutants, including carbon monoxide, hydrocarbons, and nitrogen oxides. Hydrocarbons and nitrogen oxides are involved in the formation of ozone and peroxyacetyl nitrate (PAN).

Rural Sources   Rural sources include pollutants from field spraying of chemicals, agricultural burning, forest fires, and ammonia from large animal operations. Most of these are isolated sources that tend to be seasonal.

Effects of Topography on Dispersion of Air Pollutants

Topography can interact with meteorological variables to affect the dispersion of air pollutants. Valleys can become pockets from which pollutants cannot escape, especially when there is little wind. The pollutants can be channeled along the valley, perhaps affecting areas many kilometers removed from the sources, or they may stagnate in a sheltered area for a prolonged period. Mountains can serve as a barrier over which pollutant-laden air cannot flow, thereby resulting in a buildup of the pollutants. On a smaller scale, a row of trees or buildings can act as either a barrier or a channel for air pollution.

Effects on Plants

Diagnosis of the nature and causes of suspected air pollutants' injury on plants is usually a complicated process, as other biotic or abiotic factors may resemble or mimic air pollution symptoms. The diagnosis of suspected air pollution injury to plants is based on whether there is a pollution source nearby capable of causing injury, the characteristics of the terrain or location, the symptoms, the part of the plant that is affected, the distribution of affected plants, the presence or absence of biological agents, the number of plants affected, and the crop history of the affected area, and as a result diagnosis is often difficult. Plants also generally vary in their relative sensitivity to suspected air pollutants (Table 14.2).

Symptoms

Plant responses to air pollution are helpful in establishing the early presence of airborne contaminants, determining the geographical distribution of the pollutants, and establishing the concentration of the pollutants.

Ozone   In the United States, ozone is the most injurious pollutant to plants. Ozone enters the leaf through the stomata during normal gas exchange where it disrupts cellular membranes. As these structures are destroyed, groups of cells collapse and symptoms appear on leaf surfaces. Leaves often appear shiny and then dry (Figure 14.6). The most common symptoms due to ozone on leaves are small flecks or stipples. Ozone

FIGURE 14.6   Ozone injury (bifacial necrosis) on watermelon leaves.

TABLE 14.2   *Plant relative sensitivities to air pollutants*

Ozone

Very sensitive plants

Abutilon, alder, alfalfa, apricot, ash (green and white), aspen, aster, avocado, barley, bean (green and Pinto), beet (table and sugar), begonia, bentgrass, birch, bluegrass (annual), box elder, bridal wreath, broccoli, bromegrass, Brussels sprout, carnation, carrot, catalpa, celery, chicory, chickweed, Chinese cabbage, chrysanthemum, citrus, clover (red), corn (sweet), crabapple, crabgrass, dahlia, dill, duckweed, eggplant, endive, fuchsia, gourds, grape, hemlock, honeylocust, hypericum, larch (European), lilac, linden, locust (black), maple (silver and sugar), marigold, mint, mimosa, muskmelon, oak (gambel and white), oat, onion, orchardgrass, parsley, parsnip, pea, peach, peanut, petunia, pine (ponderosa, scotch, white), potato, privet, pumpkin, radish, rye, salvia, scallion, smartweed, snowberry, spinach, squash, strawberry, sweet potato, Swiss chard, sycamore, tobacco, tomato, tulip tree, turnip, verbena, walnut, wheat, and willow (weeping).

Somewhat resistant plants

Coleus, cotton, cucumber, dogwood, euonymus, geranium, gladiolus, impatiens, juniper (Pfitzer), kalanchoe, most maples, most oaks, pepper, poinsettia, tolmiea, and yew.

Sulfur dioxide

Very sensitive plants

Alfalfa, amaranthus, apple, apricot, ash (green and white), aspen, aster, bachelor's button, barley, bean (broad and garden), beech, beet (table and sugar), begonia, bindweed, birch, blackberry, bluegrass (annual), broccoli, bromegrass, Brussels sprout, buckwheat, carrot, catalpa, centaurea, chickweed, China aster, clovers, columbine, cosmos, cotton, crabapple, curly dock, dahlia, dandelion, Douglas fir, elm, endive, fir (white), fleabane, forsythia, four o'clock, hawthorn (scarlet), larch, lettuce (garden and prickly), mallow, morning glory, mulberry, mustard, oat, okra, orchardgrass, Pacific ninebark, peach, pear, pecan, pepper (bell and chili), petunia, pine (Austrian, jack, loblolly, ponderosa, Virginia, white), plantain, polygonum, poplar, pumpkin, quince, radish, ragweed, raspberry, rhubarb, rock spirea, rose, rye, ryegrass, safflower, saltbush, smartweed, soybean, spinach, spruce, squash, strawberry, sumac, sunflower, sweet pea, sweet potato, Swiss chard, tomato, tulip tree, turnip, velvetweed, verbena, violet, wheat, and zinnia.

Somewhat resistant plants

Arborvitae, box elder, canna, castor bean, celery, chrysanthemum, citrus, corn, cucumber, gingko, gladiolus, gourds, hibiscus, honeysuckle, horseradish, iris, Johnsongrass, lilac, maple, milkweed, mock orange, muskmelon, most oaks, onion, potato, privet, purslane, shepherd's purse, snowbell, sorghum, tulip, viburnum, Virginia creeper, willow, and wisteria.

Fluorides

Very sensitive plants

Alfalfa, apple, apricot (Chinese, Moorpark, Royal, Tilton), azalea, barley, blueberry, box elder, buckwheat, canna, cattail, cherry, chickweed, citrus, corn (sweet), crabgrass, cyclamen, Douglas fir,

usually attacks nearly mature leaves first, then younger and older leaves. Young plants are generally the most sensitive to ozone, whereas mature plants are relatively resistant. Certain fungi, such as *Botrytis,* readily infect plant tissue killed by ozone. Ozone also stunts plant growth and depresses flowering and bud formation. Affected leaves of certain plants commonly wither and drop early. Many environmental factors affect ozone damage, probably by altering the opening of stomata.

gladiolus, grape (European), hypericum, iris, Jerusalem cherry, Johnsongrass, larch (western), mahonia, maple, mulberry, nettle leaf goosefoot, some oaks, oxalis, peach, peony, most pines, plum, poplar, prune (Italian), smartweed, sorghum, spruce (blue and white), sweet potato, and tulip.

Somewhat resistant plants
Ash (European and Modesto), asparagus, bean, birch (cutleaf), bridal wreath, burdock, Canterbury bell, cauliflower, celery, cherry (flowering), chrysanthemum, citrus, columbine, cotton, cucumber, currant, dandelion, dock, dogwood, eggplant, elderberry, elm (American), fir (grand), galinsoga, hemlock, most junipers, laurel, lettuce (Romaine), linden (American), lobelia, locust, marigold, Mountain ash (European), nightshade, onion, orchardgrass, parsnip, pear, pepper, petunia, pigweed, planetree, plum (flowering), plantain, privet, purslane, pyracantha, ragweed, rhododendron, rose, snapdragon, soybean, spinach, squash, spruce (Engelmann), strawberry, sweet pea, tobacco, tomato, tree-of-heaven, Virginia creeper, willow (weeping) wheat, and zinnia.

Chlorine
Very sensitive plants
Alfalfa, amaranthus, apple, ash, azalea, barberry, bean (Pinto and Scotia), birch (gray), blackberry, bluegrass (annual), box elder, bridal wreath, buckwheat, catbrier, cherry, chickweed, chokecherry, coleus, cosmos, cucumber, dandelion, dogwood, gomphrena, grape, honeysuckle, horse chestnut, hydrangea, Johnsongrass, Johnny-jump-up, juniper, lilac, mallow, maple (Norway, silver, sugar), May apple, morning glory, mulberry, mustard, oak (pin), onion, peach, peony, petunia, phlox, pine (jack, loblolly, shortleaf, slash, white), poison ivy, primrose, privet, radish, rose (tea), sassafras, spruce (Norway), sunflower, sweetgum, tomato, tree-of-heaven, tulip, Venus's looking-glass, violet, Virginia creeper, witch hazel, and zinnia.

Somewhat resistant plants
Arborvitae, begonia, ivy (Boston), day lily, eggplant, hemlock, holly (Chinese), iris, lamb's-quarters, maple (Japanese), oak (red), oxalis, pepper, pigweed, polygonum, Russian olive, soybean, and yew.

Peroxyacetyl nitrate (PAN)
Very sensitive plants
African violet, alfalfa, aster, bean (Pinto), beet (table and sugar), bluegrass (annual), carnation, celery, chickweed, dahlia, dandelion, dill, Douglas fir, endive, escarole, fennel, lettuce (Romaine), lilac, muskmelon, mustard, nettle (little-leaf), oat, certain orchids, pepper, petunia, pine (Coulter, Jeffrey, Monterey, ponderosa), poinsettia, potato, primrose, rose, salvia, snapdragon, spinach, sunflower, Swiss chard, tobacco, and tomato.

Somewhat resistant plants
Azalea, bean (lima), begonia, broccoli, cabbage, chrysanthemum, corn, cotton, cucumber, onion, pansy, periwinkle, radish, redwood, sequoia, sorghum, touch-me-not, and wheat.

Source: Modified from E. J. Sikora and A. H. Chappelka, *Air pollution damage to plants.* Auburn University IPM Publication, 1996, http://www.aces.edu/department/ipm/poldmge.htm.

Concentration   The exposure of sensitive plants for 4 hours at levels of 0.04 to 11.0 ppm of ozone will result in injury. The extent of the injury depends on the plant species and environmental conditions prior to and during exposure. Ozone susceptibility can also differ between cultivars of the same species of plant. Ozone and sulfur dioxide often combine to cause plant injury at lower levels than either of these pollutants alone.

Peroxyacetyl Nitrate (PAN)   PAN causes a collapse of tissue on the lower leaf surfaces of numerous plants. The typical leaf symptoms are a glazing, bronzing, or silvering that commonly develops in bands or blotches. In grasses, the collapsed tissue has a bleached appearance, with tan to yellow, transverse bands. Conifer needles often appear yellow. Early maturity or senescence, chlorosis, moderate to severe stunting, and premature leaf drop may also occur. PAN is most toxic to small plants and young leaves. The most mature leaves are often highly resistant to PAN injury. Peroxypropionyl nitrate and peroxybutyryl nitrate may also be present in urban air and may produce symptoms that are indistinguishable from those caused by peroxyacetyl nitrate.

Concentration   PAN injury to susceptible plants occurs at levels of 0.01 to 0.05 ppm for an hour or more. This type of plant injury requires light before, during, and after exposure and is increased by any factor increasing growth.

Sulfur Dioxide SO_2 enters the plant through open stomata where it is oxidized to highly toxic sulfite (SO_3) and then to less toxic SO_4 (sulfate). If high concentrations of SO_2 exist, SO_3 accumulations are damaging. Leaves injured by SO_2 usually have a bleached appearance (Figure 14.7). Most symptoms are marginal with some shades of orange and brown. Mature or nearly fully expanded leaves are the most sensitive to SO_2.

Concentration   The degree of plant injury due to SO_2 increases as both the concentration and the length of exposure increase. Sensitive plants are injured by exposures of 0.5 ppm for 4 hours, or 0.25 ppm for 8 to 24 hours. Plants are most sensitive to sulfur dioxide during periods of bright sun, high relative humidity, and adequate plant moisture, which occur most often during the late spring and early summer.

FIGURE 14.7   Sulfur dioxide damage on the leaves of white ash. Courtesy of Dr. John Skelley, Penn State University.

Fluorides   The typical injury by gaseous or particulate fluorides is either a yellowish mottle to a wavy, reddish brown or tan "scorching" at the margin and tips of broad-leaved plants or a "tip burn" of grasses and conifers. A narrow, chlorotic to dark brown band is often present between living and dead tissue. Citrus, poplar, sweet cherry, and corn foliage exhibit a chlorotic mottling, streaking, or blotching prior to the development of the typical "burned" area. Leaves and fruit exposed to fluorides may abscise prematurely. Injured areas on some leaves of stone fruit may become brittle and exfoliate, leaving the leaves with a shothole appearance. Young, succulent growth is often the most susceptible to fluoride injury and exposed fruit may soften or become necrotic at the blossom end.

Concentration   Accumulated leaf-fluoride concentrations of 20 to 150 ppm often injure sensitive plants, although resistant varieties and species of plants will tolerate leaf concentrations of 500 to 4,000 ppm or more without visible injury. For example, a 4-week exposure of susceptible gladiolus to an air concentration of 0.1 ppb, or less than 24 hours at 10 ppb, resulted in leaf-fluoride concentrations of 150 ppm and tissue necrosis. Susceptibility to fluorides varies tremendously among cultivars or clones of apricot, begonia, corn, gladiolus, grape, peach, ponderosa and white pines, and sweet potato.

Chlorine   Injury caused by chlorine (Cl_2) is marginal and interveinal, somewhat similar to that caused by sulfur dioxide and fluorides. On broad-leaved plants, necrotic, bleached, or tan to brown areas tend to occur near the leaf margins, on the tips, and between the principal veins. Injured grass blades develop progressive streaking toward the main vein in the region between the tip and the point where the grass blade bends. The streaking usually occurs alongside the veins. Mature leaves are often more susceptible than young leaves. Conifers may exhibit tip burn on the youngest needles. Chlorine-injured vegetation is frequently observed near swimming pools, water-purification plants, and sewage-disposal facilities.

Concentration   Very susceptible plants exhibit symptoms when exposed for 2 hours or more at concentrations of chlorine ranging from 0.1 to 4.67 ppm. Chlorides typically do not accumulate in plant tissues after exposure to chlorine.

Factors Influencing Air Pollution Injury to Plants

Plant injury caused by air pollution is most common near large cities, smelters, refineries, electric power plants, airports, highways, incinerators, refuse dumps, pulp and paper mills, and coal-, gas-, or petroleum-burning furnaces. Plant injury also occurs near industries that produce brick, pottery, cement, aluminum, copper, nickel, iron, steel, zinc, acids, ceramics, glass, phosphate fertilizers, paints and stains, rubbers, soaps and detergents, and other chemicals.

Factors that govern the extent of damage as a result of air pollution are type and concentration of pollutants, distance from the source, length of exposure, and meteorological conditions. Other important factors are city size and location, land topography, soil moisture and nutrient supply, maturity of plant tissues, time of year, and species and variety of plants. Damage in isolated areas occurs when pollutants are spread long distance by wind currents. A soil moisture deficit or extremes of temperature, humidity, and light often alter a plant's susceptibility to an air pollutant.

Damage caused by air pollution is usually most severe during warm, clear, still, humid weather when barometric pressure is high. Pollutants accumulate near the earth's surface during an air inversion when cooler air is trapped at ground level.

Types of Crop Losses Due to Air Pollutants

Crop losses due to air pollution can be either direct losses or indirect losses. Direct losses are tangible losses, which are easily estimated. This could be the reduced quantity or quality of crops as a result of the presence of air pollutants. Also, processing costs could be increased such as when leafy vegetables are injured and have to be sorted to remove unmarketable leaves before packaging.

Indirect losses are less tangible, harder to measure, and generated by or related to direct losses. Indirect losses often occur with no visible damage though yields may be reduced. An indirect loss occurs when a producer is no longer able to grow a particular crop because of its sensitivity to air pollutants and the replacement crop isn't as profitable at market. Also pollutants may predispose plants to injury from insects, diseases, nutrition, and so forth.

Photosynthesis and Dry Matter Production

Air pollutants (especially ozone and SO_2) can adversely affect photosynthesis and plant dry matter production by changing stomatal physiology, chloroplast structure, CO_2 fixation, and photosynthetic electron transport systems. In general, any factor that increases the stomata openings (e.g., high supply of water, high light intensity, high supply of nitrogen) increases the movement of pollutants into the plants and potentially increases their impact on photosynthesis. Small doses of air pollution can temporarily depress the rate of net photosynthesis. When plants are exposed to levels of pollutants for extended periods, the reductions in dry matter production and yields can be significant.

Agricultural Technologies That Affect Air Pollution

Bioindicator Plants for Air Pollutants

Plants used for monitoring environmental conditions are called bioindicators or biomonitors. Bioindicators can demonstrate the presence of air pollutants and help estimate the frequency of damaging levels of air pollution and the aerial distribution of air pollution with respect to time and geographic areas. Certain plant species or, more specifically, certain cultivars or clones of plant species have been used as sensitive bioindicators of air pollutants (Table 14.3). Their use has been particularly useful in remote locations where electrical power is not available for instrumentation. Certain types of plants may also be used to indicate the amount of fluoride that has been absorbed by the plant (even if the plant does not exhibit injury). This type of sampling followed by chemical analysis is used to determine patterns of exposure or dispersion of fluorides. The major limitations of bioindicators are their often lack of specificity and the difficulty in correctly diagnosing plant injury.

Bioindicators for air pollutants include sensitive cultivars of watermelons, beans, muskmelons, cucumbers, squash, snap beans, potatoes, as well as other crops. Bioindicator plants may be planted or transplanted in suspected polluted areas and observed periodically for specific symptoms (Figure 14.8). Accurate diagnosis of plant injury on bioindicators (Figure 14.9) requires that air pollution injury is distinguished from other stresses such as insects, diseases, drought, high temperatures, and nutrient imbalances. If air pollution injury on the bioindicators is correctly diagnosed, then similar injury should be observed on susceptible crop plants in that area (Figure 14.10).

TABLE 14.3   *List of some commonly used bioindicators for air pollutants*

For ozone

Petunia (certain cultivars only)

Pinto bean

Green bean (cv. Bush Blue Lake 290)

Tobacco (cv. Bel W-3)

White pine (certain strains only)

Watermelon (cv. Sugar Baby)

For PAN

Pinto bean

Petunia (some white-flowered cultivars)

For fluorides

Iris

Ponderosa pine

For SO$_2$

Alfalfa

White pine (certain strains only)

Bean (cv. OSU 1604)

Summer crookneck squash

White birch

Source: Modified from Purdue University Cooperative Extension Service Publication HO-195. *Using plants to monitor air pollution* by J. E. Simon, M. Simini, and D. R. Decoteau, 1986.

For a more complete listing of sensitive plants to air pollutants, use the following reference: R. B. Flagler, ed. *Recognition of air pollution injury to vegetation: A pictorial atlas*, 2nd ed. (Pittsburgh, PA: Air & Waste Management Association, 1998).

FIGURE 14.8   An example of bioindicator plants growing adjacent to a commercial vegetable field.

FIGURE 14.9   Ozone symptoms on a pinto bean bioindicator plant.

FIGURE 14.10   Ozone injury on morning glory growing near injured bioindicator plants in a field.

Summary

An air pollutant is a gaseous or particulate substance released into the atmosphere in amounts injurious to plants, animals, or humans. Air pollutants are generated from both natural and human-induced sources. The National Ambient Air Quality Standards were established by the U.S. Environmental Protection Agency as a result of the Clean Air Act to determine levels of various pollutants harmful to the public health and the environment.

Common air pollutants include photochemical oxidants (ozone and PAN), sulfur dioxide, oxides of carbon (carbon monoxide and carbon dioxide), fluorides, and chlorine. Increasing concentrations of trace gases (such as chlorofluorocarbons, carbon dioxide, and methane) can contribute to the destruction of the protective ozone layer and to global warming.

Air pollutants are affected by many variables, including the source of the air pollutant and the topography of the area. Sources of air pollutants include industrial, urban, mobile, and rural. The local topography interacts with the weather of an area to affect the dispersion of air pollutants. For example, during temperature inversions valleys become pockets in which air pollutants are trapped.

The most common symptoms on plants due to injury to ozone are small flecks or stipules initially observed on nearly mature leaves. Plants injured by sulfur dioxide often have leaves with a bleached appearance, with shades of orange and brown on the margins. Fluoride injury is exhibited as a yellowish mottle to a wavy, reddish brown or tan coloring at the margins and tips of broad-leaved plants or tip burn of grasses or conifers. Chlorine injury to plants is similar to that caused by sulfur dioxide and fluoride.

Air pollutants can adversely affect crop growth and yield by reducing photosynthesis and dry matter production. Significant reductions in yield can be observed when plants are exposed to air pollutants for extended periods. Certain plants can also be used as bioindicators to monitor for air pollutants (by observing injury) in areas where instrumentation is not available due to unavailability of electricity or funds to purchase the instrumentation. The major limitation with the use of bioindicators is the difficulty in correctly diagnosing injury induced by air pollution.

Review Questions

1. Define air pollution.
2. List four types of natural air pollution.
3. What is the difference between the ozone in the ozone layer and the ozone that is an air pollutant?
4. When (time of day, time of year) would you most likely find high levels of ozone air pollution?
5. What are four factors that affect pollution damage on plants?
6. What are the differences between direct and indirect losses?
7. What are some symptoms of ozone damage on plants?
8. List three ways that air pollution can affect photosynthesis.
9. What weather conditions lead to high levels of air pollutants?
10. Define bioindicators and explain how they are used. What are their advantages and disadvantages?
11. What did the NAAQS establish and how is it regulated?
12. Why is the diagnosis of air pollution injury on plants relatively difficult?

Selected References

Applied Science Associates. 1976. *Diagnosing vegetation injury caused by air pollutants.* Washington, DC: Environmental Protection Agency.

Flagler, R. B., ed. 1998. *Recognition of air pollution injury to vegetation: A pictorial atlas.* 2nd ed. Pittsburgh, PA: Air & Waste Management Association.

Innes, J. L., J. M. Skelly, and M. Schaub. 2001. *Ozone and broad-leaved species: A guide to the identification of ozone-induced foliar injury.* Berne, Switzerland: Paul Haupt.

Lambers, H., F. S. Stuart III, and T. L. Pons. 1998. *Plant physiological ecology.* New York: Springer-Verlag.

Sikora, E. J., and A. H. Chappelka. 1996. *Air pollution damage to plants.* Auburn University IPM Publication, http://www.aces.edu/department/ipm/poldmge. htm.

Simon, J. E., D. R. Decoteau, and M. Simini. 1986. *Identifying air pollution damage on melons.* Purdue University Cooperative Extension Service Publication HO-192.

Simon, J. E., M. Simini, and D. R. Decoteau. 1986. *Using plants to monitor air pollution.* Purdue University Cooperative Extension Service Publication HO-195.

Skelly, J. M. 2000. Tropospheric ozone and its importance to forests and natural plant communities of the northeastern United States. *Northeastern Naturalist* 7:221–236.

Selected Internet Sites

http://www.epa.gov/ Home site of the Environmental Protection Agency (EPA).

http://www.epa.gov/oar/ Government site providing information on air pollution, clean air, and air quality information. Good links for acid rain, ozone depletion, and climate change. Office of Air and Radiation, EPA.

http://www.epa.gov/air/criteria.html Information on the National Ambient Air Quality Standards. EPA.

http://www.nrcs.usda.gov/technical/airquality.html Air quality information and report of Agricultural Air Quality Task Force. Natural Resources Conservation Service, USDA.

http://www.nws.noaa.gov/ Government weather site containing observations, forecasts, maps and models, weather safety, and education. National Weather Service, National Oceanic and Atmospheric Administration.

15

Mechanical Disturbances

Mechanical disturbances to plants occur as consequences of environmental conditions that influence the movement of plant parts such as from gravity, wind, and rain. Mechanical disturbances can also occur as consequences of biological interactions between a plant and other organisms (such as diseases, insects, and other herbivores) or between competing plants. Mechanical stresses imposed on plants also can be somewhat human-induced such as when management practices purposely shake or vibrate the plant as it is growing, machinery or laborers physically contact the plant, or airflow from vents and/or water from irrigation displace portions of the plant.

To differentiate the touch from vibrational mechanical stresses, the term *thigmomorphogenesis* is often used to describe plant growth responses to contact stress and *seismomorphogenesis* to describe responses to vibrational or shaking stress. Also, mechanical disturbances on plants can be categorized further as to either physical or biological depending on the source of the stress.

This chapter begins by grouping mechanical disturbances into physical mechanical disturbances (when some type of physical pressure is applied to the plant) or biological mechanical disturbances (when other biological organisms adversely affect the plant). Examples of each grouping are presented along with discussions on how crop producers can influence whether physical mechanical disturbances or biological mechanical disturbances affect plant growth. Selected mechanical disturbances that affect the rhizosphere are presented to conclude the chapter.

Background Information

Physical Mechanical Disturbance

Physical mechanical disturbances to plants occur when some type of physical pressure is placed on the plant. Physical pressure generally is measured as the amount of force applied per unit area. Physical (abiotic) stresses on plants include gravity (due to the constant pull of the earth), winds, and fire.

Physical mechanical disturbances can be the result of symmetrical and asymmetrical pressures exerted on the plant. Symmetrical pressure exerted on plants is a pressure that is the same in intensity regardless of direction. Plants often can survive high symmetrical pressures without injury. The best examples of this are aquatic plants living at great depths in the oceans, which are exposed to increasing symmetrical pressures with increasing depth below sea level. Ocean pressure is an external, inwardly directed hydrostatic pressure, whereas turgor pressure is an internal, outwardly directed hydrostatic pressure that can lead to cell rupture. Terrestrial plants, which are exposed to these high turgor pressures, have much thicker, more lignified cell walls than freshwater aquatic plants, which have much lower turgor pressure.

Asymmetrical pressure exerted on plants is a pressure that has a directional characteristic (i.e., the intensity of the pressure is not uniform in all directions). Although plants often tolerate high symmetrical pressures without injury, slight asymmetrical pressure can cause damage to the plant and its cells. The asymmetrical pressure can produce a direct physical, shearing effect on the plant or cell. In the field, asymmetrical pressure due to wind may produce a sufficient mechanical stress to shear off plant parts. Also, plant shoots may be torn from their roots or roots damaged by heaving due to the freezing of wet soils in winter. Soil impedance encountered by roots during seedling emergence or during their growth is another common example of asymmetrical pressure in the field (Fig. 15.1).

FIGURE 15.1   Soil impedance encountered by this carrot during its growth caused its root to fork.

FIGURE 15.2   An asparagus planting stripped of much of its leaf area due to the feeding of beetles.

Biological Mechanical Disturbance

Biological mechanical disturbances to plants include adverse influences from humans and other organisms (insects, diseases, and animals). Biological disturbances differ from physical disturbances in that the plant may be able to erect barriers between itself and the source of the stress. Biological mechanical disturbances are generally measured in loss of healthy plant tissue (Fig. 15.2).

Effects on Plants

Release of Stress Ethylene by Plants

Stress ethylene refers to the increased ethylene levels in and around the plant resulting from a stress or stresses. These include physical mechanical stresses from wounding and biological mechanical stresses from insects and diseases. Stress ethylene can also be released from the plant in response to nonmechanical stresses such as from certain chemicals, chilling temperatures, drought, flooding, and radiation. Stress ethylene is of metabolic origin and is produced by stressed, living cells by the same biosynthetic pathway of ethylene produced by nonstressed cells. Severe stress may result in cell death and the cessation of ethylene production. Stress ethylene may function by communicating the effect of a stress to the plant in a way that induces a response. Stress ethylene may also be involved in the onset of stress responses such as wound healing, increased disease resistance and abscisic acid production, and enhanced senescence.

Physical Mechanical Disturbances

Gravitropism (Geotropism)   In most agricultural systems, the seeds are planted in random orientation in the field and germination is below the soil surface. To survive, a germinating seed must direct the growth of its shoot upward toward a light source for photosynthesis to occur before the stored food reserves within the seed cotyledons or elsewhere are depleted. Roots typically grow downward into the soil to absorb water and nutrients and to anchor the developing plant. The ability of a plant to orientate its growth is often in response to the earth's gravitational pull.

Gravitropism (geotropism) refers to the growth movements of plants toward (positive) or away from (negative) the earth's gravitational pull. Orthogravitropic (orthogeotropic) is vertically directed plant growth (such as main stems, tree trunks, and tap roots). Diagravitropic (diageotropic) is horizontally directed growth (such as rhizomes). Plagiogravitropic (plagiogeotropic) is growth at angles other than horizontal or vertical (such as many lateral shoots and nontap roots).

Most roots are positively orthogravitropic, with primary roots more so than secondary roots. Tertiary roots are primarily diagravitropic, and this difference between root types is important in the greater exploration of the soil environment than would occur if all roots grew parallel and straight down. Main stems or tree trunks are usually negative orthogravitropic, growing 180° away from gravitational stimulus, whereas branches, petioles, rhizomes, and stolons are usually more diagravitropic. These differences allow a plant to efficiently fill available space with photosynthetic tissue to absorb CO_2 and light, and roots to absorb water and nutrients. Stems and flowers also are generally negative orthogravitropic.

The components of the gravity-sensing guidance system in plants appear to include a sensing mechanism to detect whether the organ is growing appropriately, a reaction mechanism to respond if organs are not growing appropriately and initiate appropriate growth changes in the organ, and a communication mechanism to conduct signals from the sensory to the reaction mechanism. Plants may have some type of statocyst-like sense organ used for gravity sensing similar to those observed in animals. It has been suggested that statocysts in plants are the entire cells, and each organ has a number of statocytes. Each statocyte (cell) has within it a number of statoliths that sediment to the lower side in response to gravity. These statoliths appear to be starch grains, and most statocytes are cells characterized by the presence of sedimentable starch grains.

The presence of statocytes in plant organs determines whether the plant will respond to gravity. Evidence in support of this is that removal of the extreme apex of a root (in which statocytes are usually confined) leads to a loss of gravitropic responsiveness. Little is known beyond the sedimentation of amyloplasts. In animals, statolith sediments lie on a sensory surface and the resulting electrical signals ultimately lead to the correcting movements by the animal's limbs.

The gravity-sensing reaction mechanisms in growing plant organs depend on the fine regulation of the rate of cell expansion in the upper and lower sides of the organ. For coleoptiles, auxins may become asymmetrically distributed, with increased amounts in the lower side. This mechanism would result in a redistribution of growth. Therefore, the lower half would increase by about the amount of decrease by the upper half.

The gravity-sensing mechanism operates rapidly because plants can detect a change in their orientation with respect to gravity within a few seconds. The starch grain is the only plant cell particle with the mass and density relative to the cytosol of the cell to sediment so rapidly. It is now accepted that each statolith is not a naked starch grain but rather a group of starch grains enclosed within a membrane called an amyloplast, and each statocyte may contain 4 to 12 amyloplasts. This starch grain sedimentation mechanism in statocytes is referred to as the starch-statolith hypothesis of gravity perception.

The statocytes in plants are very localized in roots, usually in the root cap. In shoots they are not as localized, and they are observed all along the shoot and frequently in cells that surround the vascular bundle or epidermis. The reaction mechanism in roots is less well understood than in coleoptiles. An inhibitor, possibly abscisic acid (ABA) or indoleacetic acid (IAA), from the root cap may be involved in bringing about the downward gravitropic response of the primary root.

Thigmomorphogenesis and Seismomorphogenesis Thigmomorphogenesis is the response of the plant to contact stresses with solid surfaces (such as from rubbing or touching the plant). Seismomorphogenesis is the response of the plant to vibration (often imposed by shaking). Most vascular plants respond to thigmomorphogenic and seismic stresses by elongating more slowly and increasing in stem diameter slightly more rapidly, causing short, stocky plants. The removal of stimulus allows the stems to resume normal rate of growth. The decrease in stem elongation may be due to differences in auxin or gibberellin concentrations. Increased stem thickening may be due to ethylene increase around the plant. The shorter, stronger plant may have an advantage in the field in that it is less easily lodged by winds than a slender tall plant.

Although many plants respond to thigmomorphogenic (touch) stresses, two of the most interesting examples are the sensitive plant and the Venus fly trap.

The sensitive plant The sensitive plant (Fig. 15.3), *Mimosa pudica*, can respond to various mechanical stimuli applied at any point on the plant. It can be a wounding response to cutting or burning, or a potentially nonwounding response to touch. A typical response for the sensitive plant is that stimulated leaves droop ("collapse") in 1 to 2 seconds. This is followed by a slower recovery to the original leaf positions in 10 to 15 minutes. The initial fast movements may be due to release of potential energy stored within the elastic walls of the cells. The recovery response in contrast requires metabolic energy and K^+ flux to raise the leaf.

Venus fly trap The carnivorous Venus fly trap (Fig. 15.4), *Dionaea muscipula*, digests freshly trapped prey. Carnivorism is relatively rare in plants, with only a few hundred carnivorous plant species worldwide. Carnivorous plants are generally restricted in their natural habitat to sunny, wet, and nutrient-poor conditions. They are chlorophyll containing and undergo C3 photosynthesis.

The Venus fly trap has leaf modifications that consist of two-lobed structures joined at a midrib of the petiole, with spines or hairs along the periphery of the lobes. There are three trigger spines on each lobe, and the lobes quickly close together in

FIGURE 15.3 The sensitive plant can respond to various stimuli applied at any point on the plant.
Courtesy of Dr. Jay Holcomb, Penn State University.

about 300 ms (with the peripheral spines intermeshing) when one or more sensitive spines are stimulated. Insects that bend the sensitive hairs may be trapped by the lobes. If insects are trapped, the lobes move closer together, squeezing the insect. Subsequently, various digestive enzymes and acids are secreted from glands in the inside of the lobes and the products of digestion are absorbed into the plants. After 2 weeks the trap reopens. If the insect is not caught, the trap does not undergo tighter closing, and reopens in about 24 hours. The opening and closing of the trap can occur for a limited number of times.

Wind Very windy places are likely to be populated by short-growing plants. This may be due to seismic or vibrational stress plant growth responses induced by the wind. Winds can also induce plant bending, and the small asymmetrical pressure due to wind may produce a sufficient mechanical stress to break or shear off the plants (Fig. 15.5).

FIGURE 15.4 Venus fly trap with opened and unopened "traps" (modified leaves used to capture insects).
Courtesy of Dr. Jay Holcomb, Penn State University.

FIGURE 15.5 Plants were torn out of the soil by heavy winds in this field planting.

Winds also increase plant transpiration rate, and wind stress often results in more and smaller stomata per unit area of leaves. Elasticity (the ability to bend without damage) of leaves and plant respiration rate may also increase due to winds. For example, a 20% to 40% increase in respiration can occur when plants are exposed to winds as slight as 7.2 m/sec. Severe physical damage to plants can occur if winds reach 32 to 40 km/h. Lesser winds can result in blowing soil particles that can adversely affect plant growth through abrasion of leaves and stem tissue. Young plants are most susceptible to abrasion by soil particles moved by wind. Abrasion slows the growth of plants or may destroy the plant completely. Plant damage from abrasion is most severe in dry sandy soils.

Wind is the principal cause of the timberline of mountains, the line beyond which trees do not grow. Twisted timberline trees, or krummholz, are not the result of wind pressure alone. Bark and twig tissue are also blasted by snow and ice particles carried by the strong winds at these altitudes. Only a part of the scrubby tree covered by the snow pack is protected. One-sided trees of the seacoast are also at least partially caused by wind (as well as from the toxicity from salt spray).

Hurricanes Hurricane damage to large plants and trees can include physical damage due to loss or damage of leaves, stems, or tree trunks. Physiological injury due to hypoxia (occurs at low oxygen levels) and anoxia (occurs in the complete absence of oxygen) conditions can also be induced by flooding of the roots. As the O_2 content and its rate of diffusion in waterlogged soils become reduced, aerobic respiration is slowed, fermentation respiration is initiated, and as a result subsequent growth is reduced, and plants may die. A plant's age largely determines its ability to recover from hurricane injury. A young, vigorous tree is more likely to survive a hurricane than an older one. In general, a damaged older plant or tree is weakened more than a younger plant or tree receiving the same kind of damage.

The greatest threat to small plants in hurricane season is from flooding and silting. Silting occurs when soil carried by rapidly rushing water is deposited on flooded land. Crown and root disorders can result from silt damage. The severity of injury often depends on how long the water remains and/or the depth of silt deposited.

Hail Hail is the product of vigorous convection currents that occur in the warm season and is the heaviest and largest unit of solid precipitation. Hail begins as rain that is carried upward by rapidly ascending updrafts of warm, moist air into regions where temperatures are below freezing. This rain freezes, acquires a coat of frost and snow, and becomes hailstones. Eventually the hailstones enter a weaker convection current and fall, accumulating a coat of water, which also freezes. This can be repeated many times with the hailstones getting larger until they reach a point where the convection current is no longer strong enough to force them upward and they fall to the earth.

FIGURE 15.6   Hail damage to field crops.

Plant damage due to hail is primarily a shredding of the leaves or damage to the fruit (Fig. 15.6). For a leafy crop, a single hailstorm can result in the loss of the entire crop. Plant productivity after a storm for some crops such as corn is reduced in proportion to the percentage of leaf damage.

Glazing and Snow Damage   Glazing is the formation of ice on tree limbs (Fig. 15.7). Glaze damage occurs during cooler seasons when relative humidity is high and temperatures drop below freezing. The weight of the ice coating the limbs can cause them to break. Also, excessive weight from accumulated wet snow can also cause limbs to be damaged and break off (Fig. 15.8).

Fire and Lightning   Fire is a sporadic environmental factor in an ecosystem, but when it is present it causes damage to many plants (Fig. 15.9). Fires that occurred

FIGURE 15.7 Glazing of ice on branches can result in plant damage.
Courtesy Dr. Rick Bates, Penn State University.

FIGURE 15.8 Excessive weight of wet snow can cause tree limbs to break off.
Courtesy Dr. Rick Bates, Penn State University.

FIGURE 15.9   Fire may kill small trees or damage the bark on large trees and permit the wood to be invaded by decay organisms.
Courtesy Dr. Rick Bates, Penn State University.

more than a century ago affected the appearance of our forests and prairies today. Many of the aspen and jack pine forests today in the Great Lake states are the direct result of forest fires 20 to 100 years ago. Fire was one of the primary forces that created and maintained the central grasslands of the Great Plains. Before European settlement, fires on the Great Plains were ignited by lightning or Native Americans. Native Americans used fire for hunting, warfare, signaling, and reducing insect populations around villages. Fires also were set to suppress woody vegetation that would have otherwise developed and to attract wildlife to the resulting fresh, new growth. Climatic factors such as periods of drought, seasonal dryness, and nearly constant wind often permit fires to burn extensively (Fig. 15.10).

FIGURE 15.10   Weather conditions were optimal for a series of extensive forest fires to take place in the western United States in 2000. On July 27, 2000, fires were burning in Mesa Verde National Park (Colorado), Montana, Idaho, Utah, Washington, Nevada, Arizona, New Mexico, Texas, and California. This image shows smoke plumes and heat signatures (red) from many fires in the western United States on the evening of July 27. Source: Earth Observatory, NASA.

Lightning strikes on trees can cause various symptoms besides fire such as loosening bark (which hangs on the tree in strips) and cracking wood (Fig. 15.11). A tree struck by lightning may exhibit no visible symptoms until it dies unexpectedly days or years later. Vaporization of the plant cell's liquid component can occur due to a lightning strike resulting in dead plants in areas of the field near the strike (Fig. 15.12).

Sound and Ultrasound   Although the audible sound frequencies from 20 to 20,000 Hz have been reported to affect photosynthesis, transpiration, cytoplasmic properties, mitosis, and the growth and yield of some plants, the more severe injury occurs with exposure to higher energy ultrasonic waves (> 20,000 Hz) not normally found in nature rather than the ever-present audible sound waves. Indications suggest that when damage does occur it is due to an ultrasonic effect on cell structure potentially due to cavitations (bubble motion) and appear dependent on the frequency, intensity, and length of exposure.

FIGURE 15.11 Lightning damage to a tree (cracking of tree trunk). Courtesy of Dr. Larry Kuhns, Penn State University.

FIGURE 15.12 Areas in a field in which plants were damaged by lightning.

Biological Mechanical Disturbances

Plant growth is a function of the amount of photosynthetically active leaf area so that a loss of leaf area from the feeding of other organisms reduces plant growth. Photosynthetic rate can be severely reduced as a result of animal grazing or by pathogen damage. Plants can compensate for the loss of leaf area by increasing the rate of photosynthesis in surviving leaves or altering the morphology of new growth.

Plants may discourage animal feeding injury by various methods, such as developing physical barriers and having toxins within their plant tissue. Also, most plants have a low protein content and as a result might not be as appealing as other higher protein sources. Typical protein values (on a dry weight basis) include plant material, 10% to 30%; microbes, 30% to 60%; insects, 40% to 60%; and artificial insect diet, 25% to 40%. The protein content of plant material further varies according to plant part. The protein content of wood is about 1% and seed is about 30%.

Many plants have external coverings of scales, hairs, or glandular hairs to discourage other organisms from feeding on them. Beneath these coverings may be thick cuticles, thick epidermal cell walls, or deposition of substances such as silica (in grasses) or resins (in conifers). Any of these modifications may help prevent organisms from penetrating the plant tissue. In addition, the plant may produce toxins that may be constitutive (always present) or induced by insect attack. Plant toxins include alkaloids, tannins, cyanogens, or a wide range of phenolic compounds.

Insects   Plant productivity generally decreases as insect populations increase. Also, the visual plant appearance (the primary criteria used by most consumers to select fresh produce and ornamental purchases) can be affected adversely by insects, rendering the crop to be of reduced quality or even unmarketable (Fig. 15.13).

Insects categorized by feeding habit   Insects can be categorized according to their feeding habits. These include chewing, sucking, and rasping insects.

Chewing insects   Chewing insects consume plant roots, leaves, or stems as adults or larvae. Cabbage worms, armyworms, grasshoppers, Japanese beetles, Colorado potato beetles, and fall armyworms are examples of insects that cause injury by chewing.

Sucking insects   Sucking insects derive nutrients by puncturing plant cells and sucking plant sap. Only the internal liquid portions of the plant are consumed, even though the insect feeds externally on the plant. Aphids, scale insects, squash bugs, leafhoppers, and plant bugs are examples of sucking insects.

Rasping insects   Rasping insects cut plant cells and feed on exposed plant sap. Small patches of dead tissue, often not penetrating through the leaf, is indicative of injury from rasping insects. Thrips is an example of a rasping insect.

Insects categorized by plant part eaten   Insects can be grouped according to the plant type that they eat. These include insects that feed on underground plant parts,

FIGURE 15.13   Insect damage on an apple fruit, rendering it unmarketable to consumers. Courtesy of Dr. Rob Crassweller, Penn State University.

chewing insects that feed on foliage or stems, piercing and sucking insects that feed on foliage and stems, and insects that feed on seeds, pods, or fruits.

Insects that attack underground plant parts   Insects that attack underground plant parts are generally soil insects that eat seeds and/or roots and stems of young developing plants. Damage can range from reduced plant stand and vigor to death of the plants. Slow plant growth, leaf yellowing, wilting, and death of the plant are symptoms of soil insect damage. Wireworms, cutworms, mole crickets, corn stalk borers, cucumber beetle larvae, and grubs are examples of insects that attack underground plant parts.

Chewing insects that feed on foliage and stems   Foliage feeders feed on aboveground plant parts, except seeds, seed pods, and fruits. Caterpillars (larvae of *Lepidoptera*) are the primary foliage feeders. The caterpillars are widely known for consuming large amounts of plant tissue, but adult forms (moth, butterflies) do not damage the plant.

Piercing and sucking insects that feed on foliage and stems   Piercing and sucking insects have highly modified hollow mouthparts that function similar to a hypodermic needle. These needlelike mouthparts are inserted into the plant and used to remove the plant sap. Some insects in this group transmit plant viruses and pathogens from plant to plant as they feed. These insects often inject certain enzymes and toxins into the plant during feeding, causing the plant to grow abnormally. Damage from piercing and sucking insects is often underestimated because the wounds they inflict are not readily apparent. These pests are also very mobile and often fly before being observed. Overall plant symptoms caused by either severe or constant attack by piercing and sucking insects are distorted leaves, overall loss of plant vigor, leaf spotting, and, in some cases, loss of foliage and death. Aphids (plant lice), leafhoppers, stinkbugs, spider mites, and silverleaf whiteflies are examples of piercing or sucking insects.

Insects that feed on seeds, pods, or fruits   As a group, insects that feed on seeds, pods, or fruits are probably the most damaging for vegetables and fruits. Most healthy plants can overcome some root and foliage damage; however, damage to edible parts of a plant result in direct loss of seeds or fruit. To further complicate the situation, attack at times of fruiting also means that not only the fruit is lost, but also the cumulative time, money, and labor expended on the crop (which is at a maximum at this time) is lost. Corn earworms, pickle worms, and cowpea curculios are examples of insects that feed on seeds, pods, or fruits.

Nature of attack of insect feeders   Three sources of metabolites are available to plant feeders: xylem content, phloem content, and cell content. The xylem sap is a dilute solution of mineral salts and a small amount of amino acids (due to nitrate reductase activity in the roots) and is usually under tension. The phloem contains a more nutritious solution, with sucrose as its compound of greatest concentration and is usually under pressure. Aphids are successful phloem feeders. Honeydew is excreted by aphids when the sugar content of the phloem is high. Ants then "farm" the aphids for the sugar. The cell content is probably the best balanced diet for feeders, but it is not a renewable source of food.

Although metabolite feeders rarely cause severe damage to plants, tissue feeders, particularly leaf feeders, can destroy large amounts of tissue leading in some cases to total defoliation. Tissue feeders include insects, mammals, mollusks, birds, and reptiles. Fish are the dominant feeder in marine environments. The basic qualification of a plant tissue feeder is a biting or tearing mouthpart. Tissue feeders also must be able to cope with a cellulose-based diet.

Diseases   Plant disease is the process by which living or nonliving entities (pathogens) interfere, over time, with a plant's physiology and metabolism. This may result in an altered plant appearance and/or lower yields. Pathogens can be parasitic or nonparasitic.

Fungi, viruses, bacteria, and mycoplasmas cause parasitic diseases. All pathogens compete with plants by using metabolites produced by the host and most reduce the physiological efficiency of the plant by reducing photosynthate capacity and/or water and nutrient uptake, by disrupting metabolic processes at the cellular level, or by producing toxins. The pathogens that cause parasitic diseases spread from plant to plant.

Abiotic (nonparasitic) plant injury is the necrosis, mottling, and stunting caused by poor plant nutrition, air or water pollution, or weather-related factors. Diagnosing a nonparasitic injury is difficult, because many situations cause similar symptoms.

Fungi   Fungi abound in most environments and are the largest group of disease organisms. Most fungi are nonchlorophyllous plants with a threadlike (hyphae) structure. They enter plants through natural openings or wounds or penetrate directly through intact tissue. Fungi cannot obtain their own food and get their

FIGURE 15.14   Late blight disease on potatoes is caused by the fungus *Phytophthora infestans* and is responsible for the Irish potato famine and the large number of Irish immigrants coming to America in the late 1840s.
Courtesy of Dr. Bill Lamont, Penn State University.

nutrients either from living tissues, from other organisms, or from dead organic matter. Fungi produce spores, which are sexual or asexual reproductive structures capable of surviving adverse conditions and dispersing themselves. Anthracnose, bean rust, cercospora leaf spot, downy mildew, early blight, fusarium wilt, powdery mildew, and sclerotinia are examples of fungal diseases. Late blight (Fig. 15.14), caused by the fungus *Phytophthora infestans*, infects potatoes, tomatoes, and eggplants. It was late blight that destroyed the Irish potato crop in the late 1840s, causing at least a million Irish deaths and forcing the migration of an even larger number of Irish to America.

Viruses   Viruses cannot function outside a host, but most have a wide range of organisms that can serve as hosts. Viruses also depend on a second organism (a vector) to move about the environment and invade plants. Viruses interfere with the biochemical reactions in plant cells and tissues. A mosaic symptom, characterized by

FIGURE 15.15 Symptoms of mosaic virus characterized by alternating bands or patches of yellow or green colors.

alternating bands or patches of various shades of yellow or green colors, is the most typical and most frequently occurring symptom of virus diseases (Fig. 15.15). Aphids, whiteflies, and leafhoppers are the most common vectors. Cucumber mosaic virus and tomato mosaic virus are examples of viral diseases.

Bacteria Bacteria cause fewer diseases than fungi and viruses. They are single cells surrounded by a cell wall. Plant parasitic bacteria lack chlorophyll and obtain nutrients from living or nonliving sources. Bacteria enter plants through natural openings or through wounds made by insect feeding, mechanical injury, or pruning. Bacterial infections increase at an enormous rate under optimum conditions and are encouraged particularly by surface moisture from rainfall, irrigation, or high humidity and by moderate temperatures. Seeds or vegetative propagation, insects, and tillage may transmit bacteria. Bacterial blight, bacterial leaf spot, and bacterial wilt are examples of bacterial diseases.

Mycoplasma Mycoplasmas are the smallest individual organisms that are plant pathogens. They must live close to plant cells and once were classified as viruses and induce similar symptoms. Symptoms caused by mycoplasma include growth abnormalities, yellowing, very short internodes, and distortions of leaf and flower tissue. The most common mycoplasma is aster yellows.

Nematodes Nematodes are neither pathogens nor insects. They are small, mostly slender, unsegmented roundworms (0.7 to 1.7 mm in length) that are generally transparent and colorless. More information on nematodes is presented in Chapter 19.

Animals   Animals are common inhabitants above ground and below the soil surface. Animals can damage plants by eating or grazing on the foliage, uprooting plants because of their tunneling in or near the plant roots, and reducing the visual quality and physiological activity of the plant by their defacing or partially destroying leaves, stems, fruit, flowers, tree limbs, and trunks.

The following are some examples of animal injury to plants.

Birds   Birds can injure the fruits of plants by their pecking, and they can deface the foliage.

Cats and dogs   Cats and dogs can be destructive to shrubs, flowers, or vegetables by their digging, rubbing, or chewing or from their urine.

Chipmunks   Chipmunks are ground-dwelling relatives of squirrels. They are omnivores and eat insects, salamanders, acorns, seeds, berries, some grains, and fruits. Chipmunks are valuable animals for the forest ecosystems as they disperse seeds for forest regeneration. They may dig up freshly planted seeds or occasionally eat flowers, but often do little damage.

Deer   Deer feed on farmers' crops, grass, twigs, bark, and evergreens in wooded sites. Deer especially cause damage to tender garden plants, flowers, or trees.

Groundhogs   Groundhogs (woodchucks) are large rodents. They prefer to live in open fields, woodlands, and hedgerows and eat a variety of vegetables, grasses, and legumes. Groundhogs are particularly damaging to peas, beans, carrot tops, alfalfa, clover, and grasses. Early morning and evening hours are the preferred feeding times for groundhogs.

Mice and rats   Mice and rats eat the same food as humans, including fruits, nuts, seeds, and mushrooms. Most mice are nocturnal and have well-developed senses of smell, taste, and hearing. Rats are larger than mice and eat a wider variety of plant materials and other objects such as electric wires and water pipes.

Moles   Moles are seldom-seen little animals. Their tunneling may help loosen and aerate the soil, allowing better penetration of water, lime, and fertilizer. Moles typically tunnel near the surface of the soil, heaving the ground up into ridges. Their tunneling can damage plants by causing air pockets around roots and by dislodging shallow-rooted plants.

Opossum   An opossum is a large, gray, long-haired mammal about the size of a house cat. It is the only North American marsupial. It is nocturnal and will eat almost anything, including berries and other fruits.

Rabbits   Rabbits prefer dense thickets or other areas of heavy vegetation. Active year-round, rabbits often feed on young succulent plant growth of annuals during the spring and summer and on buds, twigs, and the bark of trees and shrubs. Young fruit trees, brambles, and evergreen seedlings are often damaged in their dormant season by rabbits.

Raccoon   The raccoon is a distinctly marked, stocky mammal related to the bear. They are nocturnal and eat a varied diet. In spring and early summer, they survive by eating small animals. In late summer and fall, they also eat berries and other fruits, nuts, and grains. Raccoon occasionally eat vegetables such as sweet corn.

Slugs   Slugs are close relatives of snails. They damage plants with their feeding, often completely destroying seedlings. Slugs can move upward in plants large distances, and as they move they leave an unsightly slime trail.

Squirrels   Squirrels ordinarily inhabit wooded areas, often in close association with humans. They eat a wide variety of foods, typically eating fruits and nuts in fall and early winter. Acorns, hickory nuts, and walnuts are fall foods for squirrels, and they often store nuts for later use. In late winter and early spring, they prefer tree buds. In summer squirrels eat fruits, berries, corn, mushrooms, and other plant materials. They are especially damaging to sweet corn. Squirrels will nip the twigs from trees, chew bark from various woody plants, and raid bird feeders.

Voles   Voles, also called meadow mice and meadow voles, are small stocky rodents that eat grasses and herbs. Voles also feed on seeds, bulbs, and roots, and they can damage trees and shrubs during the winter when they gnaw off tender bark.

Agricultural Technologies and Mechanical Disturbances

Physical Mechanical Disturbances

Windbreaks   Windbreaks are used to reduce the effects of winds on plant development. Reduction in wind speed reduces evaporation from the leaf surface. Also, increases in air temperatures and humidity and altered CO_2 levels can occur on the leeward side of windbreaks.

In many locations, trees and brush that completely surround fields or are located only on the windward side provide adequate wind protection. In areas that are not protected from the wind, growers often plant small grains such as wheat or rye as windbreaks. The typical situation is to seed a couple rows of grain for every three

FIGURE 15.16   Use of rye windbreaks with newly planted bean plants.

or four crop rows (Fig. 15.16). This is usually done when the plant beds are formed in the field in early spring or during the fall. The windbreak row of small grains is often mowed later in the summer and used as a drive row during harvest. Windbreaks are usually killed before the plants become senescent, to reduce the risks of insect infestations.

Pruning   Pruning is the judicious removal or reduction of certain plant parts that are not required, that are no longer effective, or that are of no use to the plant. It is done to redirect energy (carbohydrates) to the development of flowers, fruits, and limbs or shoots (shrubs) that remain on the plant. Pruning is one of the most important cultural practices for maintaining woody plants, including ornamental trees and shrubs, fruits, and nuts. Proper pruning of trees improves their strength, maintains their health, enhances their beauty, and increases their value. The reasons for pruning include to train or direct the growth of plants into a particular form (topiary/espalier) or a specified space (hedges), to control their size and shape (ornamental and fruit trees), for aesthetics and safety purposes (landscape trees and urban trees) (Fig. 15.17), and to improve overall fruit quality primarily by increasing light penetration (fruit trees). Plants damaged by insects, diseases, or freezing also may require corrective pruning.

FIGURE 15.17   These trees were pruned to prevent the limbs from disturbing power lines that pass nearby.

Pruning often removes the shoot apex, destroying apical dominance (the inhibitory effect of one bud on the other buds of the plant in terms of bud break and shoot emergence) and stimulating the growth of lateral buds into shoots. It also reduces the size of the above-ground portion of the plant in relation to the root system. As a result, the undisturbed root system services a smaller number of shoots and buds and the relative uptake of water and nutrients by the remaining shoots and buds increases, and a flush of growth (regrowth) occurs.

Generally, the more severe the pruning (greater size or number of limbs removed), the greater the resulting regrowth. In essence, the plant is regrowing in an attempt to restore a balance between the top and the root system. Pruning generally stimulates regrowth near the cut and vigorous regrowth will usually occur within 15 to 20 cm of the pruning cut. Pruning also may indirectly stimulate growth of lateral shoots by allowing more light to penetrate the plant canopy. Pruning a young plant often will stimulate vigorous shoot growth and delay the development of flowers and fruits, with the length of the delay usually depending on the species pruned and the severity of the pruning.

The recommended time of year for pruning varies with plant species. Many woody ornamentals are pruned according to their date of flowering. Spring-flowering plants, such as dogwood or forsythia, normally are pruned after they bloom. Pruning spring-flowering shrubs during the summer, fall, or dormant season will remove flower buds formed the previous summer and fall. Summer-flowering plants generally are pruned during the dormant, winter season. If plants are not grown for their flowers, the best time for pruning is during the dormant winter season before new growth begins in the spring. Heavy pruning during late summer and fall is generally avoided because regrowth may occur making the plant more susceptible to winter injury.

There are many different types of pruning cuts. A thinning (out) cut removes a branch at its point of origin on the trunk, or shortens a limb to a lateral branch large enough to resume the growth of the pruned limb. Thinning cuts should leave little or no stubs. They are used to remove damaged, dead, or weak branches; reduce the length and weight of heavy limbs; or reduce the height of a tree. Thinning cuts are placed so as to distribute ensuing growth evenly throughout a tree and retain or enhance a tree's natural shape. A heading (back) cut trims a branch back to just above a bud, or trims a branch or leader back to a small branch not large enough to assume the growth of the pruned branch. A stub cut is made indiscriminately to a point on a branch or leader where no bud or branch exists. A stub cut, like a heading cut, is used when a tree is topped.

Healing naturally follows pruning or wounding and begins in the cambium, a thin layer of cells between the wood and the bark. Two areas of the cambium, the bark ridge (at the junction of two limbs) and the branch collar (a ring of slightly raised tissue where the lateral branch joins the main limb) function to close off the wound between the plant and the pruning cut.

Topiary   Topiaries are forms of plant sculpture that originated with the ancient Romans and became popular in European formal gardens. A topiary is a plant that has been pruned to an unnatural form for the plant, such as in a geometric shape or to resemble an animal (Fig. 15.18). Some topiaries, called "stuffed" or "mock" topiaries, are vines trained to grow on the outside of wire frames filled often with peat. Other topiaries are plants trained to grow within a wire frame. Boxwood, Japanese holly, yaupon holly, ligustrum, arborvitae, juniper, yew, Benjamin fig, and podocarpus are plants often shaped into topiaries. Topiaries require frequent pruning to maintain their shape.

Bonsai   Bonsai is an ancient art form that originated in China and was further developed by the Japanese. In the 13th century, the Japanese collected and potted wild trees that had been dwarfed by nature. These naturally formed miniatures were some of the first bonsai. When demand for the small trees outgrew the supply, Japanese gardeners began to develop bonsai from native trees.

A realistic illusion of a mature dwarfed tree is accomplished by choosing a plant with the trunk, branches, twigs, leaves, flowers, fruits, buds, and roots of the plant in scale with the size of the plant (Fig. 15.19). Plants chosen for bonsai often have attractive bark with a trunk that gives the illusion of maturity. The upper one-third of the root structure of a mature bonsai is often exposed to give the appearance of age. Some plant species that are popular for bonsai are Sargent juniper, Japanese black pine, Japanese wisteria, Chinese wisteria, Japanese flowering cherry, and Japanese or sawleaf zelkova.

The three basic operations to achieve a desired bonsai shape are pruning, nipping, and wiring. Substantial pruning is often necessary initially with new plants to remove excess foliage and undesirable limbs. Often all buds except those on the outside of the trunk are removed to force the growth upward and outward. Stubs are left flush with the stems and sufficient leaves are left on the plant for necessary photosynthesis. Heavy pruning usually only takes place once in the life of the bonsai.

FIGURE 15.18 Example of a topiary.
Courtesy of Dr. Dennis Wolnick, Penn State University.

Once the basic form is established, shaping is accomplished by nipping or pinching back new growth. Nipping shapes the plant and promotes luxurious foliage. Roots are also trimmed. All fibrous roots and surface roots are maintained, and one branch for each main root is left intact. After the plant has been sufficiently pruned, the plant is often trained with thick copper wire. The wiring begins at the bottom of the tree, and the wire is turned around the branches and trunk about one quarter inch apart and spiraling upward at a 45-degree angle. The wired branches are carefully bent to achieve a desired form. The wire is usually maintained in place for at least a year.

Training Training plants is a method of orientating the plant growth in a distinct way in space. Different methods are used to develop strong, well-spaced scaffolds to

FIGURE 15.19   A Bonsai plant.
Courtesy of Dr. Dennis Wolnick, Penn State University.

bear good fruit load (in fruit trees); to promote early plant production; and to provide an aesthetically pleasing plant. Training generally allows more light to penetrate the canopy and often facilitates harvesting. Training a tree can also prevent storm damage when the tree approaches maturity.

Much of a grower's effort for the first year or two of a fruit tree's existence in an orchard may involve training the tree to develop a strong framework. Freestanding trees can be trained to a central leader or to an open center. In a central leader system, the trees are trained into a Christmas tree shape with the tops always narrower than the lower branches. Wide branch-to-branch angles (or crotch angles) may be achieved using weights or braces to help widen the branch angle (Fig. 15.20).

Trained fruit trees and vegetable crops can also be trellised (Fig. 15.21). A trellis system relies on the use of wires to serve as support and training aids. Several ultimate forms or training patterns can be used in developing the trellis.

Espalier   Espalier is a French term used to describe a plant that is trained to grow along a flat plane, such as a wall or fence (Fig. 15.22). The Romans originated the technique and the Europeans refined it into an art form still popular today. Fruit trees in England are often espaliered along walls to conserve space and to provide protection and warmth. Espaliers are often used in North American landscape designs to add interest to a bare wall, to accent a trellis or fence, or to provide plant material in areas where spreading trees and shrubs are not practical.

Almost any plant can be used for an espalier. Pyracantha is the most commonly espaliered ornamental plant. Others include Bradford pear, southern magnolia, foster holly, ligustrum, crape myrtle, loquat, creeping fig, and wisteria.

FIGURE 15.20   Wooden limb spreaders can be used to widen branch angles of fruit trees. Courtesy of Dr. Rob Crassweller, Penn State University.

The framework on which the espalier is trained to grow usually is constructed from rot-resistant wood. The framework is attached to blocks of wood, which are used as spacers to keep the framework 6 to 8 inches from the wall. This provides good air circulation around the plant and prevents the plants from staining the wall. Formal espaliers require regular pruning and training to maintain their shape. Branches that have grown outward away from the intended pattern are also often pruned, and the remaining branches are often tied to support them in place.

Staking   Some plants and trees require staking after they are planted in the field to prevent mechanical injury as a result of being blown over by wind or damage by pedestrian traffic, or to redirect growth. Some annual plants such as tomatoes are

FIGURE 15.21   Trellised apples.
Courtesy of Dr. Rick Bates, Penn State University.

FIGURE 15.22   An example of an espalier.
Courtesy of Dr. Rick Bates, Penn State University.

FIGURE 15.23 Staking tomatoes helps with air circulation, increases light exposure, and facilitates fruit harvest.

often staked (and sometimes pruned) in the field to help facilitate optimum light exposure, air circulation, and harvesting of the fruit and to save space (Fig. 15.23).

Newly planted, young trees in landscapes and orchards may be staked to keep the root ball secure until its roots are well established in the soil. Tall stakes (approximately 2.5 m in height) are often used for landscape trees that are 5 cm or less in diameter of the trunk (Fig. 15.24). The tree trunk is pulled vertical with wire, rope, or tree strap attached horizontally to the stake. The tree trunk often is allowed to have a certain amount of freedom to move because as the tree flexes it develops greater strength faster. The stakes are placed so that the tree does not rub against them during winds.

Shorter stakes are often used with larger trees (greater than 5 cm in diameter of trunk) (Fig. 15.25). Guy wires are attached to the tree at equal distances around the trunk at a crotch (junction of limb to trunk). Anchor stakes are driven in the ground at approximately 45° angles about one-half the height of the tree away from the trunk. As soon as the tree has become established in either method of staking landscape trees, the support stakes and guy wires are removed.

Wounding and Rooting   Adventitious roots arise from undifferentiated but actively dividing cells deep within the stem and not from other roots. Overlying tissue must be ruptured for adventitious roots to emerge. Wounding the outer layer of a stem often weakens the constraint of these outer tissues and produces reactions within the plant that encourage root growth. Wounding and the induction of rooting can play a significant part in the natural layering of plant parts.

FIGURE 15.24   Staking young planted trees with tall stakes prevents them from being blown over and keeps the root ball stationary until the roots are well established.

Mechanical Conditioning of Transplants/Bedding Plants   Acceptable transplants are generally short and stubby with thick, strong stems and dark green leaves. Chemical plant growth regulators provided convenient and consistent plant growth management in the past; however, the loss of daminozide (Alar, B9) from the market in 1989 resulted in no growth retardants labeled for use on vegetable transplants. An increased interest in alternative methods of transplant growth control, such as mechanical conditioning, soon developed.

FIGURE 15.25   Larger trees in the landscape that are staked often use shorter stakes and guy wires that reach up to the first branch crotch.

Mechanical conditioning of plants is the deliberate physical stressing of plants applied to manage plant growth and increase quality. Desirable transplant characteristics are necessary to produce plants that will withstand the physical rigors of handling and transplanting and also survive the stress of moving the plant from the protected environment of the greenhouse to the field environment. Mechanical stress can be applied by rubbing stems, brushing shoots, shaking potted plants and whole flats, vibrating pots or plants, and perturbing plants with water, forced air, or wind. Mechanically conditioned transplants become established and resume active growth soon after transplanting.

Mechanical conditioning can significantly reduce plant height when compared to untreated plants. The degree of growth reduction often depends on the duration or intensity of the mechanical conditioning treatment. Mechanical conditioning can improve plant appearance including higher chlorophyll concentrations. The darker, thicker leaves combined with shorter stems project a healthy, vigorous plant appearance. Greater stem and petiole strengths also often result from mechanical conditioning. For example, rubbed bean stems are frequently stronger than unrubbed stems and had less stem breakage during transplanting. Plants that were mechanically treated maintained a more upright habit when planted or transplanted to outdoor conditions. Little to no effect on yield of plants is often reported when they were only treated during the transplant production stage.

Mechanical conditioning by brushing, shaking, or rubbing is currently considered too labor intensive to be economical for commercial application, and it has the potential to physically damage transplants. Touching or shaking the plant may also increase the spread of pathogens among treated plants and reduce final crop yield.

FIGURE 15.26   Ground-level view of burning savanna grasslands in South Africa.
Source: Earth Observatory, NASA.

For mechanical conditioning of plants to be commercially successful, the technique used must be simple and flexible and must not reduce plant quality.

Prescribed Fires and Controlled Burning   Wildfires are unplanned and usually occur due to lightning, human neglect, or malice. Wildfires often arise during extended dry periods when soil moisture is low and plants are severely stressed. Prescribed fires are planned and conducted at the proper time and in a safe manner to meet specific objectives (Fig. 15.26). For prescribed fires to be effective, desirable plants should be dormant, soil moisture should be sufficient to support plant growth after the fire, and favorable environmental conditions should exist to ensure predictable fire behavior and simplify fire control. In the management of grasslands, prescribed fires can increase grass nutritive quality, palatability, availability, and yield; reduce hazardous fuels; suppress unwanted plants; and improve wildlife habitat. Burning can benefit certain wildlife species by increasing habitat diversity and the nutritive quality, availability, and yield of seeds and forage. Although most plants have no way of escaping fires, animals usually escape by running or flying away, going below ground, or moving to unburned islands of vegetation.

Foresters often set surface fires to prevent future major fires by reducing the amount of fuel present. Fires also assist some tree species to reproduce. Jack pine, for example, has serotinous cones, which won't open and release their seeds until they are exposed to a heat source such as fire. Fire assists in the germination of other seeds by killing competitive vegetation and removing dry litter on the soil surface to expose moist mineral soil.

FIGURE 15.27   An aromatic soap suspended on a stick in an attempt to keep deer out of a planting.

Deterring Animal Injury   Animal injury of plants can be controlled or partially controlled by using various crop production practices. Plantings usually are made in areas where animals are not expected to be a problem. If animals are present, protective structures such as fences or electric wires may be placed around the planting to exclude the animals. The electric wires provide a mild shock to the animal when it touches the wire and hopefully conditions the animal to not approach the area surrounded by the wire or fence.

Other protective measures may be used to deter animals, sometimes with limited success. A common method used to exclude deer from plantings is to place human hair (in mesh bags or spread on the soil) or aromatic soaps (suspended from stakes or sticks) sporadically in a planting (Fig. 15.27). It is suspected that human or humanlike scents may deter the deer from entering the area. Protective collars or wraps may also be placed around plants to reduce physical injury.

Biological Mechanical Disturbances

Using Pesticides to Reduce or Eliminate Pests   A pesticide is a substance for preventing, destroying, repelling, or mitigating any pest. Pesticides include insecticides, fungicides, herbicides, algicides, fumigants, miticides, molluscicides, and nematicides. Pesticides are chosen that are effective for use on the specific crop and against the specific pest problem, but if they are not properly applied, pesticides can injure the crop and/or farm personnel.

Pesticides can be grouped according to their chemical nature. Inorganic pesticides are made from minerals such as copper, boron, lead sulfur, tin, or zinc. An example is the Bordeaux mixture ($CuSO_4$ + lime). Synthetic pesticides typically contain carbon, hydrogen, and one or more other elements, such as chlorine, phosphorus, or nitrogen. Examples include Sevin (carbaryl), malathion, diazinon, and maneb.

Microbial pesticides are microscopic organisms, such as beneficial viruses, bacteria, and fungi that are cultured, packaged, and sold. Some of the pathogens multiply and spread after application, whereas others do not. Examples of microbial pesticides are the bacterium *Bacillus thuringiensis* and the polyhedrosis virus. Plant-derived organic pesticides are made from plants or plant parts. Examples include rotenone and pyrethrins.

Pesticides grouped according to mode of action   Pesticides can also be grouped according to their mode of action (how they work). Protectants are applied to plants, animals, structures, and products to prevent entry or damage by pests. They can prevent infection from occurring by killing fungal spores before or after they germinate, but they often do not prevent all infections.

Contact poisons kill pests when they contact the poison. Stomach poisons kill when swallowed. Systemics are taken into the blood of a host animal or sap of a host plant. Because toxic concentrations can remain in plants for more than a few days, they provide some protection against new infections.

Translocatable, systemic herbicides kill plants by being absorbed by leaves, stems, or roots and moving throughout the plant. Fumigants are gases that kill when they are inhaled or otherwise absorbed by the pest. Selective pesticides kill only certain kinds of plants or animals, whereas nonselective pesticides kill many kinds of plants or animals.

Alternatives to pesticides   Appropriate use of pesticides in an integrated pest management (IPM) plan is an important part of controlling certain diseases and insects. Pesticide use may be reduced or eliminated through the use of appropriate cultural practices combined with resistant cultivars.

Crop rotation   Growing the same crop in a field year after year encourages the buildup of pathogens and insects. By rotating crops, the pathogen or insect levels do not increase.

Sanitation   Removing infected plant material eliminates a source of infection. Alternately, infested crop residue could be buried and steam used to disinfect greenhouse, nursery pots, and containers.

Planting date   The best approach is to plant when environmental conditions favor growth of the plant but not growth and reproduction of the pathogen or insect.

Improving soil structure and drainage   Poor soil drainage can create an environment favorable to certain root pathogens. Improving soil structure and drainage can alleviate such problems.

Proper fertility   Properly fertilized crop plants are generally more resistant to diseases than underfertilized or overfertilized plants.

Transgenic plants   A relatively new practice to manage insect pests is the use of plants that have been genetically engineered to produce an insect toxin not naturally produced by plants. Insecticidal protein from *Bacillus thuringiensis* (Bt) has been transferred to selected crops as a method for protection against selected insects. Because Bt hybrids contain an exotic gene, they are commonly referred to as transgenic plants. The Bt gene in these plants produces a protein that kills selected insect larvae upon ingesting a small amount of plant tissue. An example of this currently in use is Bt hybrids of corn. Protection of Bt corn from European corn borer has been shown to be far more effective than from optimally timed insecticides.

Other practices Overcrowding of plants (high crop population density) can lead to excess humidity around leaves, favoring fungal and bacterial diseases. Timely harvest can also be important, as some diseases develop on the harvested commodity only as the crop matures. Pathogen-free seed is also important. For some crops, selected plants that repel certain pests are planted adjacent to the crop to protect the plantings from pest infestations. In general, practices that reduce plant stress can increase plant resistance to pathogens.

Integrated Pest Management (IPM) Integrated pest management (IPM) is an approach to pest control that utilizes regular monitoring of pest populations to determine if and when treatments are needed. IPM employs physical, mechanical, cultural, and biological tactics to keep pest numbers low enough to prevent intolerable damage or annoyance. In IPM programs, treatments are only made when and where monitoring has indicated that the pest will cause unacceptable economic, medical, or aesthetic damage. Treatments are chosen and timed to be most effective and least disruptive to natural pest controls.

IPM provides a process for identifying and reducing the factors causing the pest problems. It is also designed to determine whether the cost of a particular pest management action is worth the investment and to reduce reliance on pesticides that may harm the environment. Whether or not an organism is viewed as a pest is an issue of whether its damage or annoyance is tolerable.

Types of injury or damage caused by pests and determining tolerable damage or injury levels Economic damage occurs when pest activity affects the production of biological yield (i.e., fruits, flowers, and foliage) in such a way that reduces its marketability and serves as an economic good. Aesthetic damage occurs when the organism causes an undesirable change in the appearance of the product.

Injury or damage level is the level of the pest presence or pest-related damage that can be tolerated without harm to the plant as an economic good. Total eradication of pest organisms is virtually impossible and is often undesirable because it can result in the reduction or total loss of the pests' natural enemies, eliminate food of predatory or parasitic biological controls, and can upset broader ecological balance. The amount of damage that can be tolerated must be determined, as well as the critical pest population required before it causes that level of damage and a treatment level that maintains the pest below this population size.

The life cycles and seasonal variations of both the pests and the natural enemies are considered in timing pesticide treatments. Correct timing is even more crucial in the application of the many, less-toxic commercial pesticide products. Treatments are often applied only where the problem is most severe and the application will have the greatest impact.

In many pest situations there are likely to be locations where the problem is more severe or less severe. Spot treatment with pesticides is often used by treating only the critical areas. This reduces pesticides and spraying costs and minimizes the amount of chemical applied to a field.

General Control Measures for Insects and Diseases   *General insect control measures* Insects must be carefully controlled during the production of crops as they can reduce the vigor and/or visual appearance of the plant. Insects can also become embedded within leaf clusters of some edible crops and not be observed during harvest but appear after purchase or during consuming or processing. Insects are difficult to control once they are within the leaf clusters or husks. Therefore, control must be exerted before insects move into these structures.

Plant growing areas are monitored regularly for insects. Control measures are usually taken when insects reach a certain threshold amount. Several cultural practices reduce insect infestations. These cultural practices include controlling weeds in nearby areas (Fig. 15.28), burying plant debris that might harbor overwintering pests, and increasing populations of natural enemies of certain pests.

If insecticides are used, they are carefully chosen from an accepted list for use against the specific target insect. Insecticides are generally not applied in the middle of hot, dry days because coverage of wilted plants is incomplete and high temperatures will volatilize some insecticides before they reach the plant.

General disease control measures   Methods to control diseases can be broken down into five categories: resistant and tolerant cultivars, practices that promote vigorous

FIGURE 15.28   Poor weed control along the edges of a field can cause increased insect problems for the crop being grown.

plant growth, proper water management practices, practices that reduce inoculum, and fungicide sprays.

Using certified disease-free seed or planting pieces and choosing resistant and tolerant cultivars are often the least expensive, easiest, and most effective ways to reduce losses to some diseases. For some diseases, such as soil-borne vascular wilts and the viruses, the use of resistant cultivars is the only way of ensuring control. More than 85% of the vegetable acreage in the United States is planted with cultivars with tolerance or resistance to one or more disease organisms.

Vigorous growth is desired for proper plant development and prevention of disease. Stressed plants, especially those low in potash and calcium, are more vulnerable to disease. Plants that receive excessive amounts of nitrogen may develop succulent growth, which can encourage certain pathogens. Maintaining proper crop nutrition is extremely important for long-term growth and disease prevention. Selection of cultivars adapted to the location and resistant to common pathogens is also an important aspect of disease control. Following recommended planting dates could also ensure that the plants achieve vigorous growth. Also, planting dates may be adjusted so that the crops mature before disease strikes.

Proper water management is important to prevent soil waterlogging. Seeds and seedlings are likely to rot in wet soils. The foliage must also remain dry as wet leaves are more likely to support the establishment of waterborne pathogens because inoculums are spread from infected to healthy leaves by water droplets. It may be beneficial to irrigate only early in the day so that the leaves dry before evening or use trickle flooding, or other methods of subirrigation.

Inoculum levels of diseases are reduced through good sanitation and by controlling insect vectors. Only certified or treated seed to pathogens should be used in crop production systems. When harvest is complete, the remains of crops are removed or destroyed, because stems, leaves, and roots can function as disease reservoirs and can harbor insects for the following year.

Proper crop rotation can reduce disease pressures. Improving soil structure by increasing the organic matter of the soil and removing or burying diseased plant matter (which could serve as a source of inoculum) reduce the incidence of diseases. In addition, pesticide applications to the plants can be used to reduce injury due to fungal and some bacterial diseases. Chemicals that are used include seed treatment fungicides, foliar fungicides, and soil fumigants. These chemicals should be carefully chosen and properly applied.

In summary, successful control of diseases is achieved by using several steps selected on the basis of the pathogen, its life cycle, the host plants, labor availability, and cost. The degree of effectiveness that the grower achieves in controlling a particular disease will also depend on the growing season and the level loss of product that the grower can tolerate.

Special Considerations of Mechanical Disturbances of the Rhizosphere

In addition to the mechanical disturbances of the aerial environment, there are mechanical disturbances that affect the rhizosphere. Some of these include distur-

FIGURE 15.29   Example of slashing or pruning a plant root system with circling roots. Courtesy of Dr. Jim Sellmer, Penn State University.

bances to the roots as a result of root pruning, root girdling, soil compaction, and grade changes.

Root Pruning   Root pruning is often practiced with container-grown annual and herbaceous perennial plants and small shrubs to encourage new root formation and discourage root circling. This is accomplished by cutting the root ball with vertical superficial cuts when the plants are transplanted (Fig. 15.29). Large trees and shrubs may be root pruned the year preceding transplanting as the plant then has several months to develop new fibrous roots within the root ball before being transplanted. Pruning is occasionally necessary to reduce the root size in order to fit the hole at transplanting.

Root Girdling   Root girdling occurs when a plant root entwines around another large root or the base of the plant and prevents or hinders the uptake of water and nutrients (Fig. 15.30). The girdling root occurs often below ground level. Root girdling may contribute to poor plant vigor (such as small leaves, death of small branches or stems) and/or top dieback of trees (browning or drying of the outer margin of leaves or in the area between the leaf veins). Girdling roots may be the result of unfavorable conditions that prevent the roots from developing in a normal spreading manner. Roots of container-grown plants are often forced to grow in a circular fashion, and root girdling can occur if these roots are not pruned at the time of transplanting. Also, the use of proper planting techniques, such as making the hole large enough to accommodate the roots of the transplanted plant material, promotes good root growth and can discourage root girdling.

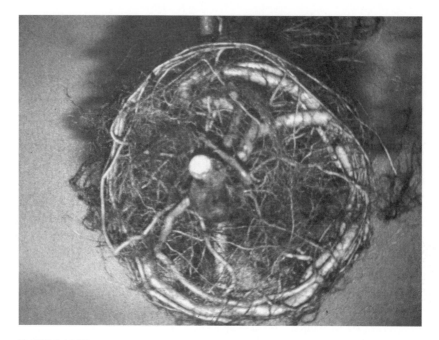

FIGURE 15.30   The root system of this tree exhibited poor root formation and girdling roots. The roots were not cut or spread at the time of planting. The roots eventually girdled themselves, killing the tree.
Courtesy of Dr. Rick Bates, Penn State University.

Soil Compaction   Soil compaction is the packing effect of a mechanical force or gravity and time on a soil. Several types of soil compaction can occur in agricultural fields. Each type generally has different causes and treatments. Surface crusting is caused by the impact of raindrops on weak soil aggregates and can result in reduced seed emergence and water infiltration. Soils with high organic matter, high biological activity, or high sand content are less likely to form crusts.

Surface compaction occurs from the surface down to the tillage depth (Fig. 15.31) and can be loosened by normal tillage, root growth, and biological activity. The degree of surface compaction is determined by the moisture in the soil at the time of compaction and the ground contact pressure of equipment or animals. Subsoil or deep compaction lies beneath the level of tillage. Ground contact pressure and the total weight on the tire (the axle load) significantly affect the amount of subsoil compaction. Deep compaction is difficult to eliminate and may permanently change the soil structure. The plow pan or tillage pan is subsoil compaction that is only a few centimeters thick and lies directly beneath the normal tillage depth. It develops when the depth of tillage is the same from year to year, or when the rear wheel of the tractor rides in the moldboard plow furrow.

Soil compaction can also occur due to natural phenomenon such as a falling tree, the result of a storm, or the movement of glaciers during the last ice age. Soil compaction decreases the soil volume occupied by pores and increases the density and strength of the soil mass. Also, the number of macropores (large pores) decreases and

FIGURE 15.31   The photo shows compaction layers resulting from moldboard plowing and secondary tillage. Relatively uncompacted soil is at the surface layer, whereas a layer reflecting compaction due to secondary tillage is below the uncompacted soil. Below the secondary tillage layer is the "plow pan," which is the compacted layer formed by moldboard plowing.
Source: Michigan State University.

micropores increases, which slows the rates of infiltration and drainage and aeration from that compacted layer. This occurs because large pores are the most effective in transporting water when the soil is saturated. Because a compacted soil has less air-filled space than before it was compacted, the exchange of gases slows down and the likelihood of aeration-related problems for plant development increase.

A moderate amount of compaction promotes good seed-soil contact and fast germination, and prevents excessive drying out around the seed. Soil compaction that exceeds useful levels can decrease the infiltration and percolation rates. Soil compaction also can adversely affect plant growth by decreasing root growth and reducing the soil volume explored by roots, which can decrease nutrient and water uptake and leaf growth. Although roots are typically smooth and cylindrical in friable soil, they become stubby and gnarled in soils that are compacted. These effects on roots probably contribute to the adverse effects of compacted soils on the plant's supply of water and nutrients.

Many urban trees and shrubs are affected by soil compaction from the paving of streets, parking lots, and sidewalks (Fig. 15.32). Upper soil layers are often required to be compacted prior to paving to provide an adequate base for the asphalt. If a tree is completely surrounded by compaction and paving, there is often little room for water and air exchange in the soil, resulting in a potential buildup of carbon dioxide. This condition is especially critical if the paving was done after the tree had established its root system. Symptoms of soil compaction and paving include long-term poor growth, marginal browning of leaves, twig dieback, summertime yellowing of leaves, and smaller leaf size. Frequently, more than one of these symptoms will exist on a tree impacted by compaction and paving.

Soil compaction can also occur due to residential or commercial construction activities in an area, resulting in detrimental effects on plants (especially trees) in that area. Construction traffic compacts soil most severely near the surface, the

FIGURE 15.32   An example of an urban tree surrounded by asphalt exhibiting reduced growth and vigor compared to the tree not surrounded by asphalt.
Courtesy of Dr. Rick Bates, Penn State University.

area where the majority of tree roots lie. Prominently marking protected areas, erecting barricades around designated trees, and avoiding vehicular traffic or parking in restricted areas can prevent injury (Fig. 15.33). Other types of plant injury due to construction include bark removal, branch breakage, and trenching injury.

Grade Changes   Soil fill on top of root systems of plants and trees can cause serious damage to the plants, at least partially due to increased soil compaction and poor air and moisture circulation to the roots. As little as 10 to 15 cm of fill can be damaging to some trees. The typical symptoms of fill damage are small leaf size, yellowing of foliage, and branch dieback. These symptoms may not be expressed until several years after the grade change was made, resulting in a long decline and often death. Soil removal around plants and trees can also be damaging, as very little soil can be removed without damaging the root system. Tree "wells" are often used to minimize grade change effects on trees. The "well" is an area surrounding the tree within the drip line that is left unmodified (Fig. 15.34).

FIGURE 15.33 Erecting barricades around designated trees can prevent injury from compaction due to construction traffic and direct physical injury from machinery and personnel.
Courtesy of Dr. Larry Kuhns, Penn State University.

FIGURE 15.34 An example of a tree well used when the soil grade was changed.
Courtesy of Dr. Rick Bates, Penn State University.

Summary

Physical mechanical disturbances to plants occur when some type of physical pressure is applied to the plant, whereas biological mechanical disturbances occur as the result of adverse influences from humans and other organisms. Gravitropism is a physical mechanical disturbance that is affected by the gravitational pull of the earth and allows plants to orientate their growth for optimum survival. Thigmomorphogenesis is the response of plants to contact stress (as illustrated by the response of the sensitive plant and the Venus fly trap to touch), and seismomorphogenesis is the response of plants to vibration stress. Winds may induce mechanical disturbances to plants by seismic or vibration stress and very windy places are often inhabited by short-growing plants. Hurricanes, hail, glazing of tree limbs, and fire and lightning are other examples of mechanical disturbances to plants.

Loss of leaf area due to biological mechanical disturbances (such as from feeding of other organisms) reduces photosynthetic area and capability and reduces plant growth. For example, as insect population and feeding increases and healthy leaf area decreases, growth and visual appearance of the plant is reduced. Plant diseases also may cause changes in plant activity (such as reduction in photosynthesis) and result in reduced growth and visual quality of the plant.

Crop producers can influence whether physical mechanical disturbances affect plant growth through the use of windbreaks in the field, selective pruning to remove plant parts or tissue, training and staking of plants to orientate their growth and provide stability, mechanical conditioning of transplants, and prescribed fires and controlled burning to control unwanted vegetation. Crop producers can influence whether biological mechanical disturbances affect plant growth through the use of pesticides, IPM practices, and animal deterrents. Some mechanical disturbances that affect the rhizosphere include root pruning, root girdling, soil compaction, and grade changes.

Review Questions

1. How do plants adjust their growth in a windy environment?
2. What is thigmomorphogenesis?
3. Explain the role of fire in maintaining long leaf pine forests.
4. How do symmetrical pressures on plants differ from asymmetrical pressures? Give two examples of each.
5. What are some advantages to the plant of being able to respond to gravity?
6. Why might the wounding of plant tissue be advantageous to plant rooting?
7. Name three methods by which plants deter animals from feeding on them.
8. What is the difference between a metabolic insect feeder and a tissue insect feeder?
9. What is the difference in nutritional content that insects receive if they feed from the xylem versus the phloem?

10. Explain why a *Mimosa pudica* plant responds within seconds to a touch stimulus, yet it may take 10 to 15 minutes for the plant to reposition itself after being touched?
11. What are statoliths?

Selected References

Bernard, K. E., C. Dennis, and W. R. Jacobi. 1999. *Protecting trees during construction.* Colorado State University Cooperative Extension Service Publication No. 7.420.

Caron, D. M. 1997. *Animals in the garden.* Delaware Cooperative Extension Service Publication HYG-63.

Elmendorf, B., and H. Gerhold. 2001. Why should trees be pruned? *Sylvan Communities,* Winter 2001:29–30.

Fishel, F. 1997. *Pesticides and the environment.* University of Missouri Cooperative Extension Service Publication G7520.

Halfacre, R. G., and J. A. Barden. 1979. *Horticulture.* New York: McGraw-Hill.

Janick, J., R. W. Schery, F. W. Woods, and V. W. Ruttan. 1974. *Plant science.* 2nd ed. San Francisco: W. H. Freeman.

Lambers, H., F. S. Stuart III, and T. L. Pons. 1998. *Plant physiological ecology.* New York: Springer-Verlag.

Latimer, J. G., and R. B. Beverly. 1993. Mechanical conditioning of greenhouse-grown transplants. *HortTechnology* 3:412–414.

Levitt, J. 1980. *Responses of plants to environmental stress: Vol. 1, Chilling, freezing, and high temperature stresses.* New York: Academic Presses.

Relf, D. 1997. *The art of bonsai.* Virginia Cooperative Extension Service Publication No. 426–601.

Salisbury, F. B., and C. W. Ross. 1978. *Plant physiology.* 2nd ed. Belmont, CA: Wadsworth.

Schrock, D. S. 1996. *Preventing construction damage to trees.* University of Missouri Agriculture Publication G6885.

Swan, J. B., J. F. Mocrief, and W. B. Voorhees. 1994. *Soil compaction: Causes, effects, and control.* University of Minnesota Extension Service Bulletin. BU-3115-GO.

Wade, G. L. and R. R. Westerfield. *Basic principles of pruning woody plants.* University of Georgia Cooperative Extension Service.

Walterscheidt, M. J. 2001. Maples: Environmental stresses. *Sylvan Communities,* Winter 2001:31–33.

Selected Internet Sites

http://www.fs.fed.us/fire/   USDA Forest Service fire reports.

http://www.hortipm.tamu.edu/   Hort IPM Web site, Texas A&M University.

http://www2.champaign.isa-arbor.com/   International Society of Arboriculture Web site. Technical information on most aspects of tree care.

http://www.ippc.orst.edu/cicp/Vegetable/veg.htm   Internet resources on vegetable pest management.

http://www.nrcs.usda.gov/technical/ECS/   Technical information on ecological sciences, including agricultural ecology, aquatic ecology, ecological climatology, forestry & agroforestry, range & grazing land ecology, understanding ecosystem, and wildlife management, Natural Resources Conservation Service, USDA.

http://www.nysipm.cornell.edu   Integrated pest management information, Cornell University.

http://www.peak.org/~mageet/tkm/ecolenv.htm   Internet resources related to the science of ecology and the state of the environment, Peak Organization.

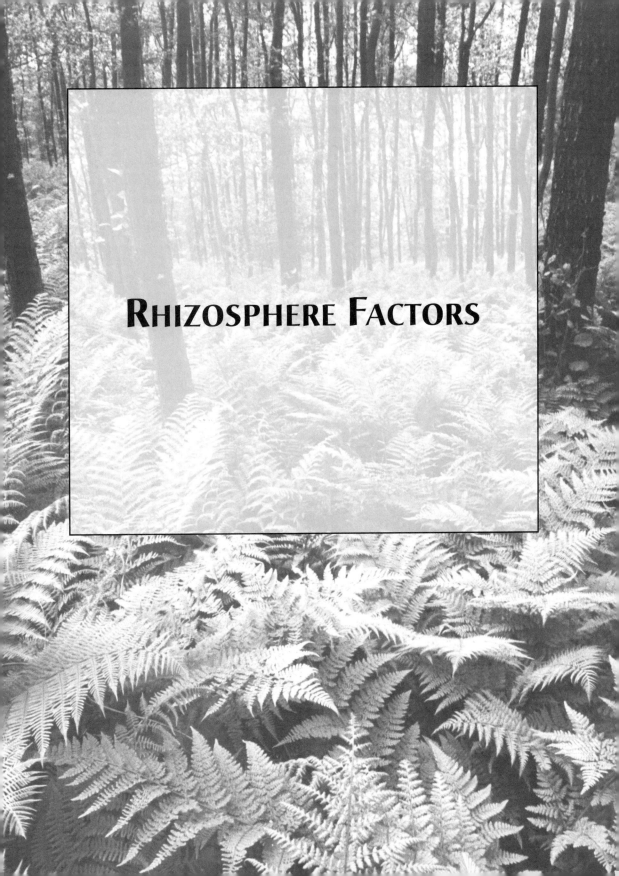

RHIZOSPHERE FACTORS

16

Overview of the Rhizosphere

The rhizosphere, often called the plant root zone, is the portion of the soil in the immediate vicinity of plant roots in which microbial populations and biochemical reactions are influenced by the presence of the roots. The soil is a living ecosystem made up of a complex of physical, chemical, and biological substrates where energy and matter are captured and transformed by plants, animals, and microbes. Water in the soil is cycled and a portion of this is stored and available to plants. Roots are important to the plant for water and nutrient uptake and for stability (by anchoring the plant in the soil).

This introductory chapter provides an overview of soils followed by brief discussions of other components of the rhizosphere such as water, plant roots, soil organisms, and allelochemicals. These subjects will be covered more in-depth in Chapters 17 to 20.

Soil

Soils are heterogeneous materials (containing solid, liquid, and gaseous phases) composed of abiotic (nonliving) and biotic (living) components. Abiotic soil components include mineral matter (clay, silt, sand), water, air, and organic matter. The soil's mineral matter comes from the breakdown of parent material (the more or less chemically weathered materials that underlie the soil and from which the soil develops), whereas organic matter comes from the breakdown of plants and animals. The soil solution makes up the liquid phase and contains dissolved minerals; it serves as a medium for ion transport to root surfaces. Oxygen, carbon dioxide, and nitrogen gases are dissolved in soil solutions, but their uptake by roots is predominantly from the air gaps between soil particles. Liquid and gaseous percentages of a soil vary significantly with soil texture and condition, weather, and plant water uptake. Biotic soil components include microorganisms, such as nematodes, protozoa, algae, fungi, and bacteria, and macroorganisms, such as earthworms and plant roots.

Textural Triangle

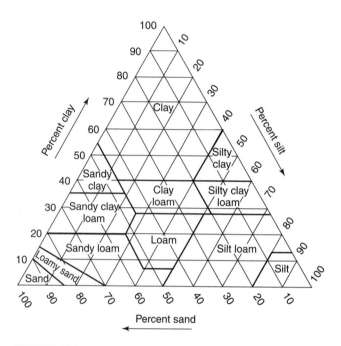

FIGURE 16.1 The soil textural triangle is a graphic representation of the various classes of soil according to the amounts of sand, silt, and clay in the various classes.
Source: Natural Resources Conservation Service, USDA.

Soil Types

Soils are classified or typed according to their texture or makeup. Soils are composed of particles of sand (2.0 to 0.02 mm diameter), silt (0.02 to 0.002 mm diameter), and clay (< 0.002 mm diameter). Soil texture is the proportion of sand, silt, and clay in a soil. The soil textural triangle (Fig. 16.1) is used to classify a soil into one of 12 categories, each of which has different physical and chemical properties.

A soil's texture greatly affects its productivity and applicability for use in agriculture. Coarse-textured soils such as sands and loamy sands have very low moisture-holding capacities (with good drainage and aeration) and are generally low in plant nutrients. Fine-textured soils such as clays, sandy clays, and silty clays have very high moisture-holding capacities and are usually high in plant nutrients, though they hold water and nutrients so tightly in some situations that availability for plant uptake and use may be limited. Also, clay soils are especially susceptible to crusting (which can result in poor seedling emergence). Loamy soils such as silts, silt loams, sandy loams, and clay loams are often excellent soils for production of many crops because of their intermediate water-holding capacity and nutrient content.

Soil Fertility

The soil is the main source of nutrients for plants. Plant macronutrients found in the soil include nitrogen (N), phosphorus (P), potassium (K), calcium (Ca), magnesium (Mg), and sulfur (S); and plant micronutrients include iron (Fe), copper (Cu), manganese (Mn), zinc (Zn), boron (B), cobalt (Co), molybdenum (Mo), and chlorine (Cl). Carbon (C), hydrogen (H), and oxygen (O) are obtained from non-mineral sources of carbon dioxide or water. Excessive amounts of nutrients can cause toxicity, and insufficient or deficient amounts can cause poor or abnormal growth. The cation exchange capacity of a soil is the degree to which a soil can absorb and exchange mineral ions and is highly dependent on soil type.

Organic Matter

The organic matter or organic composition of a topsoil often ranges from 1 to 6 percent and consists of plant and animal residue that is in various stages of decomposition. Humus is the dark organic material that is produced through the decaying action of organic residue by the action of microorganisms and chemical reactions. Increasing the amount of organic matter in a soil often improves the soil's structure, which enhances soil aeration and promotes penetration of water and roots into the soil, and increases its fertility and biological activity.

Soil Horizons

Soils typically exhibit some development of horizons. Soil horizons are horizontally stratified zones that are enriched or depleted in clay, organic matter, and nutrients relative to the material from which the soil formed. An horizon may also show some aggregating of sand, silt, and clay to give structure. Soil horizons have been given letter designations (Fig. 16.2). Although individual soils have different combinations of the horizons, they do not have to have all of the possible horizons.

The letter designations of soil horizons and their characteristics are the following:

O horizon   The O horizon is made up of organic matter. The organic matter may be recognizable as leaves, twigs, and needles, or it may be decomposed beyond recognition.

A horizon   The A horizon is the first mineral horizon. It is the zone colored dark by organic matter and soil materials are aggregated into structure.

E horizon   The E horizon is a zone of eluviation. Materials have been transported out of this zone by water. This zone is light colored and low in clay. It may be white sand.

B horizon   The B horizon is a zone of deposition. There may be an accumulation of clay or iron and aluminum oxides. Where iron or aluminum oxides have been deposited, the soil will appear redder or yellower than the soil above and below.

Soil Profile

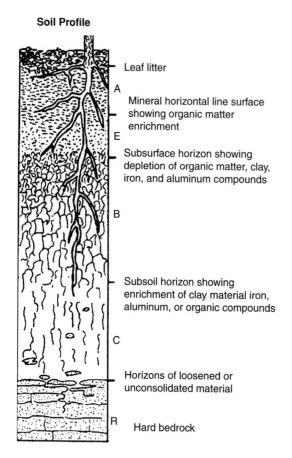

Leaf litter

A

Mineral horizontal line surface
showing organic matter
enrichment

E

Subsurface horizon showing
depletion of organic matter, clay,
iron, and aluminum compounds

B

Subsoil horizon showing
enrichment of clay material iron,
aluminum, or organic compounds

C

Horizons of loosened or
unconsolidated material

R   Hard bedrock

FIGURE 16.2   Diagram of a soil profile and location of soil horizons. Soil horizons are distinct horizontally stratified zones that have been given letter designations.
Source: University of Arizona, Cooperative Extension Service. Master Gardener Manual by L. Bradley, M. Kilby, R. E. Call, D. Kopec, J. Capizzi, D. Langston, J. D. Claridge, O. Maloy, T. DeGomez, T. Mikel, T. Doerge, N. Oebker, J. Green, J. Tipton, R. Gibson, M. Wilcox, R. Gibson, D. Young, and R. Grumbles. 1998.

C horizon   The C horizon is the parent material from which the soil was formed. It may be weathered bedrock or material deposited on the bedrock by wind or water.

R horizon   The R horizon is bedrock and not soil.

Soil-Forming Factors

The environment and the material from which the soil is derived determine the type of soil that is formed and the horizons that develop in a soil (Fig. 16.3). One or more of the following soil-forming factors can determine soil type and soil horizons:

FIGURE 16.3   An example of a soil series. The Hazelton soil series is found in Pennsylvania, Kentucky, Maryland, New Jersey, Virginia, and West Virginia, and its resulting soil profile (top left) is the product of the environment in which it is located (Pennsylvania location of Hazelton soil series, top right) and the material from which it is derived and its uses (i.e., woodland, cropland, hay, and pasture land occur in the areas) (bottom).
Source: Natural Resources Conservation Service, USDA.

Climate   The climate factors of temperature and precipitation determine what plant life can grow. Warmer and wetter climates increase the rates of biological and chemical processes. In a desert climate, very old soils show little horizon development.

Parent material   The parent material that a soil forms from will determine the properties of the soil. For example, soils in the mountains forming from granite tend to be acidic and sandy. Soils derived from sandstone are sandy, and soils derived from shale contain a lot of clay.

Topography   The topological aspects of slopes determine the amount of solar radiation that a soil receives and its average temperature. The temperature influences the rates of evaporation and whether a soil will be wet or dry. The vegetation inhabiting a soil and the rates of chemical processes are influenced by the temperature and moisture.

Biotic activity   Soil development is influenced by biological activity in or near the soil. Plant roots and microorganisms cause chemical and physical weathering of the parent material. Organic matter accumulates and improves the water-holding capacity of the soil and the retention of nutrients.

Time   The time that the other factors have to act determines how well developed the horizons become.

The Water Cycle and the Groundwater System

Water cycles throughout the ecosphere and is renewed frequently (Fig. 16.4). Solar energy drives water from the oceans, lakes, rivers, and continents by evaporation. Precipitation can then occur as rain, snow, or hail. Water can move out of the soil back to the atmosphere by a combination of evaporation from the soil surface and transpiration

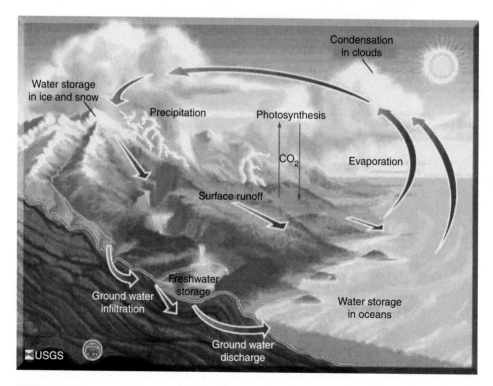

FIGURE 16.4   Graphic representation of the various processes that occur in the water cycle.
Source: U.S. Geological Service.

from the leaf surface (which removes soil water from the root zone). The combination of water loss by evaporation and transpiration is referred to as evapotranspiration. Typically in a planted field, transpiration removes more soil moisture than evaporation. Water can also run off the soil into bodies of water (streams, rivers, lakes) or drain into the groundwater system. Groundwater and water in streams, rivers, and lakes can ultimately end up in the oceans, where the cycle can begin again.

Precipitation as it lands on the earth percolates downward into the soil to form the groundwater system. A soil has a limited capacity to retain or store moisture, and additional moisture above this capacity collects as puddles and flows into ditches or runs off into nearby streams, rivers, lakes, and oceans. Precipitation intensity and the soil's infiltration capacity often determine whether precipitation enters the soil (and its subsequent leaching potential for nutrients) or runs off the soil surface (increasing possible soil erosion) (Table 16.1). The soil's infiltration rate depends on soil texture, structure, compaction, freezing, and saturation. Typically, soils that are compacted, frozen, or already saturated with water have very low infiltration rates and are susceptible to high runoff rates. Runoff and erosion in temperate regions are the highest during the early spring (snowmelt), in the fall, and during intense or long rains throughout the year. The runoff causes erosion and can contribute to water pollution.

Soil water from deeper, wetter horizons diffuses through the soil profile in response to evapotranspiration. Soil water movement from wet to dry areas through very small pores can also occur as the result of capillary flow. Water that is not taken up by plant roots continues to penetrate downward. Sandy, poorly structured and low organic matter soils have a lower field capacity (the percentage of water remaining in a soil 2 to 3 days after having been saturated and after free

TABLE 16.1   *Influence of soil hydrologic group type on relative infiltration, leaching, and runoff potential*

Soil hydrologic group type	Infiltration capacity/ permeability	Leaching potential	Runoff potential
Deep, well-drained sands and gravel	high	high	low
Moderately deep to deep, moderately drained, moderately fine to moderately coarse texture	moderate	moderate	moderate
Impeding layer, or moderately fine to fine texture	low	low	high
Clay soils, soils with high water table, shallow soils over impervious layer	very low	very low	very low

Source: Modified from M. A. Cavigelli, S. R. Deming, L. K. Probyn, and R. R. Harwood, eds. 1998. *Michigan field crop ecology: Managing biological processes for productivity and environmental quality.* Michigan State University Extension Bulletin E-2646.

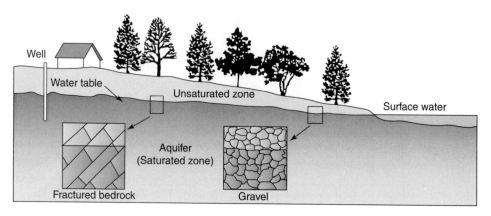

FIGURE 16.5   The groundwater system. The top of the zone of saturation is the water table. The aquifer is located below the zone of saturation and is made up of a porous rock layer that traps water between two impervious layers of rock.
Source: Penn State Cooperative Extension Service. Agrichemical Fact Sheet No. 8 *The fate of pesticides in the environment and groundwater protection* by C. L. Brown, and W. R. Hock. 1990.

drainage due to gravity has practically ceased) than clay, which is well structured and high in organic matter. When soil moisture is below field capacity, added water will be stored in the soil. When soil moisture is at or above field capacity, added water will percolate through the soil profile or run off the soil surface, potentially carrying soil nutrients such as nitrates with it. Eventually the water fills all the cracks and crevices between soil particles and bedrock. The top of this zone of saturation is the water table (Fig. 16.5). The bottom of this zone contains the aquifer, which is a porous rock layer that traps flowing water between two impervious rock layers.

Plant Roots

Roots generally anchor the plant in the soil by forming an extensive branching network that constitutes about a third of the total dry weight of some plants. The roots of most plants do not usually extend into the earth beyond a depth of 3 to 5 m. Although the roots of many herbaceous species are often located in the upper 60 to 90 cm, some plants have root systems that extend for several hundred meters into the earth.

Roots also function in water and nutrient uptake, and the bulk of roots that function for this purpose are generally confined to the upper meter of soil. Some plants have roots with specialized functions such as for food and water storage. Some aquatic plants produce roots in water and other plants produce aerial roots, but the great majority of plants develop their root systems in soils.

When a seed germinates, its radicle emerges and develops into the first root. This first root may develop either into a thickened taproot, from which thinner branch roots arise, or lead to the formation of numerous adventitious roots. A fibrous root system with numerous fine roots of similar diameter then develops from the adventitious roots. Many plants have a combination of taproot and fibrous root systems. Root hairs also may develop adding significant additional root surface area that aids in water and nutrient absorption.

Monocots often have fibrous root systems, whereas most dicot plants have taproot systems with one or, occasionally, more primary roots from which secondary roots develop. Other types of roots, such as adventitious roots, may develop in both monocots and dicots. Ivies and other vines may have lateral adventitious roots develop along the aerial stems that aid in climbing. Certain plants with specialized stems (such as rhizomes, corms, or bulbs) may produce only adventitious roots.

Root growth is highly dependent on the availability of water and minerals in the rhizosphere. If the rhizosphere is poor in nutrients or too dry, root growth is generally slow. As conditions improve in the rhizosphere, root growth increases. If fertilizer and irrigation provide abundant nutrients and water, a small root system may result that still meets the needs of the plant.

Soil Organisms and Allelochemicals

Soil organisms and allelochemicals may also be located in the rhizosphere. Soil organisms can be either microscopic or macroscopic, and their activity in degrading biomass into organic matter can greatly affect the soil. Earthworms in many soils may be one of the most influential soil organisms, and their presence often indicates a "healthy" soil.

Allelochemicals are the compounds produced by plants that may adversely affect plant growth. Although the roles of specific allelochemicals may not be entirely understood, allelochemicals generally benefit the plant producing them by reducing competition in the rhizosphere.

Summary

The rhizosphere, often called the plant root zone, is that area of the soil near the plant's roots where soil microbe activity and biochemical reactions are influenced by the roots. Soils are composed of abiotic and biotic components and are classified into 12 general categories according to the proportion of soil, silt, and clay. Soils also are primary sources of nutrients for plants, and the organic matter composition of the soil greatly influences its fertility and biological activity. Soils can be differentiated according to their horizons, which are zones in the soil that are enriched or depleted in clay, organic matter, and nutrients relative to the material from which it was formed. The environment and the material from which it was formed determine the type of soil found in a location.

The water cycle is one of the most important components of the ecosystem, providing water to biological organisms through surface containment (such as in ponds, lakes, and rivers) or contributing to the groundwater system. Whether precipitation from the water cycle enters the soil or runs off the soil surface is determined by the precipitation's intensity and soil infiltration capacity.

Plant roots provide support for the plant by forming an extensive branching network and serve as a means for water and nutrient uptake. Although the majority of a plant's roots are used for these purposes, some plant roots have specialized functions for food and water storage. The various types of plant roots include taproots, adventitious roots, root hairs, and aerial roots. Besides plant roots, soil organisms and allelochemicals are also commonly found in a soil.

Review Questions

1. Define the following:
   a. Soil
   b. Rhizosphere
2. Match the following letter of soil horizon with its appropriate title.

Letter designation	Title
A	Topsoil
B	Organic matter
C	Zone of eluviation
E	Zone of deposition
O	Parent material
R	Bedrock

3. What textural class of soils drains the quickest?
4. List four factors that help decide the type of a soil in a region.
5. What are some functions of roots?

Selected References

Cavigelli, M. A., S. R. Deming, L. K. Probyn, and R. R. Harwood, eds. 1998. *Michigan field crop ecology: Managing biological processes for productivity and environmental quality.* Michigan State University Bulletin E-2646.

Miller, G. T. Jr. 1975. *Living in the environment: Concepts, problems, and alternatives.* Belmont, CA: Wadsworth.

Stern, K. R. 1991. *Introductory plant biology.* 5th ed. Dubuque, IA: Wm. C. Brown.

Taiz, L., and E. Zeigler. 1998. *Plant physiology.* 2nd ed. Sunderland, MA: Sinauer.

Selected Internet Sites

http://www.nrcs.usda.gov/technical/   Technical resources and information including tools and data on soil science education, soil surveys and maps, soil geography, and soil use and management. Natural Resources Conservation Service, USDA.

http://www.wcc.nrcs.usda.gov/wcc.html   Government weather site containing observations, forecasts, maps and models, weather safety, and education. National Water and Climate Center, USDA.

http://www.nws.noaa.gov/   Government weather site containing observations, forecasts, maps and models, weather safety, and education. National Weather Service, National Oceanic and Atmospheric Administration.

http://www.statlab.iastate.edu/soils/nsdaf/   Technical resources and information including tools and data on soil science education, soil surveys and maps, soil geography, and soil use and management. National Soil Survey Center, Natural Resources Conservation Service, USDA.

17

Water

Most living herbaceous plants are about 80% to 95% water, so having sufficient water is a requirement for proper plant growth and development. Water is an excellent medium or solvent for transport of organic molecules (e.g., sucrose in the phloem), inorganic molecules (e.g., mineral nutrients from roots to leaves in xylem), and atmospheric gases (e.g., diffusion of oxygen to sites of respiration). Water also helps to maintain plant tissue temperature at levels suitable for proper metabolic activity. In addition, unlike most animals, plants lack a well-developed skeletal system and depend on their cells' hydrostatic or turgor pressure for overall structure and support. Hydrostatic pressure is the intercellular pressures in plant cells that occur as a consequence of normal water balance in the plant.

The total amount of water in all its forms on our planet is about 1.5 trillion liters. Of this amount about 97% is in salty oceans (Fig. 17.1). Most of the remaining water is contained in icecaps and glaciers, or in saline inland seas. Twenty-two percent of the nonocean portion of the earth's water is groundwater, some of which lies too deep under the surface of the land to access. Other water is in the atmosphere, soil environment, lakes, rivers, and streams. As a result, only 0.3% of the water on the earth is usable by humans and most of that is tied up in groundwater.

Almost 60% of all the world's fresh water withdrawals go toward irrigation. Consequently, the distribution of surface water type is important in that it at least partially determines where food production (including field agriculture) can occur (Fig. 17.2) and populations can exist (Fig. 17.3).

This chapter begins with an overview of water potential and how water is absorbed and moves through the plant. The effect of drought on plant growth is addressed, as well as methods and uses of irrigation. The effects of flooding on plant development and soil conditions and a brief introduction to acid precipitation is presented. A discussion of common methods of evaporative control of soil moisture, irrigation systems, and drainage systems follows. The chapter concludes with a summary of the theory of operations of using hydroponics and hydrophilic gels.

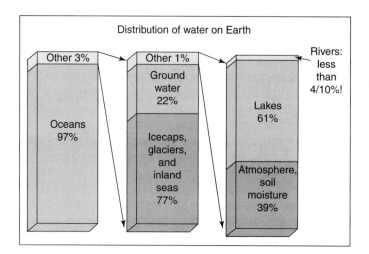

FIGURE 17.1   Distribution of water on earth. The left-side bar shows where the water on earth exists; about 97% of all water is in the oceans. The middle bar represents the 3% of the "other" part of the left-side bar (that portion of all of earth's water that is not in the oceans). Most, 77%, is locked up in glaciers and icecaps, mainly in Greenland and Antarctica, and in saline inland seas. Twenty-two percent of this portion of earth's water is groundwater. The right-side bar shows the distribution of the "other" portion of the middle bar (the remaining 1%).
Source: U.S. Geological Survey.

Background Information

Unique Properties of Water

Water is an excellent solvent, dissolving greater amounts of a wide variety of substances than do other related substances. This is due to water's small size and polar nature (i.e., water has a partial negative charge at the oxygen end and a partial positive charge at the hydrogen end of the molecule). Water provides the medium for the movement of molecules within and between plant cells and greatly influences the structure of proteins, nucleic acids, polysaccharides, and other cell constituents.

Water has the highest ability to store heat energy of any liquid or solid and has the highest ability to conduct heat of all liquids. It also has the highest heat of vaporation of any known liquid (i.e., it takes 1 calorie to warm 1 g of water 1°C, but it takes 540 calories to evaporate the same amount of water). This means that a plant loses substantial heat through the evaporation of water in transpiration. This functions in temperature regulation (cooling) of plant tissues, especially during hot and sunny weather. Also, when 1 g of water freezes (at 0°C) about 80 calories are released, a property utilized during overhead irrigation for frost protection for certain plants.

Global water dynamics also greatly influences climates and weather. Each day the sun vaporizes about 1,250 cu km of water. As this water vapor condenses, heat is slowly released over land areas and bodies of water. As a result, water is responsible

SSM/I Surface Type, 10/28/2002 4 EST

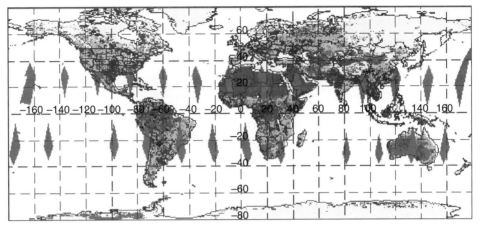

LEGEND

■ Flooding or saturated land
■ Dense vegetation
□ Vegetation
□ Moist or dry arable land
■ Semi-desert
■ Desert
■ Water / vegetation mix or wet soil
■ Rain over vegetation or soil
□ Wet snow, dry snow, glacial, or ice

FIGURE 17.2   Land surface types (descriptions based on water availability effects on the soil status and/or vegetation of an area) across the land regions of the earth.
Source: NOAA Satellite and Information Services.

for a massive circulation and distribution of heat over the entire globe. The high heat capacity of water also prevents large bodies of water from heating or cooling too rapidly contributing to "lake effects" and maritime climates.

Water has an extremely high surface tension. Surface tension can be defined as the force per unit length that can pull perpendicular to a line in a plane of the surface. It is the surface tension at the evaporative surface of leaves that generates the physical forces that can pull a stream of water through the plant's vascular system.

Water also has a definite tensile strength. A substance's tensile strength is the maximum tension that it can withstand before breaking. It is the great cohesive forces between water molecules that allows an appreciable tension to exist in an uninterrupted water column in a wettable capillary or tube (such as a xylem vessel) for capillary rise to occur. This is important for the continuous movement of water from the root through the plant to the surrounding air during transpiration.

Water is relatively transparent to visible radiation. This enables sunlight to reach chloroplasts within the cells of leaves and to reach plants submerged at appreciable depths in bodies of water.

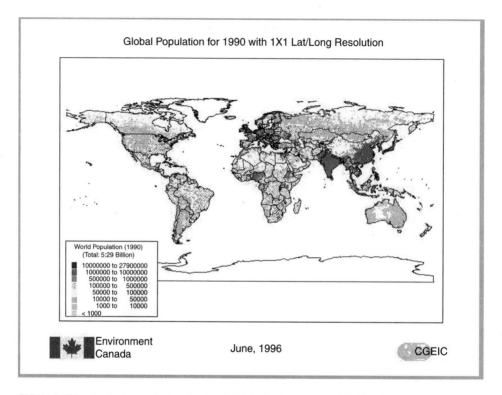

FIGURE 17.3   Global population for 1990. This depicts the worldwide distribution of population in a latitude/longitude grid system.
Source: Global Population Distribution Database developed by the Environment Canada and the United Nations Environment Programme.

Water Potential

The status of water in soils, plants, and the atmosphere is commonly described in terms of water potential (Ψw). Water potential is the chemical potential (quantitative expression of the free energy associated with the chemical) of water in a specified part of the system compared with the chemical potential of pure water at the same temperature and atmospheric pressure. Water potential is the sum of effects of the osmotic potential ($\Psi \pi$), hydrostatic pressure (Ψp), and gravitational potential (Ψg):

$$\Psi w = \Psi \pi + \Psi p + \Psi g,$$

where Ψw = water potential,
$\quad \Psi \pi$ = osmotic potential,
$\quad \Psi p$ = hydrostatic pressure,
$\quad \Psi g$ = gravitational potential

The osmotic potential is the chemical potential of water in a solution due to the pressure of dissolved materials. The osmotic potential always has a negative value because water tends to move across a semipermeable membrane from pure water into water containing solutes. The hydrostatic pressure (also called turgor pressure), which can be positive or negative, refers to the physical pressure exerted on water in the system. Gravitational potential refers to the influence of gravity on water to move vertically an appreciable distance. When there is little or no change in vertical position, such as in considerations of chemical reactions or the crossing of membranes, the gravitational term is often omitted from the equation for water potential. Total water potential can have positive or negative values, depending on the sum of the components.

When dealing with soils, matrix potential (Ψm) is another component that is included in the soil water potential equation (the equation would then be $\Psi w = \Psi\pi + \Psi p + \Psi g + \Psi m$). The matrix potential is the force with which water is adsorbed onto surfaces such as cell walls, soil particles, or colloids. The matrix potential always has a negative value because the forces tend to hold water in place, and it becomes more negative as the water film becomes thinner.

Forms of Atmospheric Moisture

Atmospheric moisture exists in several forms, including vapor, rain, snow, sleet, hail, dew, fog, and mist. Rain forms when the condensation of atmospheric moisture occurs at temperatures above the freezing point. Snow forms when the atmospheric moisture condenses at temperatures below the freezing point. Sleet forms when rain falls through a layer of subfreezing air and hail is formed if air currents carry frozen particles back into the subfreezing air and accumulates a thicker layer of ice before falling again. Hail particles may accumulate sufficient ice under certain conditions to become large and cause mechanical injury.

Dew is formed at night when atmospheric moisture condenses on cold surfaces. It can form only when the sky is relatively clear for the soil and plants to lose sufficient heat such that their surface temperatures fall below the dew point (temperature at which a given mixture of air and water vapor will reach 100% relative humidity or at which condensation will start to occur). Fogs and mists are low clouds that do not settle and cannot be measured by rain gauges.

The Supply of Water by the Soil

The absorption of soil water through the root system is the primary source of water for most plants as water absorption of dew and rain by leaves is often of negligible importance for water uptake. The quantity of water within a soil depends primarily on the climate, particularly on the excess of precipitation over evaporation in a location and the soil's physical characteristics.

The availability of water for plant uptake depends on the water-storing properties of the soil. Soils consist of mineral particles of varying diameter, expressed as relative percentages of sand (2 to 0.05 mm), silt (0.05 to 0.002 mm), and clay (smaller than 0.002 mm). These particles may be bound together into aggregates by organic matter and clay particles. The actual spaces in the soil are only a few centimeters in size (often due to drying cracks and earthworm or termite channels). Smaller size spaces exist between soil aggregates and account for the majority of pore space and soil particles. These spaces, regardless of size, are called pore space.

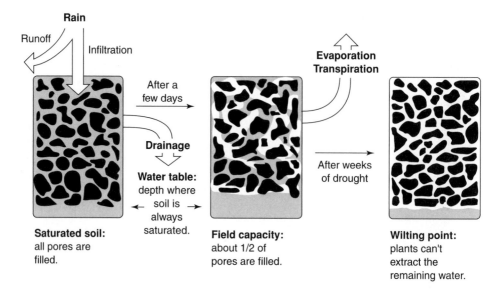

FIGURE 17.4   Diagrammatic presentation of the changes in soil water during and following a rain. Field capacity minus wilting point is the amount of water available to plants. Source: University of Minnesota Cooperative Extension Service Publication BU-07501-GO, *Soil Management* by Ann Lewandowski, 1995.

When the soil is saturated with water, the pore space becomes temporarily water filled (Fig. 17.4). Because gravity exerts a suction of about 5 kPa, all pores larger than 60 μm drain spontaneously after a soil has become saturated with water. In freely draining soil, this drainage can take 2 to 3 days. After the water drains, the soil is considered to be at soil water capacity. The gravitational water normally lost in drainage is not available to plants unless drainage is impeded. If a drought subsequently occurs, more water will be lost and the moisture in the soil will reach a point where plants are not able to extract the remaining water and the plants wilt.

The amount of soil water available for uptake by a plant depends on the size distribution of the soil pores, which is influenced by soil texture and soil structure (extent and type of aggregation). In general, medium- to fine-textured soils tend to hold more water for plant use than coarse-textured soils.

Effects on Plants

Classification of Plants Based on Water Use

Plants can be classified on the basis of their water requirements. Hydrophytes grow in water and are frequently characterized by having a low oxygen requirement and poorly developed root systems. Mesophytes are intermediate in water requirements, have well-developed root systems, and are characteristic of mature, well-developed soils. Xerophytes grow in dry habitats and have adaptations that enable them to grow and reproduce in very rigorous environments.

Water Acquisition via Roots

Water absorption via roots is the primary mechanism for water uptake by most plants, though each plant may develop different types of root systems. Native plants from ecosystems in humid areas often do not require deep and widely spreading root systems for water uptake, and their root:shoot ratio (i.e., root dry wt/shoot dry wt) tends to be low. For example, roots account for only 21% to 25% of the total biomass of coniferous forests, whereas the proportion rises to 30% to 40% in drier, tropical Savannah woodlands. Plant species from water-limiting desert and prairie ecosystems often have deep root systems that can represent 60% to 90% of the plant's total biomass. Deep rooting is an important feature of some established perennials. Also, rapid root growth at the start of the growing season may provide perennial plants an additional competitive advantage over annual plants.

Water contacts roots in two ways: the water may move to the root via capillary action or gravity, or the root may grow and intercept moist soil. Water may move along the surface of a soil particle, vaporize into a pore space, condense onto the surface of another soil particle, and move long distances by repeating the process. As extraction of water adjacent to a root proceeds, a depletion zone develops causing water in adjacent regions of the soil to move toward the root. Thus, the root-surface interface is a dynamic region.

Water absorption by roots is affected by the number of roots within a given volume of soil, the size of the roots, the rate of root length growth, the distance the water must travel in the root from the site of absorption to the site of its loss or utilization, and the age of the root at the absorbing site. The amount of water absorbed is also affected by root age, whether the roots are perennial or annual, and the rate of new root production. Root hairs assist in water absorption by traversing air gaps between the root and moist soil particles. Water moves many times more rapidly along a root hair than it does as water vapor diffusing across air gaps.

Water Movement through Plants

Water transport from the soil, through the plant, and to the atmosphere takes place in a soil-plant-air continuum that is often interconnected by a continuous film of liquid. Water moves through the plant along a gradient, either from high to low water potential (if transport occurs across a semipermeable membrane), from high to low hydrostatic pressure (if no membrane is involved), or from high to low partial water vapor pressure. The low partial pressure of water vapor in the air, compared with that inside the leaves, is the major driving force for water loss as transpiration from leaves. This drives water transport along the gradient in hydrostatic pressure between the xylem in roots and leaves, and down a gradient in water potential between the soil and the cells in the roots. As soils lose moisture and become drier, soil water potential and plant water potential decrease.

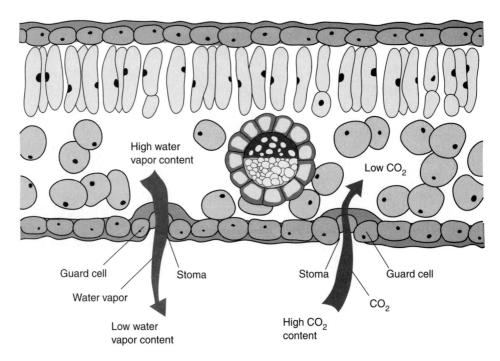

FIGURE 17.5   Evaporation from the leaf sets up a water potential gradient between the outside air and the leaf's air spaces. The gradient is transmitted into the photosynthetic cells and on to the water-filled xylem in the leaf vein.
Source: Oregon State University Extension Service, *Master Gardener Handbook,* Chapter 1, Botany Basics, Ann Marie VanDerZanden, 1999.

Transpiration

Transpiration is the process of water loss from the internal leaf atmosphere. More than 90% of the water entering a plant evaporates primarily into the leaf air spaces and then passes through the stomata into the atmosphere (Fig. 17.5). Usually less than 5% of the water escaping from plants is through the cuticle.

Transpirational pull is a method that water travels through the plant. This path that water travels may begin in the roots 3 to 6 m or more beneath the surface and travel up the trunk of the tree to the topmost leaves that may be more than 90 m above the soil. This occurs because there are continuous tubular pathways of xylem that run throughout the plant, extending from the young roots up through the stem and branches to the tiny veinlets of the leaves. The water is raised through the columns of a plant through a combination of factors. Capillary action will allow water to rise in narrow tubes, and the heights attained are inversely proportional to the diameter of the tube. Plants also have root pressure, but this is considerably less than needed to raise water to the top of trees.

The pulling force due to evaporation of water from leaves and stems provides the primary mechanism for water to reach to the top of trees (this is often referred to as the cohesion-tension theory). Water molecules are electrically neutral, but are asymmetrical in shape. This results in the molecules having a very slight positive charge at one end and a very slight negative charge at the other end. As a result, the molecules are polar. When the negatively charged end of one water molecule comes close to the positively charged end of another water molecule, weak hydrogen bonds hold the molecules together. This permits a certain amount of tension to exist.

When water evaporates from the mesophyll cells in a leaf and diffuses out of the stomata (transpires), the cells involved develop lower water potential than the adjacent cells. Because the adjacent cells then have correspondingly higher water potential, replacement water moves into the first cells through osmosis. This continues across the rows of mesophyll cells until a small vein is reached. As transpiration occurs it creates a "pull" or tension on water columns, drawing water from one molecule to another all the way through an entire span of xylem cells. The cohesion strength of the water in the columns is usually more than adequate to move the water to the top of a tree. The passage of water is partly through cell protoplasm and partly through spaces between cells, between cellulose fibers in the walls, and in the centers of dead cells.

Measuring Soil Moisture

To optimize crop production, water must be available and irrigation, when needed, must be efficiently used. Monitoring soil moisture is therefore important, and a number of methods have been developed and used by growers for monitoring soil moisture to determine when irrigation may be needed. These monitoring methods can be broadly categorized as plant-based and soil-based monitoring.

Plant-Based Monitoring   The plant's appearance can be used by growers to determine when to water. Although this may not be the best way to determine water needs, it is the easiest to recognize. Visible signs of water stress include wilting and a change in leaf color. Young leaves of water-stressed plants may become dull, darken, or turn grayish.

Visible signs may not always indicate what is best for the plant at a certain stage. For instance, allowing the plant to undergo mild wilting may help "harden" the plant, making it more resistant to environmental stresses. At other times, plant injury from water deprivation may occur before signs of stress are noticeable.

A pressure chamber is an instrument that measures the water potential in the xylem of the plant. Xylem pressure is nearly equivalent to leaf water pressure and indicates plant water status. The higher the gauge reading on the pressure chamber the greater the plant water stress. Because of the diurnal nature of water relations in a plant, critical values of water pressure are usually needed for each crop at different times of the day. Pressure chambers only indicate when a plant needs water, but not how much it needs.

Infrared thermometers measure leaf temperature without making direct contact with the leaves by measuring the amount of infrared wavelengths (or heat) given off by the leaf. Transpirational or evaporative loss of water through the stomata results in reduced leaf surface temperatures due to evaporative cooling effects on the leaf tissue. A rise in leaf temperature may indicate water stress due to reduced evaporative cooling occurring at the leaf surface.

Soil-Based Monitoring *Appearance of soil* The appearance of the soil can provide an indication of the soil water status (at least near the soil surface). The following are some of the characteristics of the various soil types and suggestions on when irrigation is needed.

Sandy soil dries rapidly and should be irrigated when it appears dry.
Sandy loam dries moderately fast and should be irrigated when soil appears dry.
Silt loam dries moderately slowly and should be irrigated when soil assumes a pale color.
Clay dries slowly and should be irrigated when soil is still pliable.

Soil moisture feel Feeling the soil is a simple, effective method to determine when to irrigate and how much water to apply at an irrigation. A soil has a specific feel characteristic according to the amount of available soil moisture it contains. The following are some characteristics of how some soil classes may feel when they have various amounts of available moisture within them.

Below 20% available moisture Fine soils (loam, silt loams, clay loams) will be powdery dry and will not form a ball if a handful of the soil is squeezed. Coarse soils (sandy and loamy sands) will also not form a ball when squeezed, and single grains of soil will flow through fingers with ease.
35% to 40% available moisture Fine soils are almost powdery and a ball can be formed under pressure. Some soil will fall or flake away when the hand is opened after the ball is made. Coarse soils will form a weak ball and soil grains will stick to the hand.
50% available moisture Fine soils will form a ball readily when squeezed, but the ball is very brittle and breaks readily. The soil falls or crumbles into small granules when the ball is broken. Coarse soils form very weak balls and soil grains will stick to hands.
60% to 65% available moisture Fine soils will form a firm, pliable ball and the soil doesn't stick to hand. These soils also will form ribbons when a handful of soil is pushed through the thumb and forefinger. Coarse soils form weak, brittle balls and soil particles will stick to the hand in a patchy layer. These soils will not ribbon at this available soil moisture.
70% to 80% Fine soils will form a damp and heavy ball when squeezed. The ball will shatter into large particles when broken. Fine soils will form a distinct ribbon when squeezed between the thumb and forefinger. Coarse soils will form a weak ball and soil particles will stick to the palm of the hand. No ribbon will be formed when squeezed.
100% available moisture Fine soils produce a wet, sticky, doughy, and slick ball when squeezed. These soils handle similarly to stiff bread dough or modeling clay. After squeezing, water will be left on hands and these soils will ribbon readily. Coarse soils form a ball upon squeezing that has some stickiness and

sharp fingerprint outlines will be left on it. These soils will not ribbon but will form a smooth layer on the thumb.

Gravimetric method   The gravimetric method requires that a portion of soil be removed from the field and weighed. The soil sample is dried in an oven and reweighed. Percent water content is determined by dividing the weight of water (initial − oven-dry soil weight) by oven-dry soil weight. This method provides an estimate of water content of the soil but not the water that is available to the plant.

Tensiometers   A tensiometer is a sealed, water-filled tube with a porous ceramic tip on the lower end and a vacuum gauge on the upper end. The tube is installed in the soil with the ceramic tip placed at the desired root zone depth and with the gauge above the ground (Fig. 17.6). Water passes into and out of a chamber in the tensiometer to and from the surrounding soil through a membrane, depending on the amount of water in the soil.

Tensiometers are generally placed at several locations in the field and at various depths to determine soil moisture throughout the root zone. Tensiometers are most effective when used for scheduling irrigation in shallow-rooted, water-stress-sensitive, and frequently irrigated crops. As the soil dries, water leaves the tensiometer and vice versa, as the soil is wetted. The flow of water from the tensiometer creates a vacuum that is measured as pressure. The standard unit of measure of soil water tension or soil suction is the "bar," which is a unit of pressure. Most tensiometers are calibrated in hundredths of a bar (centibar) and can operate in a range of 0 to 80 centibars.

The pressure changes in the tensiometer are related to soil moisture or the relative wetness of the soil, and minimum values have been established for many crops

FIGURE 17.6   A tensiometer measuring soil moisture in the root zone.

TABLE 17.1   *Influence of soil type on when to begin irrigation*

Soil type	Soil moisture to begin irrigating (centibar)[1]
Loamy sands, sandy loams, very fine sandy loams	25
Silt loams	40
Clay loams, silt clay loams	60

[1] Readings will vary according to crop, rooting depth, type of irrigation system, and soil type.

Source: Adapted from K. Harrison and A. Tyson. 1993. *Irrigation scheduling methods.* University of Georgia Cooperative Extension Service Bulletin 974.

to determine when irrigation should be used (Table 17.1). The following are some general guidelines for interpreting tensiometer readings (in centibars) under many conditions:

Readings 0 to 5   This range indicates a nearly saturated soil and often occurs for one or two days following a rain or irrigation. Plant roots may suffer from lack of oxygen if readings in this range persist.

Readings 6 to 20   This range indicates field capacity. Irrigation is discontinued in this range to prevent waste of water by percolation and also to prevent leaching of nutrients below the root zone.

Readings 21 to 60   This is the usual range for starting irrigation. Most field plants having root systems 0.5 m deep or more will suffer until readings reach the 40 to 50 range. Irrigation is started in this range to insure maintaining readily available soil water at all times. It also provides a safety factor to compensate for practical problems such as delayed irrigation or inability to obtain uniform distribution of water to all portions of the field.

Readings 61 and higher   This is the stress range for most soil and crops. Deeper-rooted crops in medium-textured soils may not show signs of stress before readings reach 61. A reading of 61 does not necessarily indicate that all available soil water is used up, but that readily available water is below that required for maximum growth.

The need for irrigation will also be affected by environmental factors such as soil type, air temperature, wind speed, and relative humidity. Irrigation is often begun sooner on sandy soils than heavier clay soils because sandy soils hold less water than clay soils at the same meter reading. Irrigation is also often initiated at higher soil moisture levels during hot weather, high winds, or low humidity.

Gypsum Blocks   Gypsum blocks measure changes in electrical conductivity of soils. The wetter the soil the greater the conductivity. Similar to tensiometers, gypsum blocks can be buried at different depths to obtain a soil profile characterization of irrigation needs.

Gypsum blocks indicate when to irrigate but not how much to apply. Depending on soil type, gypsum blocks are generally replaced every 2 to 5 years. Gypsum blocks lose their effectiveness in determining soil water status when they're used in saline or alkali soils where salts affect conductivity.

Thermal dissipation sensors   Thermal dissipation sensors monitor the dissipation of heat in a porous ceramic block in contact with the soil. Water flows into and out of the ceramic block depending on the water status of the soil. The wetter the soil, the more moisture in the block. The ceramic block is a good heat conductor when wet and a poor conductor when dry. An electric current quantifies heat conduction. The sensor can detect small changes in soil moisture and can be linked by computer to an automatic irrigation system. Once calibrated, the sensor can remain in place for long periods.

Efficient Water Use by Plants

The efficiency of water use in crop production is influenced by climate, soil, and nutrient factors. Efficiency can be measured using the transpiration ratio, the amount of water transpired (in kg) per kilogram of plant biomass produced. This can be about 500 for C3 plants in humid regions, about 250 for plants with C4 photosynthesis, and 50 for desert plants adapted with CAM photosynthesis. Certain variables that will increase the transpiration ratio include intense sunshine, increased temperatures, lower humidity, and greater wind velocity.

Drought Stress

Drought is a meteorological term that is defined as a period of time without sufficient rainfall. Drought stress occurs when insufficient water is available in the soil because water input (by rain or irrigation) has been reduced and the atmospheric conditions existed that caused continued loss of water by transpiration and evaporation. Drought stress may occur on a daily basis or over a prolonged period. If the stress continues, the plants may die of desiccation unless they possess mechanisms of resistance that either prevent or slow water loss in certain tissues or organs, or unless they are able to increase rates of absorption and translocation of water.

Drought has profound effects on plant growth, yield, and quality. The early deleterious effects due to drought stress are often a loss of turgor (including wilting) that affects the rate of cell expansion and ultimate cell size. Loss of turgor appears to be the plant process most sensitive to drought stress. The reduction in cell expansion rates and cell size contribute to reduced rates of stem elongation and leaf expansion, regulation of stomatal aperture, and an overall decrease in the photosynthetic and growth rate. Drought can also enhance leaf senescence and abscission, reducing total leaf area and potentially improving the plant's fitness for a water-limited environment, and root extension into deeper, moist soil areas.

Most plants avoid drought in natural ecosystems by growing where there is adequate moisture to sustain their needs. For example, marsh species are not found in upland areas and upland species are not found in deserts. Within their range of adaptability, most plants avoid drought stress by either conserving available moisture by enhancing their uptake of limited soil moisture or timing their life cycle so that it is completed during wet growing seasons (before onset of drought).

A water conservation mechanism practiced by most plants involves regulation of guard cells. Because 99% of the water leaving a plant is lost through the stomata, the rapid closure of stomata at the first sign of drought is an effective means of conserving moisture when the water supply is low. Other features to enhance water conservation include the development of small thick leaves with a thick waxy coating. Plants with small thick leaves lose water more slowly than plants with large thin leaves, because smaller leaves provide less surface area from which water can evaporate. Small hairs on a leaf surface are also effective in reflecting light to shade the leaf and reducing air movement directly above the leaf surface. Many succulents have adapted to survive conditions of extreme water stress by storing water in thick fleshy leaves and by opening stomata for entry of carbon dioxide only during the cool of night. A modification of normal photosynthesis (CAM metabolism) permits these plants to take in and store carbon dioxide at night for use in photosynthesis the following day.

A major factor determining the amount of water absorbed by a plant is the depth and concentration of its root system. Deep taproots in combination with shallow surface roots allow plants to capture moisture from light rains, as well as water from lower in the soil profile. Turfgrasses often form very dense mats of roots through which little water is able to pass. By developing these highly fibrous root systems, a large amount of moisture is collected before it can penetrate to more deeply rooted species.

Flooding and Anaerobiosis

Flooding occurs in a soil when water within it is in excess of field capacity. During flooding the gas (air) phase of the soil is replaced by the liquid (water) phase. The diffusion rate of oxygen in water is about 10^{-4} that of air and consequently the rate of replenishment to plant surfaces that are submerged is extremely low. Under most nonflooded conditions the oxygen supply in the air-filled spaces of the soil is replenished from the atmosphere at rates more than sufficient to maintain oxygen in the range of 15% to 20%.

In waterlogged soils, plant tissue may be subjected to low (hypoxic) or zero (anoxic) concentrations of ambient oxygen. As a result, plants may undergo fermentation metabolism (anaerobic respiration). Initial response in some plants, such as corn, is lactic acid fermentation, but subsequent response is often alcoholic fermentation. Some adenine triphosphate (ATP) is generated slowly by fermentation, but the energy efficiency (the energy stored as ATP relative to the energy potential available in glucose) of fermentation is considerably less than in aerobic respiration. The decreased energy produced by this type of respiration is often insufficient to maintain proper plant metabolism. Other plant effects due to low O_2 include premature senescence and increased production of the ethylene precursor ACC (1-aminocyclopropane-1-carboxylic acid) and/or abscisic acid. Roots that are deficient in O_2 lack sufficient energy to support many physiological processes of the shoots including nutrition absorption, water uptake, and photosynthesis, and death of the plant may result.

The formation of aerenchyma is a mechanism that some plants have to renew the supply of oxygen to flooded root tissues. Aerenchyma are gas-filled chambers that create an internal gas exchange channel from the aerobic shoot to the hypoxic or anoxic root. Air enters through the stomata of leaves or lenticels of the stem and passes through the network of aerenchyma channels to the submerged root. Oxygen consumption for respiration in the root results in a negative pressure gradient that draws air by mass flow to the root. In some species that form aerenchyma at hypoxic conditions, more than 50% of the total root respiration can be supported by internal aeration.

Flooding of soils can also affect the soil characteristics, such as increasing the pH of acid soils and decreasing the pH of alkaline soils. The rate of decomposition of organic matter in flooded soils tends to be only half that in nonflooded soils. The major end products of decomposition of organic matter in flooded soils are carbon dioxide, methane, and humic materials. In addition, high concentrations of ethanol and hydrogen sulfide, which can damage plants, are produced in waterlogged soils.

Sedimentation as a result of flooding can also be a problem to plants and trees. Deposits of silt or sand as shallow as several centimeters may injure tree roots by limiting the supply of oxygen. Strong currents, waves, or suspended particulates may cause soil around the base of trees to be eroded, exposing tree roots. Exposed roots also can make the trees more susceptible to mechanical dislodging by winds.

The degree of flood-induced injury to plants is affected by several factors. Flooding during the growing season often is more harmful to woody plants than flooding during the dormant season. Mature trees often are most susceptible to flooding in late spring just after the first flush of their annual growth. Also, the longer the time period that the trees are exposed to flooding, the greater the potential for injury. Short periods of flooding during the growing season can be tolerated by most trees. However, if flooding is recurrent and keeps the soil saturated or prevents recovery from the previous flooding, injuries will accumulate and serious damage may occur.

Water temperature can also affect the degree of injury due to flooding. Cold water is often less injurious than warm water due to cold water's capacity to hold more dissolved oxygen. When plants are dormant and temperatures are low, oxygen depletion in water-saturated soils is very slow and few effects on plants are observed, but when temperatures are warmer ($> 20°C$), oxygen can be rapidly depleted from the soil water by plant roots, soil fauna, and soil microorganisms, and growth and survival by the plant is reduced. Rapidly flowing water (with higher oxygen content) generally is less harmful than stagnant water. Water from floods may also carry various chemicals that have been picked up as runoff from agricultural fields and other areas or from sewage released when treatment facilities become unable to handle large volumes of water. The impact generally depends on the type and dosage of chemicals.

Acid Deposition

Acid deposition is the process by which acidic components of atmospheric air pollutants are deposited to the earth by means of rain, sleet, snow, fog, or as dry particles

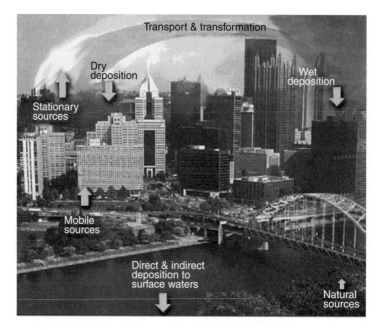

FIGURE 17.7   Processes involved in acid deposition.
Source: Clean Air Status and Trends Network, EPA.

(Fig. 17.7). Prevailing winds from west to east cause air pollutants emitted in the Midwest to be deposited in New England (Fig. 17.8) and Canada. Over the last quarter of a century, acid precipitation has emerged as a critical environmental stress that affects forested landscapes, aquatic ecosystems, and human-made structures in North America, Europe, and Asia.

The composition of acid precipitation includes gases and particles derived from the gaseous emissions of sulfur dioxide (SO_2), nitrogen oxides (NO_x), and ammonia (NH_3), and particulate emissions of acidifying and neutralizing compounds. During atmospheric transport, some of the SO_2 and NO_x will be converted to sulfuric acid and nitric acids. Unpolluted rain is often slightly acidic with a pH of about 5.6 (this is because the CO_2 dissolved in the rain produces a weak acid, H_2CO_3). When SO_2 and NO_x are dissolved in water droplets, the pH of the rain can decrease to 3 to 4 (or lower in some cases).

Acid deposition can increase the concentrations of protons (H^+) and strong anions (SO_4^{2-} and NO_3^-) in soils, especially in the northeastern United States. This has led to increased rates of leaching of base cations and to the associated acidification of soils. If the supply of base cations is sufficient, the acidity of soil water will be effectively neutralized. Because mineral weathering is the primary source of base cations, the rate of mineralization may not be great enough to keep pace with leaching rates accelerated by acid deposition in acid sensitive areas.

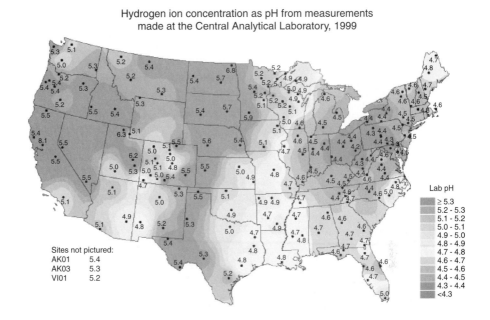

FIGURE 17.8   Hydrogen ion concentration as pH from measurements made at the Central Analytical Laboratory, 1999.
Source: National Atmospheric Deposition Program, EPA.

The amount of sulfur in surface waters often is also closely correlated with the amount of sulfur from acid deposition. Inputs of acidic precipitation to regions with base-poor soils have resulted in the acidification of soil waters, shallow groundwaters, streams, and lakes. The most severe acidification of surface water in the northeast generally occurs during spring snowmelt. Short-term acid episodes also occur during midwinter snowmelts, and large precipitation events in summer and fall. Acidification can have marked effects on the chemical composition of surface waters. Decreases in pH and increased aluminum (Al) concentrations contribute to declines in the number of macroanimals and zooplankton. High concentrations of both H^+ and inorganic Al are toxic to fish.

Acid precipitation has caused reduced growth of certain tree species (Fig. 17.9). The mechanisms by which acid deposition causes stress to trees are only partially understood, but they generally involve interference with calcium (Ca) nutrition and Ca-dependent cellular processes. Strong evidence suggests that acid precipitation causes dieback of red spruce by decreasing cold tolerance. Results of controlled exposure studies suggest that acidic precipitation reduces the cold tolerance of current year red spruce needles by 3 to 10° C. Calcium depletion and Al mobilization may also be affecting red spruce in the Northeast. Acidic precipitation may also

FIGURE 17.9   Acid-rain damage to a forest at Slamba Poremba, Poland.
Source: The Environmental Picture Library, U.S. Geological Survey (photo courtesy of C. Martin).

contribute to episodic dieback of sugar maple by causing depletion of nutrient cations from marginal soils. On leafy plants, acid deposition can increase the leaching of cations from the leaf surface. This occurs because hydrogen ions in the precipitation exchange with cations held on the cuticular exchange surface and because acidity changes the chemical nature of the cuticle so that it is more susceptible to diffusion and mass flow of nutrients to the leaf surface.

Agricultural Technologies that Affect Water

Methods of Evaporative Control

Mulches   A mulch is any material used at the surface of a soil primarily to prevent the loss of water by evaporation or to reduce weed establishment and growth. Using mulches can also alter soil temperature and affect plant photosynthetic rates and pest management through the material's reflective properties. Sawdust, manure, straw, leaves, paper, bark, or nonbiodegradable materials such as plastics can be used. Mulches are most practical for home garden use and for high-value crops (such as some small fruits, vegetables, and ornamentals).

The mulches used for many crop production systems are often made of plastic or paper sheeted material. The mulch cover is spread and fastened down either between the rows or over the rows of plants. As long as the ground is covered, evaporation and weeds are reduced or eliminated.

Cover Crop   A cover crop is a crop that is not harvested for sale but is grown to benefit the soil and/or other crops in a number of ways (Fig. 17.10). Cover crop benefits include reduced evaporation of water from the soil surface, reduced soil erosion, improved soil quality, reduced weed pressure, and reduced insect, nematode, and other pest problems. Cover crops are grown during or between primary cropping seasons. They are versatile and easily adapted to conventional, low input, and organic field crop ecosystems.

FIGURE 17.10   Use of a cover crop to improve soil structure. This crop would be plowed under well before field planting with a new crop.

In many areas, groundwater contamination and soil loss is greatest during the winter. Fall plowing exposes soil to weathering and disrupts soil organisms' habitats. An efficient way to minimize soil deterioration is to allow cover crop residues to remain on the surface and/or to plant a cover crop. Both dead and living cover crops physically protect the soil from degradation that occurs from direct exposure to rainfall. In addition, the roots of live cover crops can absorb or utilize excess nitrogen, providing a favorable environment for soil organisms, and supply crop residue for new plantings.

Contour Planting and Terraces   Contour planting or planting terraces constructed on the contour is done to conserve precipitation and prevent erosion. Terracing slows down runoff and increases the time for water from precipitation to infiltrate the soil and facilitates water storage and conservation.

Weed Control and Tillage   Transpiration from weeds can extract soil moisture. Herbicides and tillage operations are often used to eliminate unwanted plants. Tillage influences erosion by exposing the soil to the effects of wind and water. Tillage also increases organic matter decomposition by burying residues so they are exposed to greater microbial activity, increasing soil temperature and aeration (both factors that increase decomposition rate), physically breaking up soil

FIGURE 17.11   A no-till conservation tillage field planted in peppers.
Courtesy of Dr. Mike Orzolek, Penn State University.

aggregates, and exposing soil organic matter to microbial activity. In summer, cultivation may provide an insulation layer that decreases underlying temperatures. Tillage methods employed to reduce wind erosion include producing a rough furrow at right angles to the prevailing high winds, strip cropping, and the use of stubble mulch. Tillage, if not properly done, can result in root pruning.

Conservation tillage   Conservation tillage is practiced in some production systems to help reduce soil erosion and soil water loss by leaving crop residues on the soil surface and by decreasing or eliminating tillage. There are many forms of conservation tillage, but no-till (Fig. 17.11) appears to be the most effective at reducing erosion and increasing soil organic matter. Soil organic matter may increase 5% to 20% under a no-till management system. Soil organic matter that accumulates under no-till tends to concentrate in the top 2 to 5 cm of soil.

Summer Fallow   The soil profile is generally a large reservoir of water. Summer fallow is a common procedure for water conservation in the soil profile in which the crop stubble stays on the land until spring of the next year. Weeds are controlled during summer fallow by occasional cultivation or herbicides. Evaporative loss, as a result, is usually limited to the soil surface. The soil is tilled the following spring before appreciable weed growth has taken place. The overall effectiveness of summer fallow in maintaining soil moisture levels is somewhat variable.

Types of Farming Systems Based on Water Use

Two general types of farming are dryland farming and irrigation farming, depending on how water is handled. Dryland farming involves growing crops in the field without irrigation. The average mean total precipitation varies across the United States with greatest amounts along the East Coast, in portions of the mid-Atlantic states, and in the upper West Coast (Fig. 17.12). Although total amounts of precipitation in an area are important in dryland farming, equally important is the timing and distribution of the precipitation in the life cycle of the growing crop to a marketable product. As a result, areas that have high mean annual precipitation rates

FIGURE 17.12   Annual mean total precipitation for areas across the continental United States (1 inch = 2.5 cm).
Source: National Climatic Data Center, NOAA.

but with poor timing and/or distribution of rain during the critical crop growing periods, such as may occur in the southeastern and mid-Atlantic states, often require supplemental irrigation for satisfactory production of many crops.

Dryland farming usually consists of using special cultivars of certain crops, reducing the rate of seeding to correspond to the limited amounts of moisture that may be available, and employing tillage practices that conserve and economize water. Irrigation farming utilizes the management practice of application of water (irrigation) to the land. Typically the surface of the soil is maintained only as moist as needed for good crop production. Irrigation is scheduled so as to keep soil wet throughout the root zone to encourage good root growth.

States in the western and southern United States tend to use more water for irrigation than states in the northeast, mid-Atlantic, and upper Midwest (Fig. 17.13). Crops that tend to receive irrigation also varies (Fig. 17.14) as 100% of the rice, almost 70% of the orchards, and approximately 65% of the vegetables planted in the United States are typically irrigated. In contrast, less than 10% of the planted acreage for soybeans, oats, wheat, and tobacco are typically irrigated. These differences in irrigation use among crops may be due to differing water requirements, the climate of the available growing environment for the desired crop, and cost effectiveness and the availability of irrigation for the particular crop.

Irrigation

Watering or irrigating supplements natural rainfall when soil moisture is insufficient to meet plant needs. Even brief periods of water stress can reduce crop yield and quality. Crops must receive adequate moisture, especially during critical periods in their development. Proper irrigation practices are important to ensure good seed

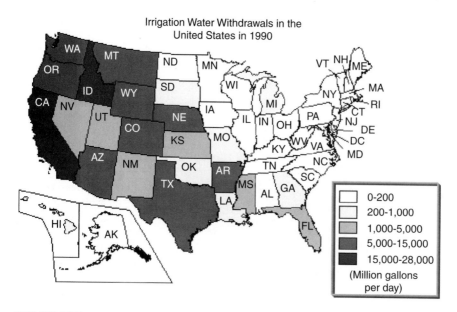

FIGURE 17.13 Irrigation water withdrawals in the United States in 1990 (1 gallon = 3.78 liters). Source: U.S. Geological Survey.

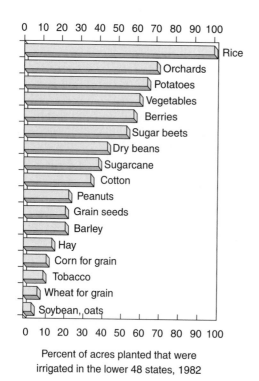

FIGURE 17.14 Percent of acres planted that were irrigated in the lower 48 states in 1982 (1 acre = 0.405 hectares).
Source: U.S. Geological Survey.

germination and seedling establishment, control diseases and insects, reduce soil erosion, conserve water, and protect water quality. Irrigation can also be used to cool plants during warm temperatures and for frost control during freezing temperatures.

Many factors affect the loss of moisture from the soils, including surface runoff, evaporation, percolation to the water table, and plant transpiration. The largest proportion of water loss from the field is through transpiration, with the rate of loss fluctuating with duration and intensity of solar radiation, wind speed, relative humidity, and plant canopy area and species.

Incipient wilting occurs when plants wilt under weather conditions favoring rapid transpiration, even when field moisture is adequate. Such stress increases in frequency and intensity as soils dry. The plant's permanent wilting point is the moisture content of the soil at which plants wilt, water uptake ceases, plant tissue and cells are damaged, and the plant fails to recover even when placed in a humid atmosphere or with abundant soil moisture.

Types of Irrigation Systems

Three basic types of irrigation systems used for most crop production systems are overhead systems, surface systems, and subsurface systems. The type of system used by a grower will depend on the characteristic of the field, crop, and available resources. Each system will have its advantages and disadvantages depending on the grower situation.

Overhead Systems *Portable aluminum pipe sprinkler system* A portable aluminum pipe sprinkler system is designed to cover only a portion of the acreage to be covered at one time. This system can be moved one or more times per day. It may use a variety of sprinkler sizes. It has the highest labor requirement of any sprinkler system and is only used on relatively small acreages.

Solid set irrigation system A solid set irrigation system is an aluminum pipe system that is placed in the field at the start of the irrigation season and left in place throughout the season (Fig. 17.15). It is adapted to most soil types, topography, field sizes, and shape. It has a low labor requirement once it is placed in the field and the system can be automated. It is normally used for crops with a high cash value.

Permanent sprinkler system A permanent sprinkler system is an underground pipe system with only a portion of the risers and sprinklers above ground. This system has the same characteristics as a solid set, except labor requirements are lower and the system cannot be moved to another location.

Traveling gun A traveling gun is a self-propelled, continuous-move sprinkler mounted on a two-, three-, or four-wheel trailer with water being supplied through a flexible hose (Fig. 17.16). The unit follows a steel cable that has one end anchored at one end of the field and the other end is attached to the machine. The cable winds

FIGURE 17.15 Solid set irrigation system.

FIGURE 17.16 A traveling gun irrigation system.

onto a cable drum on the machine as the machine moves through the field. It is best suited to square or rectangular fields, straight rows, and flat to rolling topography. Uniformity of water application will vary from excellent to fair. It is easily transported from field to field and farm to farm.

Center pivot system A center pivot system is an electrically or hydraulically powered, self-propelled lateral line on which sprinklers or spray nozzles are mounted (Fig. 17.17). Towers that are spaced 34 to 85 m apart support the lateral line. Center

FIGURE 17.17   A center pivot system.

pivots are available as fixed-point machines and as towable machines that can be used on two or more pivots. The pattern of irrigation is a circle; however, with an end gun, it is possible to irrigate corners or odd shapes of fields. The slope of land must be less than 10%. Center pivots have a high initial cost and are usually custom designed to fit the area to be used.

Lateral move system   A lateral move system is a self-propelled, electric drive lateral line on which low or very low pressure nozzles are mounted. The system moves in a straight line down the field. Towers that are spaced 34 to 55 m apart support the lateral line. The base tower controls the speed of travel and is located at the end of the lateral line or at some point in the lateral line. It follows a buried guidance or above-ground guidance cable. A lateral move system is designed for flat fields with widths up to 1,220 m and lengths up to 2,745 m. It has a medium to high initial cost and the operational costs are low.

Surface Systems   *Drip or trickle irrigation*   A drip or trickle irrigation system is a low-pressure system where a lateral line is placed along side every crop row adjacent to the plant (Fig. 17.18). Water is discharged through the line through emitters, through orifices in the pipe, or through microsprinklers. Trickle tubing or tapes are designed to be operated at low water pressure (such as $0.84\,\text{kg cm}^{-2}$ or 12 psi). A drip or trickle system can be used on most soil types and works best on flat terrain. The filter system is very important. With the use of drip systems the amount of water used is reduced because of placement near the root zone. Also limited water sources and low pressure can be used. Field operations can be performed during irrigation. A disadvantage with the use of drip systems is that the emitters can become plugged with mineral deposits and they require a high level of management.

FIGURE 17.18   Trickle irrigation on tomatoes. The black plastic line is the trickle tubing that emits small amounts of water through holes in the tubing.

Furrow irrigation   Furrow irrigation is the application of water in furrows that have a continuous slope in the direction of water movement to crops (Fig. 17.19). The system is limited to fields that have natural slopes of 2% or less and to soils that are not excessively sandy. Fields must be graded to provide uniform slope and water distribution. Furrow irrigation is not designed for small applications of water. Labor requirements are fairly high and the initial costs can be high depending on the amount of grading. Operational costs are low and it does have a good uniformity of water distribution.

Subsurface Systems   Subsurface irrigation systems are used in areas where there is a naturally occurring shallow water table (such as in selected areas in Florida) and adjustment of water level is possible (Fig. 17.20). Pesticides are applied very carefully with subsurface irrigation systems so as not to contaminate the water source.

Drainage

Most crops will not grow where surface water accumulates for long periods of time. Seeds planted in wet soils are more susceptible to decay, and emerging seedlings in wet soils become more susceptible to soil-borne diseases such as damping off. It is also difficult to make satisfactory seed beds in poorly drained soils. Drainage is the removal of excess water from soil. It can generally be achieved in two ways: surface drainage and subsurface drainage.

FIGURE 17.19 Furrow irrigation involves flooding the furrows between the planted raised rows to irrigate the crop.

FIGURE 17.20 Subsurface irrigation requires raising the water table until it reaches the plant root zone. This picture shows the rising water table in the drainage ditch.

FIGURE 17.21   Field drains are used to remove excess water from the fields. This is a field drain running parallel to the planted rows. This field drain will lead into a lateral drain that is often perpendicular to the field drains.

Surface Drainage   Ditches are the most common type of surface drainage in agriculture. Field drains are ditches as much as 46 cm deep that collect water from the row crops and land surfaces and carry it to a larger drain (Fig. 17.21). Lateral drains or collecting ditches receive water from field drains. They typically are 0.3 m deeper than the field drains and have side slopes with a low angle so that they can easily be traversed with motorized farm equipment. Main drains collect water from laterals and carry it to drainage outlets.

The slope of the land or topography often dictates the type of drainage used in a field. A parallel system of surface drains is usually used when topography is uniform. The field drains are spaced uniformly and are at right angles to collecting drains. Random systems are used when topography is not uniform and when only some areas require drainage. Each problem area is considered separately and enough field drains are installed to provide necessary outlets.

Many field-grown crops such as vegetables and small fruits are grown on raised beds (Fig. 17.22) that serve to facilitate drainage and also to increase soil temperatures. Raised beds are high ridges formed by turned furrows that are flanked by ditches to carry excess water. Beds are often 15 to 20 cm high with a flat surface. Raised beds are essential with furrow irrigation or in areas with excessive rainfall. Raised beds encourage wet soils to dry quicker and can help to prevent waterlogging of plant roots.

Subsurface Drainage   Subsurface drainage is most often accomplished by using tile drains or perforated plastic pipe. Tile drains or pipe are installed below ground to remove excess water from high-value agricultural land. Tile drains and pipes occupy

FIGURE 17.22 Raised beds prepared in the field prior to planting or seeding.

no arable land and eliminate maintenance problems associated with open ditches, such as weed control. Water enters through cracks between the individual tiles or through the perforations in the pipe and flows by gravity from the root zone down to the water table.

Wind Screens

Winds blowing over leaf surfaces can increase transpiration rates and result in a considerable loss of moisture from the plant. There is a relatively constant water vapor layer near the leaf surface caused by natural plant transpiration. The water vapor layer near the leaf tends to slow the movement of water from the leaf. However, as the water vapor is removed by wind action through increases in transpiration and evaporation, it is replaced with more water vapor from the leaf. Wind blowing across a crop can also carry with it small particles that can damage plant tissue by abrasion.

Installation of windscreens or barriers near plantings can be effective in reducing plant exposure to winds. Most windbreaks or screens modify air movement for a distance of about two times their height. Shrub or tree borders of drought tolerant plants are often used for long-term wind protection. Alternately, fabric materials placed on upward rising poles or grain crop plantings are also used.

Hydroponics

Hydroponics is the science of growing plants without using soil. This is accomplished by providing the plant a solution of water containing mineral nutrients. The term *hydroponics* is derived from two Greek words, '*hudor*,' which means water, and '*pomos*,' which means work. Hydroponics is used in numerous ways around the world, ranging from large-scale production of commercial crops (flowers, vegetables, and fruits) through medium-sized office and community growing units to smaller pots of indoor plants.

Hydroponics is an intensive growing method requiring optimum light, temperature, and humidity. It can be used to grow almost any herbaceous plant; however, it is more difficult and exacting than traditional methods. The primary advantage of using hydroponics for commercial crop production is to increase yields and improve plant quality, most often in greenhouses. Hydroponics also allows the growth and production of plants in areas where suitable soil is not available for cultivation. The nutrient solution of a hydroponic system needs to contain all of the essential elements necessary for plant growth. Although weeds and soil pathogens are usually not a major problem when using this method, a hydroponics system has many drawbacks, including the need for specialized equipment and relatively high maintenance costs.

Common hydroponic systems include the nutrient film technique, aeroponics, the aeration method, and the float system. The nutrient film technique uses a plastic trough or tube as the container through which a constant, thin film of nutrient solution flows. Plants are suspended through holes in the top of the trough. The trough is gently sloped so gravity pulls the solution back to the nutrient reservoir.

Aeroponics is the growing of plants in a container in which the roots are suspended in a nutrient mist rather than in a solution. The most popular container for aeroponics is an enclosed A-frame constructed of Styrofoam boards. The plants are placed in holes along the sloped sides of the frame. The nutrient mist is delivered to the roots. Any excess runs down the inside of the frame, is collected at the bottom, and is recycled back to the nutrient reservoir.

The aeration method, one of the first systems to be developed, uses an aquarium air pump to bubble oxygen to the roots of plants immersed in the nutrient solution (Fig. 17.23). Plants are suspended above the solution by tray or other means that is set into the container of nutrient solution. A layer of inert material, such as gravel, rock wool, clay pebbles, or vermiculite, provides stability for the plants while allowing the roots to grow into the nutrient solution. The float system requires the plants to be placed through a flat-surfaced material that floats on a nutrient solution with its roots submerged into the solution.

Hydrophilic Gels

Hydrophilic gels are moisture-holding polymer compounds that absorb many times their weight in water. These gels release their available moisture as the environment around them becomes dry. Hydrophilic gels may be able to increase available moisture, improve media aeration, reduce compaction, improve drainage, and increase plant survival and growth. These products are used in fluid drilling of vegetable seeds and as packing material for bare-root plants during shipping. They also can be used as an amendment for nursery and greenhouse media, and as a transplant aid for turf, bedding plants, and woody ornamentals.

There are two types of absorbent polymers. The first contains starch and is rapidly decomposed by bacteria and fungi. The life span of the starch-containing prod-

FIGURE 17.23   A hydroponic system.
Courtesy of Mike Orzolek, Penn State University.

ucts in the soil varies from a few days to a few weeks. The second contains synthetic polymers. Synthetic polymers can be divided into cross-linked polyacrylamides or cross-linked polyacrylates. Synthetic polymers may remain active for 2 years or more. Polyacrylamide polymers are the most durable.

Summary

Water is utilized by plants for the transport of organic molecules, inorganic molecules, and atmospheric gases; tissue temperature regulation; and overall structural support. Unique properties of water include its high ability to store heat and its high heat of vaporation. It also has a high surface tension and tensile strength and is relatively transparent to visible radiation. The status of water in soils, plants, and the atmosphere is often described in terms of water potential, which is the sum of the osmotic potential, the hydrostatic potential, the gravitational potential, and the matrix potential (for situations involving soils).

The primary source of water for most plants is through the absorption of soil water through the root system. Water may move to the root via capillary action or roots may grow to intercept moist soil. The water absorbed by the root is affected by the number of roots, the size of the roots, the rate of root growth, the age of the root, and the presence of root hairs. Water transport from the soil, through the plant, and to the atmosphere takes place in a soil-plant-air continuum that is interconnected by a continuous film of water. Transpiration (or the evaporative loss of water from leaves and stems) is the primary driving mechanism for water to reach the top of plants and trees in this continuum.

For optimum plant growth in the field it is often necessary to monitor soil moisture. Plant-based soil moisture monitoring methods include evaluation of plant appearance, measuring plant water potential, and measuring leaf temperature. Soil-based soil moisture monitoring methods include evaluating the visual appearance and feel of the soil, determining the weight of the soil, and measuring soil water potential.

Drought is a meteorological term that occurs when there has been a period of time without sufficient rainfall. This often results in water stress to the field-grown plants. The early deleterious effect of drought (or water stress) on plants is a loss of turgor that subsequently affects the rate of cell expansion and cell size. Mechanisms developed by certain plants to avoid water stress include the ability to regulate the opening of the guard cells of the stomata, having thick leaves with thick waxy coverings and/or small hairs on the leaf surface to reflect light, and utilizing storage capabilities for water in fleshy leaves.

Flooding is the result of excessive soil moisture and can harm plant growth through the development of anaerobic (no available oxygen) conditions. Insufficient oxygen can result in the death of the plant. Flooding can also affect the soil characteristics of pH, the rate of organic matter decomposition, and the production of toxic levels of compounds.

Acid precipitation can adversely affect forest landscapes and aquatic ecosystems by increasing the concentration of protons and strong anions in the soil and groundwater reservoirs. Acid precipitation may interfere with plant Ca metabolism and Ca-dependent cellular mechanisms, decrease cold tolerance of some tree species, and leach cations from leaf surfaces of some plants.

Methods to control evaporative loss of soil water practiced by crop producers include the use of soil mulches, cover crops, contour plantings and terraces, tillage practices, and summer fallow. Irrigation is often used when rainfall in an area is insufficient to maintain soil water levels at an adequate level for optimum plant growth, to cool plants during high temperatures, and for frost protection during freezing temperatures. Other systems that regulate the amount of moisture that the plant receives include utilizing special drainage systems to remove excess water, using wind screens to reduce water loss due to transpiration, growing plants in hydroponic systems, and adding hydrophilic gels to the soil to enhance moisture content and retention.

Review Questions

1. List four factors that affect root effectiveness.
2. What is the ecological significance of the high heat capacity of water?
3. List four soil-based techniques for measuring soil water.
4. How does an infrared thermometer determine the water needs of a plant?
5. Describe the probable soil water situation and some grower concerns when a tensiometer in the field is reading the following:
   a. 0 to 5
   b. 6 to 20

      c. 21 to 60

      d. 61+

  6. What are some of the principles of dryland farming?

  7. List three factors that influence the efficiency of water use in cropping systems.

  8. Define the following:

      a. Soil saturation

      b. Field capacity

      c. Available soil moisture

  9. What is the importance of root hairs in water uptake by plants?

 10. How does the following help conserve soil moisture?

      a. Plastic mulches

      b. Stubble mulch

      c. Cultivation and weed control

 11. Describe the following irrigation systems and their advantages and disadvantages in their use over other types of available systems:

      a. Permanent sprinkler system

      b. Traveling gun

      c. Furrow irrigation

      d. Center pivot system

      e. Portable aluminum pipe

Selected References

Cavigelli, M. A., S. R. Deming, L. K. Probyn, and R. R. Harwood, eds. 1998. *Michigan field crop ecology: Managing biological processes for productivity and environmental quality.* Michigan State University Extension Bulletin E-2646.

Driscoll, C. T., G. B. Lawrence, A. J. Bulger, T. J. Butler, C. S. Cronan, C. Eagar, K. F. Lambert, G. E. Likens, J. L. Stoddard, and K. C. Weathers. 2001. Acidic deposition in the northeastern United States: Sources and inputs, ecosystem effects, and management strategies. *Bioscience* 51:180–198.

Hensley, D. 1993. *Take a hard look at water-holding compounds.* University of Hawaii Horticulture Digest #99.

Janick, J., R. W. Schery, F. W. Woods, and V. W. Ruttan. 1974. *Plant science.* 2nd ed. San Francisco: W. H. Freeman.

Kramer, P. J., and J. S. Boyer. 1995. *Water relations of plants and soils.* San Diego: Academic Press.

Lambers, H., F. S. Stuart III, and T. L. Pons. 1998. *Plant physiological ecology.* New York: Springer-Verlag.

Larsen, D. C., and M. K. Thornton. 1990. Irrigation on crop important during growth. *The Potato Grower of Idaho* (August):18–19.

Niles, M. A. 2000. *Atmospheric deposition program of the U.S. geological survey.* U.S. Geological Survey Fact Sheet FS-112–00.

Nuss, J. R. 1996. *Drought & the landscape garden.* Penn State University PENNpages, Document No. 2940156.

Sorenson, R., and D. Relf. 1996. *Home hydroponics.* Virginia Cooperative Extension Service Publication No. 426–084.

Stern, K. R. 1991. *Introductory plant biology.* 5th ed. Dubuque, IA: Wm. C. Brown.

Taiz, L., and E. Zeigler. 1998. *Plant physiology.* 2nd ed. Sunderland, MA: Sinauer.

Tanino. K. K., and B. Baldwin. 1996. *Physiology of drought in stressed plants.* University of Saskatchewan, Department of Horticulture Publication.

Selected Internet Sites

http://www.nrcs.usda.gov/technical/water.html   Water resources information on water and climate data, water management documents, and water supply forecasts. Natural Resources Conservation Service, USDA.

http://www.usgs.gov/   Water resources of the United States, including water data, publications and products, technical resources, and programs. U.S. Geological Survey.

http://water.usgs.gov/pubs/acidrain/   Trends in precipitation chemistry in the United States, 1983 to 1994, U.S. Geological Survey.

18

Nutrients

Plants require 16 essential chemical elements (nutrients) for growth. The soils in which many plants are grown provide many of these needed chemical elements. The chemical elements provided by the soil are often referred to as mineral elements. Hydrogen, carbon, and oxygen are obtained by plants from carbon dioxide and water and are not considered mineral elements. Of the essential elements, usually only a few (often nitrogen, phosphorus, and potassium) have a high probability of being deficient in most soil-based crop production systems. Plant nutrients come from many different sources and exist in many different chemical forms in the soil (Table 18.1). Soil pH also exerts a major influence on nutrient availability to plants.

This chapter describes the essential plant elements and their general role in plant growth; presents overviews of the important nutrient cycles of carbon, nitrogen, phosphorus, and potassium; discusses the concept of critical nutrient concentration; and illustrates how crop producers can affect elemental availability.

Background Information

Methods of Nutrient Acquisition by Plants

Although most terrestrial plants absorb nutrients via their roots (as discussed in Chapter 17), other methods of nutrient acquisition exist. For example, leaves are capable of absorbing gaseous forms of nutrients (such as carbon dioxide, water vapor, and nitrogenous and sulfur compounds) through the stomata. Nutrients in the water on wet leaves are also available for absorption by leaves to a limited extent, and foliar application of micronutrients may be economically important. Other mechanisms of nutrient acquisition include those found in carnivorous plants, symbiotic associations with microorganism associations with host plants.

TABLE 18.1 *Nutrients essential for plant growth and forms taken up by the plant*

Element	Chemical symbol	Form(s) taken up by plant
Macronutrients		
Carbon	C	CO_2
Hydrogen	H	H_2O
Oxygen	O	H_2O, O_2
Nitrogen	N	NH_4^+, NO_3^-
Phosphorus	P	$H_2PO_4^-$, HPO_4^{2-}
Potassium	K	K^+
Calcium	Ca	Ca^{2+}
Magnesium	Mg	Mg^{2+}
Sulfur	S	SO_4^{2-}
Iron	Fe	Fe^{2+}, Fe^{3+}
Micronutrients		
Zinc	Zn	Zn^{2+}, $Zn(OH)_2$
Manganese	Mn	Mn^{2+}
Copper	Cu	Cu^{2+}
Boron	B	$B(OH)_3$
Molybdenum	Mo	MoO_4^{2-}
Chlorine	Cl	Cl^-

Essential Plant Nutrients

It was once accepted that the only substance needed by plants for satisfactory plant growth in a field was water. It was subsequently discovered that certain "earth substances" were beneficial for plant growth. In the 1800s, Von Liebig suggested that certain elements had an essential role in plant growth and that growth was limited to the extent that an essential element was lacking.

Each of the 16 accepted essential elements has at least one specifically defined role in plant growth so that the plants fail to grow and reproduce normally in the absence of the element. In general, essential elements function in plants as constituents of compounds, in activation of enzymes, in osmoregulation of plant cells, or in modification of plant membrane permeability. In actuality, most essential elements have multiple roles in plants.

Essential elements are often divided into macronutrients and micronutrients, according to their relative concentration in plant tissue.

Macronutrients   Macronutrients are those elements needed by plants in relatively large quantities. The following elements are considered macronutrients:

Carbon (C)   Carbon is available from CO_2 and is assimilated by plants during photosynthesis. It is a component of organic compounds such as sugars, proteins, and organic acids. These compounds are used in structural components, enzymatic reactions, and genetic material, among others.

Hydrogen (H)   Hydrogen is derived from H_2O and is incorporated into organic compounds during photosynthesis. Hydrogen ions are also involved in electrochemical reactions and maintaining electrical charge balance.

Oxygen (O)   Oxygen is derived from CO_2 and is also a part of organic compounds, such as simple sugars. Oxygen is necessary for all oxygen-requiring reactions in plants including nutrient uptake by roots.

Phosphorus (P)   Phosphorus is a component of several energy transfer compounds in plants. Phosphorus is also important in the structure of nucleic acids (which serve as the building blocks for the genetic code material in plant cells). Phosphorus promotes early maturity and fruit quality.

Potassium (K)   Potassium is an activator in many enzymatic reactions in the plant. K movement in and out of these cells also controls turgor in the guard cells of the stomata. Potassium is also important in cell growth primarily through its effect on cell extension. With adequate K, cell walls are thicker and provide more tissue stability.

Nitrogen (N)   Nitrogen is a component in many compounds including chlorophylls, amino acids, proteins, nucleic acids, and organic acids. A large part of the plant body is composed of N-containing compounds. Because N is contained in the chlorophyll molecule, a deficiency of N will result in a chlorotic condition of the plant.

Sulfur (S)   Sulfur is a component of sulfur-containing amino acids such as methionine and in the sulfhydryl group of certain enzymes. Sulfur is present in glycosides, which give the odors characteristic of onions, mustard, and garlic.

Calcium (Ca)   Calcium is a component of calcium pectate, a constituent of cell walls. In addition, Ca is a cofactor of certain enzymatic reactions and is involved in cell elongation and cell division.

Iron (Fe)   Iron is used in the biochemical reactions that form chlorophyll and is a part of one of the enzymes that is responsible for the reduction of nitrate nitrogen to ammoniacal-nitrogen. Other enzymes such as catalase and peroxidase also require Fe.

Magnesium (Mg)   Magnesium plays an important role in plant cells as it appears in the center of the chlorophyll molecule. Certain enzymatic reactions require Mg as a cofactor. Magnesium aids in the formation of sugars, oils, and fats.

Micronutrients   Micronutrients are essential nutrients that are needed by plants in smaller amounts than macronutrients. The amount of micronutrients assimilated by crops is also small in comparison to the total quantity of these nutrients that may be present in the soil. The availability of micronutrients is influenced by soil conditions such as pH, moisture content, aeration, and the presence and amounts of other elements. The following elements are considered micronutrients:

Molybdenum (Mo)   Molybdenum is a constituent of two enzymes involved in N metabolism. The most important of these is nitrate reductase, the enzyme involved in the reduction of nitrate-nitrogen to ammoniacal-nitrogen. It is also a structural component of nitrogenase, which is involved in the fixation of N_2 into the ammonium form in a symbiotic relationship with legumes.

Boron (B)   Boron appears to be important for meristem development in young plant parts such as root tips. Boron is involved in the transport of sugars across cell membranes and in the synthesis of cell wall material. Because of its impact on cell development and on sugar and starch formation and translocation, a deficiency of B will retard new growth and development.

Manganese (Mn)   Manganese functions in several enzymatic reactions that involve the energy compound adenosine triphosphate (ATP). Manganese also activates several enzymes and is involved in the process of the electron transport system in photosynthesis.

Copper (Cu)   Copper is involved in several enzyme systems, cell wall formation, electron transport, and oxidative reactions.

Zinc (Zn)   Zinc is involved in the activation of several enzymes and is required for the synthesis of indoleacetic acid, a plant growth regulator.

Chlorine (Cl)   Chlorine has a possible role in photosynthesis and may function as a counter ion for K fluxes involved in cell turgor. Chlorine is involved in the capture and storage of light energy through its involvement in the photophosphorylation reaction in photosynthesis.

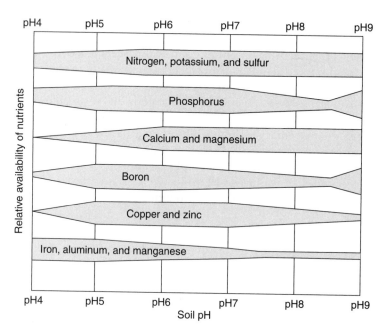

FIGURE 18.1   Relationship among pH, alkalinity, acidity, and plant growth.
Source: Texas A&M Cooperative Extension Service Vegetable Growers Handbook. 2002.

Soil pH

Soil pH measures the concentration of hydrogen ions, H+, in the soil and is a major factor in determining the availability of nutrients from the soil for the plant. Soils are acidic if they have a soil pH below 7 (Fig. 18.1). Neutral soils have a soil pH of 7. Soils are alkaline if they have a soil pH above 7. Soils that are either acidic or alkaline typically have their soil pH adjusted to near neutral to ensure proper plant growth because most essential elements reach near maximal availability and most plant toxic elements become nontoxic in this pH range (Fig. 18.2).

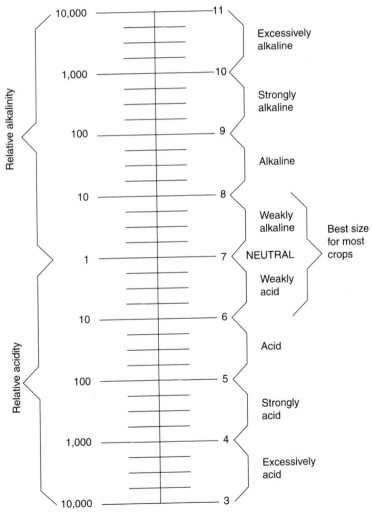

FIGURE 18.2   Availability of plant nutrients is affected by soil pH.
Source: Purdue University Cooperative Extension Publication AY-267, *Soil Acidity and Liming of Indiana Soils*, C. D. Spies and C. L. Harms. 1968.

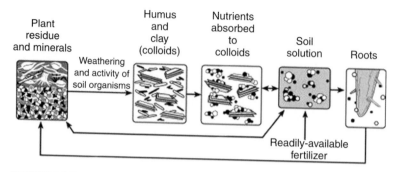

FIGURE 18.3 Nutrients in the soil exist in several forms that affect availability to the plant. Source: University of Minnesota Cooperative Extension Service Publication BU-07501-GO, *Organic Matter Management,* Ann Lewandowski. 2002.

Biogeochemistry

Biogeochemistry is the study of how, when, and in what form soil nutrients become available to plants, microbes, and other organisms. Nutrients in the environment exist in several forms (Fig. 18.3) and are typically cycled; several processes affect their availability.

Important Nutrient Cycles   The ecological principle of the Law of Conservation of Matter states that matter is neither created nor destroyed, but it may change chemical form. As a result, chemical elements, including the essential elements of plants and animals, tend to circulate in characteristic paths from environment to organisms and back to the environment. These somewhat circular paths are known as nutrient cycles. The more important nutrient cycles for plants include the carbon cycle, the nitrogen cycle, the phosphorus cycle, and the potassium cycle.

The carbon cycle   The primary source of carbon for plants is the CO_2 in the atmosphere. During the photosynthesis, CO_2 from the atmosphere is incorporated into plant biomass as organic compounds of starches and sugars (Fig. 18.4). These energy-rich compounds are transferred through the food chain to herbivores and carnivores.

A significant portion of the carbon in many crops is harvested and removed from the field or production area. After harvest, both above-ground (shoot) and below-ground (root) residues that remain in the field enter the soil. Some root residues also enter the soil system while the plant is alive. When roots die, portions are sloughed off or carbon compounds leak out of them. Most microbial activity in soils occurs in the rhizosphere, that portion of the soil affected directly by the root.

In addition, some plant tissue (either before or after harvest) may be consumed by humans or animals. The carbon in this tissue may be subsequently released as CO_2 as a result of human and animal respiration or returned directly to the soil as excrement. Carbon in the soil is a food source for soil microorganisms such as bacteria, fungi, and actinomycetes. Thus, some of the carbon that enters the soil as residue or manure becomes part of the microorganisms (5% to 15% of the original

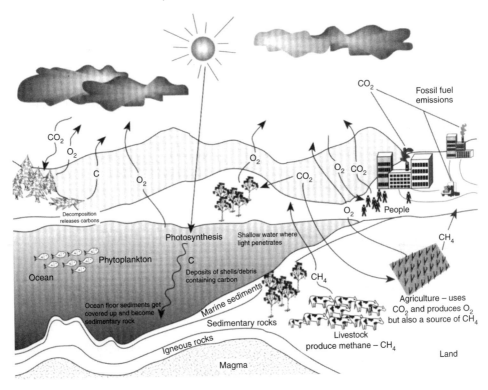

FIGURE 18.4   The carbon cycle.
Source: Office of Mission to Planet Earth's Education Office Goddard Space Center, NASA.

plant residues) and is called microbial biomass. The largest proportion of residue carbon, though, is released to the atmosphere as CO_2 when soil organisms respire (60% to 75% of original plant residues). The conversion of organic carbon to CO_2 is called carbon mineralization.

As a result of microbial activity, carbon undergoes many complex chemical transformations that are collectively known as decomposition. Decomposition rates are influenced by factors that influence microbial activity such as temperature, moisture, aeration, pH, amount and quality of residue, residue particle size, and degree of burial in the soil.

The nitrogen cycle   Although gaseous nitrogen (N_2) makes up approximately 80% by volume of the earth's atmosphere, it cannot be used in this form by most plants and animals. Plants can only take up nitrogen in the organic forms of ammonium (NH_4+) and nitrate (NO_3-) (Fig. 18.5). Animals primarily receive nitrogen from amino acids within plants and other animals.

Only certain kinds of bacteria and some blue-green algae can convert or fix N_2 directly into usable forms. Nitrogen fixation converts N_2 to ammonium by either bacteria (biological nitrogen fixation) or chemical processes (chemical nitrogen fixation). Some free-living bacteria fix N_2, but most biological nitrogen fixation comes

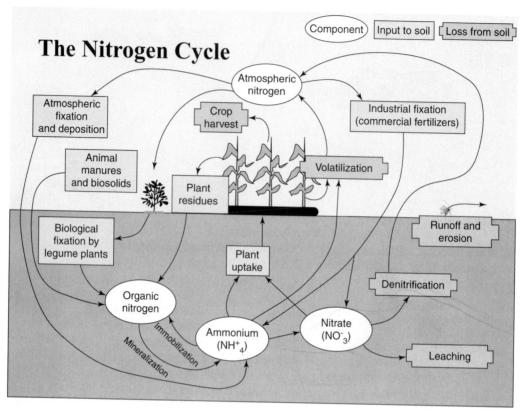

FIGURE 18.5 The nitrogen cycle.
Used with permission from the Potash & Phosphate Institute.

from symbiotic nitrogen-fixing bacteria in legume root nodules. Denitrification is the conversion of nitrate to gaseous N_2, nitrous oxide (N_2O), and/or nitric oxide (NO) by specialized bacteria (denitrifiers) under aerobic conditions.

Synthetic fertilizers are important sources of nitrogen in agricultural systems and contain one of three nitrogen forms: urea [$CO(NH_2)_2$], ammonia/ammonium, or nitrate. All nonnitrate forms commonly used in agriculture are readily converted to soil nitrate by mineralization and/or nitrification. In addition, some nitrogen is available to plants in crop residues and green manures. Unfortunately, almost all the nitrogen in crop residues and green manures, and about half of that in animal manures, is in organic forms not immediately available for crop uptake. These forms of nitrogen are naturally slow-released at a rate that depends on the factors that influence mineralization, immobilization, and nitrification.

Organic nitrogen is converted to ammonium through nitrogen mineralization. Nitrogen mineralization is conducted by a wide array of soil organisms. When organic carbon is consumed by the microbial biomass, some organic nitrogen may also be consumed and then become part of short-term or long-term organic matter pools.

Ammonium is generally not very mobile in soils. Because of its positive charge, ammonium is attracted to negatively charged soil colloids (clay and organic matter).

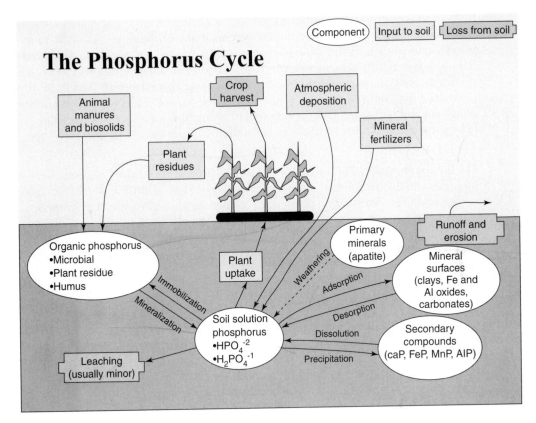

FIGURE 18.6   The phosphorus cycle.
Used with permission from the Potash & Phosphate Institute.

Ammonium can be lost from the system to the atmosphere when ammonium is transformed to ammonia via ammonia volatilization. This chemical process occurs when urea or manure (which contains a large proportion of nitrogen as urea and uric acid) is surface applied. Volatilization rates are highest at high pH (> 7.9) and under dry and warm conditions.

Nitrification occurs when ammonium is converted to nitrate by specialized bacteria called nitrifiers. Nitrate is very mobile in soils because of its negative charge. If nitrate is not taken up by the crop, it is very susceptible to loss by both leaching and denitrification. Nitrate leaching occurs whenever soil nitrate is present and water moves down through the soil profile. Denitrification is the conversion of nitrates to N_2, N_2O, or NO by specialized bacteria under anaerobic conditions. Soil oxygen is depleted when soils are saturated and/or microbial respiration is very high. Large amounts of soil nitrogen are also lost via runoff and erosion.

The phosphorus cycle   The phosphorus cycle is one of the sedimentary cycles by which materials travel from the land to the sea and back again (Fig. 18.6). Elements in such cycles usually take millions of years to cycle. The great reservoir of phosphorus is the rocks or other deposits that were formed in past geological ages. These are gradually eroded, releasing phosphates to the ecosystems, some of which are available to the plant. Plants can absorb available phosphorus from the soil and pass it on to animals. It is eventually returned to the soil, rivers, and oceans as animal excretion or as a result of decomposition after death or through atmospheric deposits. Much of this phosphate with time can escape to the sea, where part of it is deposited in shallow sediments and part is lost to deep sediments. Fishing by humans and the excretions of fish-eating guano birds return a large amount of phosphorus to the land each year.

The potassium cycle   The return of potassium from crop residues and manures is important in maintaining soil potash (Fig. 18.7). The annual losses of available potassium by leaching and erosion greatly exceed those of nitrogen and phosphorus. Also, some potassium may be converted or "fixed" into readily nonavailable, nonexchangeable potassium. The fixed potassium can be subsequently released to exchangeable potassium, which is available for plant uptake. Often supplemental

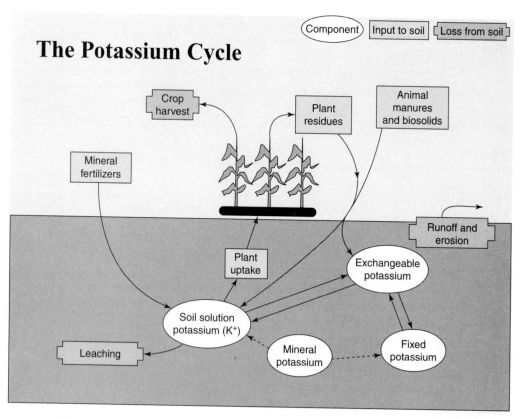

FIGURE 18.7   The potassium cycle.
Used with permission from the Potash & Phosphate Institute.

potassium, through the use of fertilizers, supplies much of the potassium needed for crop production. This is essentially true in cash-crop areas and in regions where sandy soils are prominent. The most common form of potassium for fertilizers is that mined from naturally occurring beds of potassium ore, which developed as sea salts evaporated and potassium salts crystallized as deposits.

Effects on Plants

Critical Nutrient Concentrations

Normal plant growth is achieved when all the essential elements are provided in a suitable nutrient concentration range. This is called the adequate or sufficient nutritional concentration range. A critical concentration for a nutrient occurs at the point where plant growth is reduced by 10% because of a shortage of the element in question. The critical concentration is the borderline between elemental sufficiency and deficiency.

A deficiency of an element essential for plant growth will result in a decrease in the normal growth of the plant and will affect the yield of a crop. The deficient range occurs at elemental concentrations lower than those in the transitional zone and is accompanied by a drastic restriction in growth at which time the plants may begin to show deficiency symptoms (Table 18.2). Nutrient deficiency symptoms in a plant are the expression of metabolic disorders resulting from the insufficient supply of an essential element. Symptoms of nutrient deficiencies are often used to diagnose growth problems.

Essential elements can be toxic when they are in excess and detrimentally affect growth and yield of the crop. The toxic zone is where tissue elemental concentrations are greater than those in the adequate zone. A gradual decrease in plant growth occurs in the toxic zone. As the tissue concentration increases further, toxicity symptoms, often necrosis, can occur. Such symptoms are often used to diagnose the problem.

Most deficiency or toxicity symptoms can be categorized into one of five types: (1) chlorosis, which is a yellowing of plant tissue due to a reduction in chlorophyll formation; (2) necrosis, or death of plant tissue; (3) lack of new growth or terminal growth resulting in rosetting; (4) an accumulation of anthocyanin and an appearance of a reddish color; and (5) stunting and reduced growth with either normal or dark green coloring or yellowing. Deficiency symptoms can be used to help determine nutritional needs (see Soil and Tissue Analysis later in this chapter). Symptoms plus plant and soil analyses together with a general knowledge of crop needs and the chemistry of the soil provide the best information for determining crop nutrient needs.

Many deficiency symptoms can be induced by lack of several nutrients or by growing conditions. Knowledge of soil pH and general soil conditions must be used to accurately diagnose nutrient imbalances. Typical nutrient deficiency symptoms can also be caused by many other conditions. Certain herbicides, diseases,

TABLE 18.2   *Nutrient deficiency symptoms in plants and typical situations where symptoms may occur*

Nutrient	Deficiency symptoms	Occurrence
Nitrogen (N)	Stems thin, erect, hard. Leaves small, yellow; on some crops (tomatoes) undersides are reddish. Lower leaves affected first.	On sandy soils especially after heavy rain or after overirrigation. Also on organic soils during cool growing seasons.
Phosphorus (P)	Stems thin and shortened. Leaves develop purple color. Older leaves affected first. Plants stunted and maturity delayed.	On acid soils or very alkaline soils. Also when soils are cool and wet.
Potassium (K)	Older leaves develop gray or tan areas on leaf margins. Eventually a scorch appears on the entire margin.	On sandy soils following leaching rains or overirrigation.
Boron (B)	Growing tips die and leaves are distorted. Specific diseases caused by boron deficiency include brown curd and hollow stem of cauliflower, cracked stem of celery, blackheart of beet, and internal browning of turnip.	On soils with pH above 6.8 or on sandy, leached soils. Also on crops with very high demand such as cole crops.
Calcium (Ca)	Growing-point growth restricted on shoots and roots. Specific deficiencies include blossom-end rot of tomato, pepper, and watermelon; brownheart of escarole; celery blackheart, and cauliflower or cabbage tipburn.	On strongly acid soils or during severe droughts.

and insects can cause chlorosis in plants. Waterlogged or droughty soils, salinity, and mechanical or wind damage can often create problems that may resemble nutrient deficiencies.

Crop Removal Values

Crop removal values are estimated by analyzing plants and fruits for their nutrient content and then expressing the results on an hectare basis. Crop removal values can be used to estimate fertilizer. This is generally not an accurate method of determining fertilizer needs by the plants because these values are usually determined on crops that receive plenty of fertilizer, and crops in this situation will continue to take up nutrients in excess of their needs. As a result, crop removal values may overestimate the true nutrient content of the crops.

Nutrient	Deficiency symptoms	Occurrence
Copper (Cu)	Yellowing of leaves, stunting of plants. Onion bulbs are soft with thin, pale scales.	On organic soils or occasionally new mineral soils.
Iron (Fe)	Distinct yellow or white areas between veins on youngest leaves.	On soils with pH above 6.8.
Magnesium (Mg)	Initially older leaves show yellowing between veins, followed by yellowing of young leaves. Older leaves soon fall.	On strongly acid soils or on leached sandy soils.
Manganese (Mn)	Yellow mottled areas between veins on youngest leaves, not as intense as iron deficiency.	On soils with pH above 6.4.
Molybdenum (Mo)	Pale, distorted, narrow leaves with some interveinal yellowing of older leaves, such as whiptail disease of cauliflower.	On very acid soils.
Zinc (Zn)	Small reddish spots on cotyledon leaves of beans; light areas (white bud) of corn leaves.	On wet, cold soils in early spring or where excessive phosphorus is present.
Sulfur (S)	General yellowing of younger leaves and reduced growth.	On very sandy soils, low in organic matter, especially following continued use of sulfur-free fertilizers and especially in areas that receive little atmospheric sulfur.
Chlorine	Deficiencies very rare.	Usually only under laboratory conditions.

Source: G. J. Hochmuth. 1996. *Soil and fertilizer management for vegetable production in Florida.* University of Florida Cooperative Extension Service Circular HS-711.

Agricultural Technologies That Affect Nutrients

Adjusting Soil pH

Many crops will not grow in highly acidic soils. Growers generally apply lime when the soil pH is below 5.5 to raise the pH to a range of 6 to 7. Several types of materials that vary in their ability to neutralize soil acidity are used for liming.

Although most crops are somewhat tolerant of alkaline soil, reducing the soil pH is still beneficial to avoid mineral deficiencies and toxicities. Methods to reduce pH of alkaline soil involve the field application of sulfur-containing compounds. The amount of sulfur to reduce soil pH depends on soil type and desired soil pH.

Nutrient Replacement

Nutrient supply rates in the soil ultimately govern the rates of nutrient acquisition by plants. To maintain long-term plant productivity some or all of the nutrients that may have been previously utilized by the plant or lost by other pathways (such as leaching) must be periodically replaced. These nutrients may be replaced by using synthetic fertilizers, manures, or specific cover crops.

Fertilizers   Fertilizers are applied to a field or production system to increase the fertility of the soil so as to avoid nutrient deficiency levels in the plant and attain optimum plant growth and production. Fertilizers from inorganic and organic sources are available for use by growers and producers.

Inorganic fertilizers   Commercial inorganic fertilizers (Table 18.3) are added to a soil to directly increase the amounts of specific nutrients available to the plant. They are not used to improve the physical condition or make soil nutrient reserves available. Commercial fertilizers furnish limited elements in the most economic manner.

Most inorganic fertilizers contain salts of the macronutrients nitrogen, phosphorus, and potassium. Straight fertilizers (such as superphosphate, ammonium ni-

TABLE 18.3   *Some common nutrient fertilizer sources*

Nutrient	Source	Content	
Nitrogen		% N	
	Ammonium sulfate	21	
	Ammonium nitrate	33	
	Ammonium phosphate	10–18	
	Urea	46	
	Sodium nitrate	16	
	Blood meal	13–15	
Phosphates		% P	% P_2O_5
	Superphosphates	7–22	16–50
	Ammonium phosphate	21	48
	Diammonium phosphate	20–23	46–53
	Steamed bone meal	10–13	23–30
	Rock phosphate	11–13	25–30
	Phosphoric acid	24	54
Potassium		% K	% K_2O
	Potassium chloride	40–50	48–60
	Potassium sulfate	40–42	48–50
	Sul-Po-Mag	19–25	25–30
	Potassium nitrate	37	44

trate, and muriate of potash) contain only one of these three nutrients. Compound or mixed fertilizers contain two or more mineral nutrients. Numbers (such as 10-10-20) refer to the effective percentage of nitrogen, phosphorus, and potassium, respectively, in the fertilizer. For example, a fertilizer labeled 5-10-10 means the bag contains 5% by weight of nitrogen, 10% of phosphate (P_2O_5), and 10% of potash (K_2O). The percentage of elemental phosphorus in the fertilizer is obtained by multiplying the percentage of P_2O_5 by 0.44, and the percentage of elemental potassium is obtained by multiplying the percentage of K_2O by 0.83. Therefore, a 5-10-10 fertilizer is 5% nitrogen, 4.4% phosphorus, and 8.3% potassium. The fertilizer label also indicates what kinds of materials are used to make the units of available fertilizers. As an example, a 5-10-5 fertilizer may indicate 2.5% nitrogen as 20% ammonium sulfate, 2.5% nitrogen as 15% sodium nitrate, 10% phosphorus as 20% triple superphosphate, and 5% potassium as 40% potassium chloride.

Organic fertilizers   Organic fertilizers originate from residues of plants or animals or from natural rock deposits. The nutrient elements from residues must be released from the organic compounds through mineralization, a process usually involving the action of soil organisms. Mineralization depends on many factors including temperature, water, oxygen, and type of microorganisms in the soil.

Organic fertilizers generally used in commercial production include animal manures, green manures, and compost. Organic fertilizers improve soil organic matter, promote microorganisms in the soil, and supply many essential nutrients. Nutrient composition and decomposition rates of manures and compost vary according to source. Proper application of manure to fields takes into account the nutrient composition and decomposition rate for the manure used. The carbon to nitrogen ratio (C:N) of the manure is also important as those with a low C:N decompose more quickly than those with a high C:N. Because many manures do not supply adequate nitrogen at the time of application (animal manure) or at planting (green manure), a side dressing of nitrogen is also often required to provide the proper amount of nutrients for early plant growth.

Animal manures   Animal manures are important sources of organic nutrients and can be used as fertilizers (Table 18.4). Many crops generally show a favorable

TABLE 18.4   *Nutrient content of some commonly used manure sources**

Material	% Nitrogen	% Phosphate	% Potash	C:N	Availability
Cattle manure	0.25–2.0	0.15–0.9	0.25–1.5	8	med.
Horse manure	0.3–2.5	0.15–2.5	0.5–3.0	2	med.
Sheep manure	0.6–4.0	0.3–2.5	0.75–3.0	1	med.
Swine manure	0.3	0.3	0.3	4	med.
Poultry manure	1.1–2.8	0.5–2.8	0.5–1.5	-	med.–fast

*The nutrient content of most organic materials is quite variable and depends on the specific source and how the material has been handled and stored.

Source: Adapted from Colorado State University Cooperative Extension Service Fact Sheet No. 7.217, *Fertilizing the organic garden*, C. W. Basham and J. E. Ells.

response to animal manures. Disadvantages of manures may include the presence of excessive amounts of salt and introduction of weed seeds. Also, some manure may need to be properly fermented or aged before applying to the fields.

Green manures and cover crops   Green manures supply organic matter, prevent erosion, and aid in conserving soluble nutrients in the soil. Green manures can either be nitrogen-fixing legumes or nonlegume crops. The added organic matter in the soil retains more moisture and stores plant nutrients for crops and microorganisms. A 6- to 8-week period between plowing down of green cover crops and crop establishment is recommended to allow decay of the green cover crop plant material. Also freshly in-corporated plant material can encourage certain plant diseases such as damping off.

Cover crops are used to decrease runoff, erosion, and leaching between crop-ping seasons. A single crop can serve both purposes, so the terms *green manure* and *cover crops* are often used interchangeably.

Legumes   Certain legumes (members of the Fabaceae family) are usually used as green manures because of their ability to host nitrogen-fixing bacteria in root nodules. The bacteria provide the plant with readily available nitrogen by fixing at-mospheric nitrogen. In return, the plant provides the bacteria with energy in the form of carbon. The amount of nitrogen fixed in legume nodules that will be avail-able to a succeeding crop depends on the legume species, variety, age, growth, and soil conditions (Table 18.5).

Composts   Composts have relatively low amounts of nitrogen, phosphorus, and potassium as compared to inorganic fertilizers. Compost can be made from domes-tic sources (such as food scraps) or from municipal or commercial sources. Munic-ipal sources of compost include yard wastes, and common types of commercial

TABLE 18.5   *Some residual nitrogen contributions from legumes*

Previous crop	% stand	Nitrogen credit (kg/ha)
Alfalfa		
First year after alfalfa	> 50	180–270
	25–49	130–180
	< 25	90
Second year after alfalfa	> 50	130
Red clover and trefoil		
First year after clover or trefoil	> 50	130–200
	25–49	110–130
	< 25	90

Growing a legume in a rotation preceding a nitrogen-requiring crop may result in residual N in the soil that can be utilized by the following crop. Nitrogen credit is an estimate of this residual nitrogen. This value depends on soil type and other environmental factors of the location. The credit can be deducted from the overall N fertilizer recommendation for a crop.

Source: Modified from Penn State Cooperative Extension Service Publication, Agronomy Guide, 2002 (Table1-2-8).

composts include chicken and turkey litter. The advantages of compost are that the composting process allows the safe use of materials such as sawdust that normally depletes soil nitrogen if applied without undergoing the composting process, and that composting can reduce pathogen populations within the materials when temperatures reach 65° C and can destroy weed seeds when temperatures reach 80° C. The disadvantages of compost are that it is generally not as effective as raw organic matter in improving soil structure and can lose nitrogen as NH_3 when the compost is turned.

Timing and Application of Fertilizers   *Preplant Fertilizer or Broadcast Application*   A preplant fertilizer or a broadcast application uniformly applies fertilizer with a spreader prior to planting. The fertilizer is broadcast or spread on the soil surface and then incorporated with a plow, disk, or power tiller. Fertilizer is generally not broadcast to and incorporated into a field too far in advance of planting because considerable loss of nitrogen and potassium could occur as a result of leaching due to rains that may occur before the field is planted.

Band application   Band application is often used with direct seeded crops such as beans and corn. In banding, fertilizer is placed 5 cm to the side and 5 cm below the level of the seed in a row at planting time. This method is especially effective when placing phosphorus in either cold or calcareous soils or soils high in hydrous oxides of iron and aluminum. Because banding places fertilizer where it is needed and likely to be taken up by the plant, smaller amounts of nitrogen and potassium may be lost due to leaching.

Starter solution   Starter fertilizer solutions are usually relatively high in P and K and low in N and are commonly used with transplanted crops. Water is often applied to the recently planted transplants and a small amount of fertilizer can be included in this water to encourage early plant growth and development. Starter fertilizer can also be applied at planting in a band to the side and below the seed or transplant. Although starter fertilizers provide small amounts of the actual fertilizer requirements of the plants, they are important in establishing crops in cool, damp soils. Transplanted crops produce little above-ground growth during the first 2 weeks after planting, often because they undergo shock at transplanting and have a limited root system. An application of starter solution provides phosphorus as well as nitrogen and potassium directly to the plant roots and could encourage early growth and increase the early yield, although it rarely affects the total yield.

Side dressing   Although fertilizer is needed for early growth, the greatest quantities of nutrients for many annual crops are assimilated into the plants during the second half of their growing period. Also, much of the fertilizer applied at planting time may be leached out of sandy soils before being utilized by the plant. To correct this problem, a side dressing of fertilizer (usually nitrogen) can be applied 15 to 25 cm from the base of the plant several weeks after planting and then lightly incorporated into the soil. N and K are the nutrients of the most concern because of their potential to be leached through the soil. Side dressings of fertilizer are often used on sandy soils or with long-season crops.

Foliar applications or foliar sprays   Foliar applications are used to rapidly enhance the green color in plants. Nutrient uptake by plant leaves is most effective when nutrient solutions remain on the leaves as a thin film. This is often accomplished by supplementing the nutrient solution with surfactant chemicals that remove or reduce surface tension.

Nitrogen foliar fertilizers usually contain low-biuret urea because uncharged urea molecules more easily penetrate the cuticle of the leaf. Foliar sprays are widely used to furnish trace elements to crops. Although foliar applications are usually considered as a last resort for correcting nutrient deficiency problems in some crops, for citrus and some other crops it is one of the most useful methods of micronutrient application (especially in cool soils).

Slow-release fertilizers   Slow-release fertilizers include sulfur-coated urea or isobutylidene-diurea that supply a portion of the N that the plant requires. Although they are more expensive, slow-release fertilizers can be useful in reducing fertilizer nutrient leaching and in supplying adequate fertilizer for long-term crops. In ornamentals, encapsulated forms (such as Osmocote) are important.

Nutrient/chemical application through irrigation systems   Most plant nutrients can be applied through trickle or sprinkler irrigation systems using a siphon injector or with a metering pump. Nitrogen is soluble and moves readily into the soil. Although phosphorous and potassium move more slowly through the soil than nitrogen, their movement is greater with trickle irrigation than with other types of application techniques. Micronutrients are often insoluble in water but can become soluble if they are applied with chelates, thereby reducing the possibility of clogging of the irrigation emitters. Clogging can be a problem when insoluble or slightly soluble compounds lodge in the small openings of trickle lines.

Chemicals such as fungicides or insecticides can be applied through a trickle system in the same manner as nutrients. However, only soil-borne organisms are affected by the application. Regulations regarding applications of chemicals may prohibit application through an irrigation system.

Soil and Tissue Analysis

Soil analysis is the determination of the nutrient content in a soil sample from the root zone. Soil analysis quantifies the levels of nutrients potentially available to the plant roots, but does not attempt to evaluate uptake condition and amount of nutrients actually absorbed by plants. Additional information is made available through tissue analysis.

Regular tissue analysis (determination of nutrient content in plants or plant parts) can be a vital management tool for growers. Plant tissue analysis has been applied to many different plant tissues and organs, including leaves of different ages and positions, stems, roots, seeds, fruits, grain, and sap residue. Growers may depend on soil and tissue analysis to make fertilizer decisions or to determine the effectiveness of a fertilizer management program. Timely tissue sampling, along with soil or media testing and a general knowledge of the specific crop's needs can also be used to diagnose suspected nutritional problems. Plant analysis provides a more accurate indicator of nutrient disorder than diagnosing by visual appearance or presence of injury.

Rotations

Crop rotation is the growing of two or more crops in a sequence, on the same land, during a period of time. Crop rotations increase soil sanitation, fertility, and structure. Proper crop rotations can also reduce disease and insect pressures. Field-grown crops often are rotated within the year (in warm areas where multiple crops are grown) or either every year or every other year (in cooler areas) depending on the climate and previous field history.

A variety of rotations can be used as long as the ensuing planted crops are not susceptible to the same root diseases and insects as the crop that is currently being grown. Certain crops and families of crops can be grouped together according to their susceptibility to the same diseases. Crops within individual groups are rotated with crops from other groups.

Salt Injury

Salt injury can be observed on field-grown plants when fertilizers and/or irrigation have been used for many years without adequate soil drainage for necessary nutrient leaching to occur. As water flows through the soil profile, it dissolves salts of sodium, calcium, and magnesium and other substances. If these excess dissolved salts are not drained from a water basin as quickly as they enter, the water can evaporate and allow salt to build up in the soil (Fig. 18.8) and groundwater. Salt content of irrigation water may also be a major contributor to salinity problems. The presence of salts, especially calcium, magnesium, and sodium carbonates, results in a preponderance of hydroxyl ions over hydrogen ions and the soil pH is often alkaline with pH values of 9 to 10.

Salt injury on bushes and trees can also occur as a result of the use of deicing salts in areas where snow and ice is present during the winter (Fig. 18.9). Bushes and trees near intersections or on major streets where greater amounts of salt are applied, or low areas where runoff water collects, will often show the most injury. Symptoms can vary from marginal leaf browning and yellowing of leaves to branch dieback. Windblown spray from automobiles causes damage to the lower branches of the bush or tree, whereas salt uptake by roots from runoff water is often evident in the upper portion of the bush or tree. The only solution to deicing salt damage

FIGURE 18.8 Salt build up (illustrated by the whitish layer) in California soils.

FIGURE 18.9 Road deicing salt that is splashed onto the foliage or is absorbed through the roots will cause a browning of the tree, especially on the side facing the roadway. Courtesy Dr. Larry Kuhns, Penn State University.

FIGURE 18.10   Field site contaminated with heavy metals.
Source: "Phytoremediation: Using Plants To Clean Up Soils" published in the June 2000 issue of *Agricultural Research* magazine by Rufus L. Chaney, ARS, USDA.

to bushes and trees is to plant salt-tolerant species instead of more susceptible species or to reduce the amount of deicing salts used.

Phytoremediation

Soils contaminated with heavy or toxic metals or other pollutants (Fig. 18.10) pose major environmental, agricultural, and human health concerns worldwide. Current engineering-based technologies used to clean up soils, such as the removal of topsoil for storage in landfills, are often very costly and can dramatically disturb the landscape. These problems may be partially solved by using phytoremediation (the use of plants to clean up contaminated soils) (Fig. 18.11).

Certain plant species known as hyperaccumulators have the ability to extract elements from the soil and concentrate them in the stems, shoots, and leaves. An example of a hyperaccumulator is *Thlaspi caerulescens*, commonly known as alpine pennycress. *Thlaspi* is a small, weedy member of the cabbage family and thrives on soils having high levels of zinc and cadmium.

Hyperaccumulators possess genes that govern processes that can increase the solubility of metals in the soil surrounding the roots as well as the transport proteins that move metals into root cells. From there, the metals enter the plant's vascular system for further transport to other parts of the plant and are ultimately deposited in leaf cells. *Thlaspi* can accumulate up to 30,000 ppm zinc and 1,500 ppm cadmium in its shoots, while exhibiting few or no toxic symptoms. A normal plant can be "poisoned" with as little as 1,000 ppm of zinc or 20 to 50 ppm of cadmium in its shoots. The metals can be extracted from the roots and shoots at harvest, if desired.

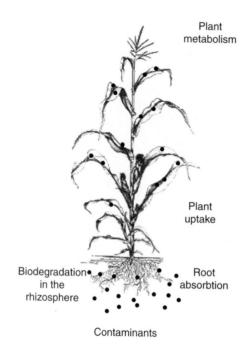

Plant
metabolism

Plant
uptake

Biodegradation
in the
rhizosphere

Root
absorbtion

Contaminants

FIGURE 18.11   Phytoremediation uses plants to remove heavy or toxic metals in soils (as indicated by the dots in the diagram). Several processes can act together to remove or stabilize contaminants in planted soils, including biodegradation of contaminants in the root zone, plant uptake followed by transformation or volatilization, and contaminant stabilization/immobilization.

Summary

Plant growth requires 16 essential elements (nutrients), and each of these has at least one specific role in plant growth so that plants fail to grow and reproduce normally in the absence of the element. These essential elements are often divided into macronutrients or micronutrients, depending on the relative amount in which each nutrient is required by plants. The availability of elements from a soil for plant uptake is often further determined by pH.

The essential elements of plants tend to circulate in characteristic circular paths from environment to organisms back to the environment. These are often called nutrient cycles and the more important nutrient cycles for plants include the carbon cycle, the nitrogen cycle, the phosphorus cycle, and the potassium cycle.

When all the essential elements are provided to the plant in a suitable general nutrient concentration range, normal growth is often achieved. When plant growth is reduced by 10% because of a shortage of an element, a critical concentration occurs (this is borderline between elemental sufficiency and deficiency). Plants often begin to show deficiency symptoms when elemental concentrations fall below the critical concentration range. Essential elements can also be detrimental to the plant when they are in excess and become toxic.

Crop producers can affect elemental availability from a soil by monitoring and adjusting the soil pH to a level that provides optimum nutrient availability (generally between 6 and 7) and by nutrient replacement. Nutrient replacement is often accomplished by using synthetic fertilizers, manures, or cover crops. Phytoremediation is the use nutrient uptake of plants (often cereals) to clean up soils contaminated with heavy or toxic metals or other pollutants.

Review Questions

1. List five general nutrient deficiency symptoms in plants.
2. List the essential elements for plant growth (include the abbreviation for the each element and the name of the element).
   a. Macronutrients
   b. Micronutrients
3. What are the disadvantages of using animal manures as fertilizers?
4. List and explain the four ranges of critical nutrient concentrations.
5. List the function in the plant of the following elements:
   a. C
   b. P
   c. K
   d. S
   e. Ca
6. How does soil pH affect soil nutrient availability?
7. What does the "3-8-18" mean on a fertilizer label?
8. When and why are starter fertilizers used in field production of crops?
9. What are crop removal values? How are they determined, and how can they be used?
10. Why is fertilizer often necessary in agriculture?

Selected References

Becker, H. 2000. Using plants to clean up soils. *Agricultural Research* 48:4–9.

Bennett, W. F. 1993. Plant nutrient utilization and diagnosing plant symptoms. In *Nutrient deficiencies & toxicities in crop plants*, ed. W. F. Bennett, 1–7. St. Paul, MN: APS Press.

Brady, N. C. 1974. *The nature and properties of soils*. 8th ed. New York: Macmillan.

Cavigelli, M. A., S. R. Deming, L. K. Probyn, and R. R. Harwood (eds.). 1998. *Michigan field crop ecology: Managing biological processes for productivity and environmental quality*. Michigan State University Extension Bulletin E-2646.

Decoteau, D. R., D. Ranwala, M. J. McMahon, and S. B. Wilson. 1995. *The lettuce growing handbook: Botany, field procedures, growing problems, and postharvest handling*. Oak Brook, IL: McDonald's International.

Gerber, J. M., J. W. Courter, H. J. Hopen, B. J. Jacobsen, and R. Randell. 1985. *Vegetable production handbook for fresh market growers*. University of Illinois Extension Publication.

Hochmuth, G. J. 1996. *Soil and fertilizer management for vegetable production in Florida.* University of Florida Cooperative Extension Service Circular HS-711.

Johnson, K. E. 1993. *Crop rotation in vegetable production.* University of Tennessee, Agricultural Extension Service. Vegetable and Small Fruit Facts #3.

Maynard, D. N., and G. J. Hochmuth. 1997. *Knott's handbook for vegetable growers.* 4th ed. New York: John Wiley.

Mengel, K., and E. A. Kirby. 1987. *Principles of plant nutrition.* International Potash Institute, Bern.

Miller, G. T., Jr. 1975. *Living in the environment: Concepts, problems, and alternatives.* Belmont, CA: Wadsworth.

Parnes, R. 1990. *Fertile soil: A grower's guide to organic and inorganic fertilizers.* Davis, CA: AgAccess.

Selected Internet Sites

http://minerals.usgs.gov/ Current information on the occurrence, quality, quantity, and availability of mineral resources. U.S. Geological Survey.

http://www.nrcs.usda.gov/technical/nutrient.html Nutrient management information, including ecological sciences nutrient management, animal waste management, manure management planner, nutrient and pest management, and phosphorus index. Natural Resources Conservation Service, USDA.

19

Soil Organisms

A large number of organisms live in the soil, and their activity strongly influences soil properties and resulting plant growth (Fig. 19.1). A "healthy" soil often supports a large number of organisms. These may include macroscopic organisms (such as earthworms) or microorganisms (such as fungus, actinomycetes, and bacteria) (Fig. 19.2). Herbivores are the soil organisms that feed on plants and are involved in degradation of higher plant tissue into organic matter. Examples of herbivores include parasitic nematodes, snails, slugs, and larvae of some insects.

Soil organic matter is composed of organic compounds. These organic compounds can be divided into three forms: water-soluble fulvic acid, humic acid that is soluble at high pH but not at low pH, and water-insoluble humin. Fulvic acid is readily used by microorganisms for building tissue and for energy. Humic acid can also be used by microorganisms, but it contains compounds that cannot be used by all organisms. Humin is very resistant to use by the microorganisms.

This chapter describes the more common and important soil microorganisms and macroorganisms, discusses some of the effects these organisms have on plant growth, and highlights some effects that agricultural systems have on soil organisms.

Background Information

Microorganisms

Nematodes   Nematodes, also called threadworms or eelworms, reside in most soils. They are round in cross section and spindle-shaped and are mostly microscopic and slender. The caudal (or tail) end of the nematode is usually pointed. Nematodes reproduce by eggs that are deposited in the soil or in plant tissue or retained inside the female's body in a jellylike mass, which becomes a tough protective capsule called a cyst

Functions of Soil Organisms

Type of Soil Organism		Major Functions
Photosynthesizers	•Plants •Algae •Bacteria	**Capture energy** • Use solar energy to fix CO_2. • Add organic matter to soil (biomass such as dead cells, plant litter, and secondary metabolites).
Decomposers	•Bacteria •Fungi	**Break down residue** • Immobilize (retain) nutrients in their biomass. • Create new organic compounds (cell constituents, waste products) that are sources of energy and nutrients for other organisms. • Produce compounds that help bind soil into aggregates. • Bind soil aggregates with fungal hyphae. • Nitrifying and denitrifying bacteria convert forms of nitrogen. • Compete with or inhibit disease-causing organisms.
Mutualists	•Bacteria •Fungi	**Enhance plant growth** • Protect plant roots from disease-causing organisms. • Some bacteria fix N_2. • Some fungi form mycorrhizal associations with roots and deliver nutrients (such as P) and water to the plant.
Pathogens Parasites	•Bacteria •Fungi •Nematodes •Microarthropods	**Promote disease** • Consume roots and other plant parts, causing disease. • Parasitize nematodes or insects, including disease-causing organisms.
Root-feeders	•Nematodes •Microarthropods (e.g., cutworm, weevil larvae, & symphylans)	**Consume plant roots** • Potentially cause significant crop yield losses.
Baterial-feeders	•Protozoa •Nematodes	**Graze** • Release plant available nitrogen (NH_4^+) and other nutrients when feeding on bacteria. • Control many root-feeding or disease-causing pests. • Stimulate and control the activity of bacterial populations.
Fungal-feeders	•Nematodes •Microarthropods	**Graze** • Release plant available nitrogen (NH_4^+) and other nutrients when feeding on fungi. • Control many root-feeding or disease-causing pests. • Stimulate and control the activity of fungal populations.
Shredders	•Earthworms •Microarthropods	**Break down residue and enhance soil structure** • Shred plant litter as they feed on bacteria and fungi. • Provide habitat for bacteria in their guts and fecal pellets. • Enhance soil structure as they produce fecal pellets and burrow through soil.
Higher-level predators	•Nematodes-feeding nematodes •Larger arthropods, mice, voles, shrews, birds, other above-ground animals	**Control populations** • Control the populations of lower trophic-level predators. • Larger organisms improve soil structure by burrowing and by passing soil through their guts. • Larger organisms carry smaller organisms long distances.

FIGURE 19.1   Soil organisms and their functions.

Source: Soil Quality Institute, Natural Resources Conservation Service, USDA. Tugel, A. J., and A. M. Lewandowski, eds. (February 2001, last update). *Soil Biology Primer* [online]. Available: http://soils.usda.gov/sqi/soil_quality/soil_biology/soil_biology_primer.html

or gall (Fig. 19.3). Under ideal conditions (temperature between 27°C and 30°C), many of the types of nematodes complete their life cycle in as little as 4 weeks.

Nematodes are generally characterized on the basis of their food demands. These include those that live on decaying organic matter; those that are predatory on other nematodes, bacteria, algae, and protozoa; and those that are parasitic, attacking the roots of higher plants to pass at least a part of their life cycle attached to such tissue. The parasitic nematodes tend to have the most deleterious effects on crops, though the other groups are often more numerous in the average soil.

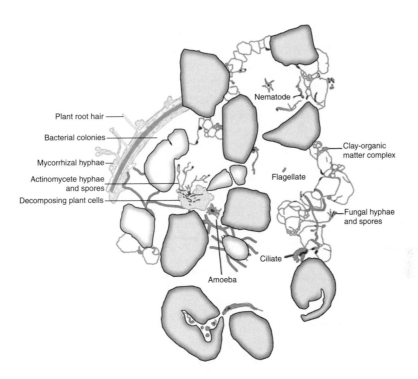

FIGURE 19.2   Many soil organisms live in microscale environments within and between soil particles.
Source: Soil Quality Institute, Natural Resources Conservation Service, USDA. Tugel, A. J., and A. M. Lewandowski, eds. (February 2001, last update). *Soil Biology Primer* [online]. Available: http://soils.usda.gov/sqi/soil_quality/soil_biology/soil_biology_primer.html (Credit: S. Rose and E. T. Elliott).

Plants seriously affected by nematodes appear stunted and wilted (Fig. 19.4). They may also exhibit nonspecific symptoms common to damaged root systems, which can easily be confused with damage due to disease, salinity, drought, and so on. Leaf chlorosis and other symptoms of nutrient deficiency may also be present. Plants exhibiting stunted growth due to nematodes usually occur in patches of the planted field rather than across the entire field.

Protozoa   Protozoa are the simplest forms of animal life and are one-celled organisms that are considerably larger than bacteria. Soil protozoa are divided into three groups: amoebae, ciliates, and flagellates (the most numerous in soils). Protozoa are the most varied and numerous organisms of the microanimal populations in the soil. More than 250 species have been isolated, with sometimes as many as 40 to 50 in a single sample of soil.

The number of protozoa in a soil is affected by aeration because available food supply is usually confined to the surface horizons. Their numbers are usually highest

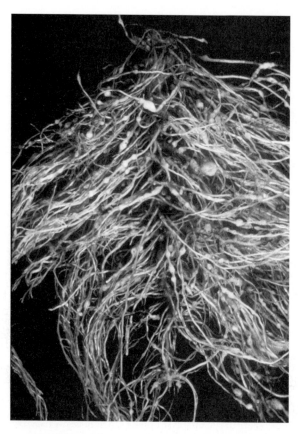

FIGURE 19.3 Nematode-infected roots exhibiting cysts (small circular masses on roots).

FIGURE 19.4 Cucumber plants infected with nematodes (smaller plants on the left) and without nematodes (larger plants on the right).

in spring and fall. Protozoa ingest bacteria and to a lesser extent other microflora. It is thought that protozoan predation on bacteria may hasten the turnover of readily available nutrients.

Algae   Most algae are chlorophyll-containing organisms that are capable of photosynthesis. As a result, soil algae are usually located near the soil surface so that they may be exposed to sunlight. A few forms, though, obtain their energy largely from organic matter. These forms live within and below the surface horizon. Algae benefit the soil by contributing organic matter to the soil.

Several hundred species of algae have been isolated from soils, and these can be grouped into three general groups: the blue-green algae, the green algae, and diatoms. Grasslands seem especially favorable for the blue-green forms, whereas old gardens favor diatoms. Populations of all types are stimulated by application of farm manure.

Fungi   Soil fungi are important soil constituents, with over 690 species of soil fungi identified, representing approximately 170 genera. Fungi contain no chlorophyll and depend on organic matter in the soil for their energy and carbon. Their importance is in the chemical transformation of soil constituents and as pathogens. Important groups of fungi for plants include molds and mycorrhiza.

Molds   Molds are distinctly filamentous, microscopic, or semimacroscopic. They respond to soil aeration with increasing numbers and are important in decomposing organic residues. The greatest numbers of molds are found in surface layers, where organic matter is ample and aeration is adequate. The four most common genera of molds are *Penicillium, Mucor, Fusarium,* and *Aspergillus.*

Mycorrhiza   Mycorrhizae are symbiotic (mutually beneficial) associations of certain fungi with the roots of higher plants. Specific fungi grow on and vigorously invade portions of the root that are primarily responsible for nutrient absorption (Fig. 19.5). The term *mycorrhiza* means "fungus-root" and is used to denote these specific associations of roots and fungi.

Eighty-three percent of dicots, 79% of monocots, and essentially all gymnosperms regularly form mycorrhizal associations. Mycorrhizae are typically absent from roots in very dry, saline, or flooded soils, or where soil fertility is extreme (either high or low). An additional key factor in the extent of mycorrhizal associations with a plant is the nutritional status of the host plant. Deficiency of a nutrient such as phosphorus tends to promote mycorrhizal associations, whereas plants under nutrient-rich conditions tend to suppress these associations.

Types of mycorrhizal fungi   The physical appearance or morphology of mycorrhizae varies among plant species, and each plant species tends to have characteristic groups of fungi capable of producing mycorrhizae. On the basis of their morphology, these associations are divided into two major groups: ectomycorrhizae and endomycorrhizae. Of the two, endomycorrhizae are the most common, but ectomycorrhizae are formed on some very important families of forest trees.

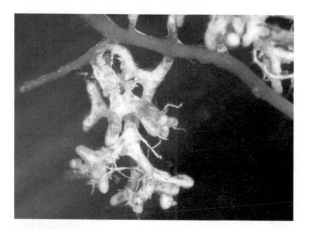

FIGURE 19.5   Ectomycorrhizae are important for nutrient absorption by tree and grape roots. The fungus does not actually invade root cells but forms a sheath that penetrates between plant cells. The sheath in this photo is white, but they may be black, orange, pink, or yellow.
Courtesy of Dr. Roger Koide, Penn State University.

Ectomycorrhizae   Ectomycorrhizae are formed by fungi belonging to the higher Basidiomycetes (mushrooms and puffballs), Ascomycetes (cup fungi and truffles), and Phycomycetes in the family Endogonaceae. The host plants of these fungi are predominantly trees such as pine, hemlock, spruce, fir, oak, birch, beech, eucalyptus, willow, and poplar. Many species of fungi may be involved in the ectomycorrhizal association of a forest, a single tree species, an individual tree seedling, or even a small segment of lateral root. As many as three species of fungi have been isolated from an individual ectomycorrhizal root cluster.

Ectomycorrhizal infection is initiated from spores (reproductive structures) or hyphae (vegetative growth structures) of the fungi that are in the soil around feeder roots. Growth of the mycorrhizal fungi on the surface of short roots is stimulated by exudates from the roots. Fungal mycelia (strands of hyphae growing together) grow over the feeder root surfaces and form an external mantle or sheath (Fig. 19.6, right). Following mantle development, hyphae grow intercellularly, or between the cells, forming a network of hyphae (called the Hartig net) around the root cortical cells. Physical or chemical properties of roots restrict the hyphae of all mycorrhizal fungi to the cortex and meristematic (actively dividing) cells of the root tip. The exact mechanism for this resistance to hyphal penetration by other parts of the plant is not known. The Hartig net, which may completely replace the tissue between cortical cells, is the major distinguishing feature of ectomycorrhizae.

Endomycorrhizae   The endomycorrhizal or vesicular-arbuscular (VA) mycorrhizal fungi are the most widespread and important root symbionts. They are found throughout the world in both agricultural and forest soils. They occur on most families of hardwood and on some conifer tree species and on many agronomic and horticultural crops, such as corn, onion, red clover, and strawberries. Trees that form

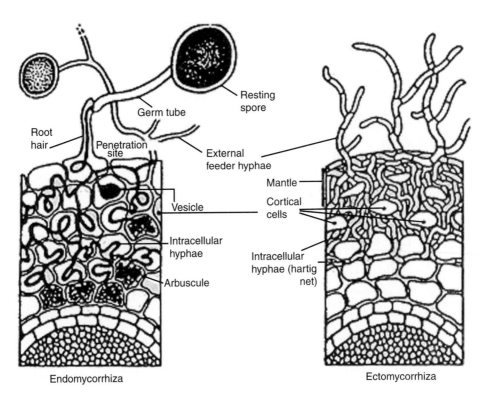

Endomycorrhiza

Ectomycorrhiza

FIGURE 19.6   Diagrams of a root infected with endotrophic mycorrhizal fungi (left) and ectotrophic mycorrhizal fungi (right).
Source: USDA Forest Service, Northeastern area.

endomycorrhizal associations include maples, elms, ash, walnut, sweetgum, yellow poplar, sycamore, cottonwood, black locust, and willow.

The endomycorrhizal fungi invade the cortex, but not the endodermis or stele of the feeder roots of plants (Fig. 19.6, left). The colonization does not alter root morphology, and the sheath of fungus mycelium common to ectomycorrhizae is lacking. Under a microscope, endomycorrhizae are diagnosed by the presence of vesicles (terminal, spherical structures that contained oil droplets) and arbuscules (complex structures formed by repeated dichotomous branching of hyphae) in the cortical cells of differentially stained feeder roots. Mycelia emanate from the infected root to form a loose network in the rhizosphere and adjacent soil. The fungi are more commonly spread by growing from feeder root to feeder root and, at times, are disseminated by moving water, soil, insects, and animals.

Actinomycetes   Actinomycetes resemble molds and are similar to bacteria. They develop best in moist, well-aerated soils. Actinomycetes are sensitive to soil pH, with no growth below pH 5.0 and optimum development between 6.0 and 7.5. Actinomycetes are numerous in soils high in humus, such as old meadows or pastures. Addition of farm manure stimulates their activities, and it is thought that the aroma of fresh plowed land is probably due to actinomycetes as well as certain molds. Actinomycetes are important in the liberation of nutrients of soil organic matter as a result of decomposition.

Bacteria   Bacteria are single-celled organisms, one of the simplest and smallest forms of life, with their greatest population in surface horizons. Bacteria are involved in all of the organic transformations of the soil, as well as the three basic enzyme transformations of nitrification, sulfur oxidation, and nitrogen fixation.

Macroorganisms

Earthworms   The earthworm is probably the most important soil macroanimal, and high earthworm activity is often associated with healthy, productive soils. There are thousands of species of earthworms in the world. Those that live in the soil can be grouped into three major behavioral classes: the litter dwellers, shallow soil dwellers, and deep burrowers. The litter-dwelling worms live in the litter layer of a forest and are generally absent from agricultural fields. Typical agricultural fields may have one to five different shallow-dwelling species and one deep-burrowing species (Fig. 19.7).

The shallow-dwelling worms (known as redworms, grayworms, fishworms, and many other names) comprise many species and live primarily in the top 0.3 m of soil. The adult is usually 8 to 13 cm long. Shallow-dwelling worms typically do not

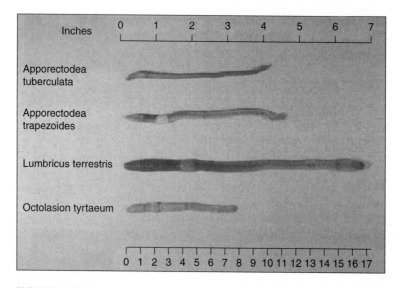

FIGURE 19.7   Examples of earthworms found in the northern midwest section of the United States.
Source: Northern Prairie Wildlife Research Center, USGS.

build permanent burrows, but instead randomly burrow throughout the topsoil, ingesting residues and mineral soil as they go. Because they do not require residues at the soil surface, they are not as sensitive to soil surface debris management as are the nightcrawlers. However, surface-dwelling worms are affected by the amount of surface mulch because of the impact of the mulch on soil temperature and moisture extremes.

The deep-burrowing worms ("nightcrawlers") build large, vertical, permanent burrows that may extend 1.5 to 1.8 m or more deep (Fig. 19.8). They move plant residues down into the mouth of their burrow, where the residues soften and can be eaten at a later time. Nightcrawlers construct middens over the mouth of their burrows (Fig. 19.9). Middens are a mixture of plant residues and castings (worm feces) and probably serve as protection as well as a food reserve. Because nightcrawlers require residue at the surface to move down their burrows, they are not often found in agricultural fields where there is no surface residue cover. A common specie of nightcrawler is *Lumbricus terrestris*. The length of a nightcrawler is 10 to 20 cm or more.

Earthworms prefer a moist habitat that is reasonably well aerated, and they are typically found in medium-textured upland soils where moisture capacity is high. Earthworms also must have organic matter as a source of food. Consequently they thrive where manure or plant residue is added to the soil.

FIGURE 19.8   Nightcrawlers build large, vertical, permanent burrows that may extend a couple meters deep.
Source: Purdue Cooperative Extension Service Publication AY279, *Earthworms and Crop Management* by Eileen J. Kladivko, 1993.

FIGURE 19.9   A mound of organic middens was moved aside to expose the entrance to a burrow. *Lumbricus terrestris* will quickly replug its burrow if its mound is removed.
Source: North Appalachian Experimental Watershed, USDA, Agricultural Research Service, Coshocton, Ohio.

Earthworms are seasonal in their activity. The shallow-dwellers are active in spring and fall but generally enter a resting state in summer and winter. The seasonal effects on earthworm activity may be due in part to the soil temperature effects on earthworm numbers and distribution. For example, a temperature of 10°C appears optimum for *Lumbricus,* with their numbers decreasing with increasing and decreasing temperatures. As the soil starts to heat up and dry out in late spring, the shallow-dwellers burrow a little deeper (perhaps 4.5 cm), curl up in a ball, and secrete a mucus to try to keep from drying out. They spend most of the summer in that state. In fall, when the soil starts to cool and become wetter, they become active again, but then often enter into a hibernation state for the winter. The nightcrawlers also tend to be more active in spring and fall, but they often do not go into a complete resting state in summer or winter because they can retreat to the bottom of their burrows during extremes of heat or cold.

Earthworms can penetrate as deep as 1 to 2 m in temperate regions. In barren soils a sudden heavy frost in the fall may kill the organisms before they can move lower in the profile. Soil cover is important in maintaining a high earthworm population (especially under such circumstances).

Roots of Higher Plants   Higher plants are the primary producers of organic matter and the storer of the sun's energy. Their roots grow and often die in the soil and as a result supply the soil microorganisms with food and energy. Living roots physically

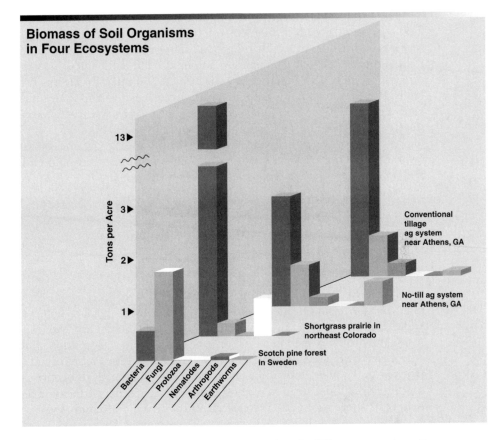

FIGURE 19.10   The relative biomass of soil organisms in different ecosystems (1 ton per acre = 2.24 metric tons per hectare).
Source: Soil Quality Institute, Natural Resources Conservation Service, USDA. Tugel, A. J., and A. M. Lewandowski, eds. (February 2001, last update). *Soil Biology Primer* [online]. Available: http://statlab.iastate.edu/survey/SQI/soil_biology_primer.htm

modify the soil as they penetrate through cracks and create new soil openings. Tiny channels are increased in size as the root swells and grows. Roots also provide a mass of living organic matter, which helps stabilize soil aggregates. As roots decompose, they provide building material for humus not only in the top few centimeters, but also in greater soil depths.

Comparative Organism Activity

Microorganisms are generally found in greater numbers in the soil than the macroorganisms (Fig. 19.10). Microorganisms are responsible for 60% to 80% of the total soil metabolism. They destroy plant residues, function in the digestive tracts of animals, and eventually decompose the dead bodies of all organisms. Soil humus is the end product of these activities.

The specific microorganisms inhabiting soils depend on many factors. Climate and vegetation greatly influence which organisms are dominant. For example, species composition in a grass prairie will differ from that in a pine forest area. Soil temperature, acidity, and moisture relations also affect the microorganisms. Vegetation in forests generally supports more diverse soil microorganisms than do grasslands. However, grasslands are metabolically more active and their total weight per acre is greater. Cultivated fields as compared to virgin soils are generally lower in numbers and weights of soil organisms, especially the soil fauna.

Macroorganisms are also important to the soil. By living in the soil many animals favorably affect its physical condition. Rodents pulverize, mix, and granulate soils, and incorporate organic molecules into lower horizons.

Effects on Crops

Microorganisms

Benefits Derived from Mycorrhizae   Infection of tree seedling roots with certain types of ectomycorrhizae can increase seedling growth by as much as 400% (Fig. 19.11). Some tree species require the mycorrhizae on their root systems to develop "normally." The observed benefit of ectomycorrhizae in the growth and development of trees has been ascribed to several factors. Among these factors are (1) increase in nutrient and water absorption by virtue of an increased absorptive surface area resulting from the formation of short roots and by mycelia permeating the soil in the vicinity of short roots; (2) increase in nutrient mineralization through biological

FIGURE 19.11   Ectomycorrhizae infections of tree seedling roots can influence seedling growth. Redwood seedlings without (left) and with (right) mycorrhizae.
Source: U.S. National Science & Technology Center, Bureau of Land Management.
Photo: Mike Amaranthus, USDA.

weathering; and (3) increase in feeder root longevity by providing a biological deterrent to root infection by soil pathogens. In general, mycorrhizae increase the capacity of infected plants to absorb nutrients, which can be especially important on infertile or adverse sites. Carbon compounds synthesized in the green tissue of the host plant not only nourish the host itself but are a source of carbon for fungal mycelia. In turn, soil-derived nutrients absorbed by the mycelia in the soil pass into host tissue. Ectomycorrhizae are able to absorb and accumulate various elements such as nitrogen, phosphorous, potassium, and calcium in the fungus mantle and then translocate these elements to host plant tissue.

Ectomycorrhizal fungi have been reported to afford protection to delicate root tissue from attack from pathogenic fungi. This protection apparently results from the fungal mantle serving as a physical barrier to infection. Even without the mantle, root cortex cells surrounded by the Hartig net also are resistant to pathogens. In addition to a physical barrier, there is an antibiotic mechanism of resistance. This resistance is derived from a chemical substance identified as diatretyne nitrile. Evidence also suggests that although the symbiotic fungus supplies the host plant with inorganic nutrients from the soil, it may also provide growth hormones, including auxins, cytokinins, gibberellins, and growth-regulating B vitamins.

Endomycorrhizae significantly increase the growth of plants in soils deficient in readily available phosphorous. The main effect of plant response to these symbionts is increased efficiency of nutrient uptake. Thus, if plants are colonized with the appropriate endomycorrhizal fungi, the estimate of production potential of a soil and its fertilizer requirement might change. However, the importance of mineral solubilization by mycorrhizal fungi to the nutrition of the host plant is not well understood. There seems to be little evidence that VA mycorrhizae can exploit less soluble forms of phosphates; rather they increase the efficiency of available phosphorous absorbed from the soil.

Nitrogen–Fixing Bacteria   Atmospheric nitrogen (N_2) cannot be used directly by plants and must be in combined form (NH_4^+, NO_3^-) for plant uptake. The process by which N_2 is reduced to NH_4^+ is called nitrogen fixation. Certain soil bacteria are needed for nitrogen fixation. Most plants that form symbiotic relationships with nitrogen-fixing bacteria are in the legume family.

The legume-colonizing bacteria are species of the genus *Rhizobium*. A different *Rhizobium* is generally effective on each legume species. The bacteria live in root nodules, which is the result of the irritation of the root surface. These bacteria invade the plant root, use the carbohydrates of their host plants as the energy source, fix the nitrogen, and pass part of the nitrogen to the infected host.

The amount of nitrogen fixed by the *Rhizobium* depends on the aeration, drainage, moisture, pH, and amount of active calcium of the soil. The nitrogen fixed may be used by the host plant, passed into the soil by sloughing off of roots and nodules, or made available to the soil when the crop dies or is killed and the plant material is incorporated into the soil.

A number of nonlegume crops also form symbiotic relations with bacteria and fix atmospheric nitrogen. These species are mostly trees and shrubs that are often found under conditions of low nitrogen. These plants include members of the genus

Alnus (alder), *Myrica* (including the big myrtle), *Shepherdia, Hippophae, Ceanothus, Elaeagnus,* and *Casuarina.* In general, the nitrogen-fixing capability of these plants has been less studied and as a result is less understood.

Plant Diseases   Plants growing in fields are under constant attack by many soil organisms, and the soil is a natural reservoir for many plant pathogens. If conditions become favorable for infection and disease, the plant often suffers. Once a soil is infested with disease organisms it is likely to remain that way for an extended period of time. Some injurious organisms disappear in a few years if host plants are not grown, whereas others are able to maintain an existence on almost any organic substrate. Infestation occurs easily and organisms from infested fields may be carried on plants, farm machinery, or animals (including humans). Also manure and erosion can lead to infestations.

Other Microorganism Effects   Plant roots require a certain amount of O_2 for normal growth and nutrient uptake. Under conditions of somewhat restricted drainage, active soil microflora may deplete the already limited oxygen supply of the soil. Oxidized forms of several elements, including nitrogen, sulfur, iron, and manganese will be further reduced by microbial action. Therefore nutrient deficiencies and toxicities, both microbiologically induced, can result.

Macroorganisms

Influence of Earthworms on Soil Fertility and Crop Productivity   Earthworms have tremendous potential to affect soil fertility and productivity. During the passage through the worms, the organic matter (which serves as food for the earthworm) and mineral constituents of the soil are subjected to digestive enzymes and grinding action. Earthworm casts are the organic materials and mineral soils that have been partially decomposed during passage through the earthworm gut, converting the organic nutrients to more available forms for the plant (Fig. 19.12). Casts are higher in bacteria and organic matter, total and nitrate nitrogen, exchangeable calcium and magnesium, available phosphorus and potassium, pH and percentage base saturation, and cation exchange capacity than the soil itself.

Earthworms can have a significant impact on soil properties and processes through their feeding, casting, and burrowing activity. The worms create channels in the soil, which can aid water and air flow as well as root development. The shallow-dwelling worms create numerous small channels throughout the topsoil, which increases overall porosity. Nightcrawlers create large vertical channels, which can greatly increase water infiltration under very intense rainfall or ponded conditions. Nightcrawler channels can also aid root proliferation in the subsoil, due both to the ease of root growth in a preformed channel and the higher nutrient availability in the cast material that lines portions of the burrow.

Earthworms improve soil structure and tilth. Their casts are quite stable after initial drying. The burrowing action of the worms moves soil particles closer together

FIGURE 19.12 Earthworm casts at the soil surface.
Source: Soil and Water Management Research Unit, USDA, Agricultural Research Service, St. Paul, Minnesota.

near burrow walls, and the mucus secreted by the worms as they burrow can also help bind the soil particles together. Increased porosity, plus mixing of residues and soil, are additional ways that earthworms improve soil structure.

The mixing of organic materials and nutrients in the soil by earthworms may be an important benefit of earthworms in reduced tillage systems, especially no-till systems. The earthworms may, in effect, partially replace the work of tillage implements in mixing materials and making them available for subsequent crops. In natural ecosystems such as forests, organisms recycle the previous year's leaf litter into the soil for release of nutrients. This process can be detrimental in some northern forests where introduced species of earthworms may cause too-rapid degradation of forest litter.

Competition for Nutrients

Soil organisms can compete with plants for available nutrients. N is the soil element most vigorously contested for by soil organisms and is an important nutrient requirement for decomposers. Soil organisms may also utilize appreciable quantities of P, K, and Ca. There is also some competition for trace elements. Soil organisms usually exact their nutrient quota first and higher plants subsist on what remains.

Agricultural Technologies That Affect Soil Organisms

Some Generalizations on the Effect of Agricultural Systems on Soil Organisms

Agricultural systems tend to reduce the species diversity and total organism population of the soil (organic farming systems may increase organism populations over conventional systems) as compared to nonagricultural ecosystems (Fig. 19.13). At the same time agricultural systems can greatly increase or decrease the number of a given species depending on the situation. Adding lime, fertilizer, and manure to infertile soil will increase the activity of certain bacteria and actinomycetes. Pesticides (especially fumigants) can reduce organism number but may also increase populations of species able to use organic pesticides as a carbon source. Monoculture systems tend to reduce species numbers but may increase the organism count of remaining species.

Typical Numbers of Soil Organisms in Healthy Ecosystems

		Agricultural Soils	Prairie Soils	Forest Soils
Bacteria	Per teaspoon of soil (one gram dry)	100 million to 1 billion.	100 million to 1 billion.	100 million to 1 billion.
Fungi		Several yards. (Dominated by vesicular-arbuscular mycorrhizal (VAM) fungi).	Tens to hundreds of yards. (Dominated by vesicular-arbuscular mycorrhizal (VAM) fungi).	Several hundred yards in deciduous forests. One to forty miles in coniferous forests (dominated by ectomycorrhizal fungi).
Protozoa		Several thousand flagellates and amoebae, one hundred to several hundred ciliates.	Several thousand flagellates and amoebae, one hundred to several hundred ciliates.	Several hundred thousand amoebae, fewer flagellates.
Nematodes		Ten to twenty bacterial-feeders. A few fungal-feeders, few predatory nematodes.	Tens to several hundred.	Several hundred bacterial- and fungal-feeders. Many predatory nematodes.
Arthropods	Per square foot	Up to one hundred.	Five hundred to two thousand.	Ten to twenty-five thousand. Many more species than in agricultural soils.
Earthworms		Five to thirty. More in soils with high organic matter.	Ten to fifty. Arid or semi-arid areas may have none.	Ten to fifty in deciduous woodlands. Very few in coniferous forests.

FIGURE 19.13   Typical numbers of soil organisms in healthy ecosystems (1 yard = 0.914 meters).

Source: Soil Quality Institute, Natural Resources Conservation Service, USDA. Tugel, A. J., and A. M. Lewandowski, eds. (February 2001, last update). *Soil Biology Primer* [online]. Available: http://statlab.iastate.edu/survey/SQI/soil_biology_primer.htm

Effect of Planting New Ground

Placing either forested or grassland areas under cultivation can result in dramatic changes in the soil environment. Species of higher plants are changed and are generally less numerous in agricultural fields and the amount of plant residue is often reduced. A monoculture or even rotation in absence of weeds generally provides a much narrower range of plant materials than in natural ecosystems. Tillage of the soil and applications of lime and fertilizer present a different environment to the soil organism. Drainage and irrigation affect soil moisture and aeration and populations of soil organisms. Also, activities associated with commercial agriculture transports species from one location to another, making possible their introduction into new areas.

Disease Control by Soil Management

Prevention is considered the best defense crop producers have against plant diseases produced by soil organisms. Once established, diseases are often difficult to eradicate. Rotation of crops in the field and the absence of the host plant is often adequate for controlling some diseases. Regulation of pH is effective to a certain extent with diseases such as potato scab and club root of cabbage. Wet, cold soils favor some seed rots and seedling diseases known as damping off. Good drainage and ridging help control these diseases. Breeding of plant immunity or resistance to particular diseases has been successful in a number of other crops.

Management of Soil Pathogens   Soil-borne pathogens (or disease-producing agents) including bacteria, fungi, nematodes, insects, and some viruses can be reduced or eliminated from soil by appropriate treatments. Cultural practices that reduce the negative impact of soil pathogens include soil solarization, composting, heat, soil fumigants, and treatment of containers and equipment.

Soil solarization   Solarization, the use of the sun's energy to generate high soil temperatures, has been shown to be effective for disinfecting soil. This can be accomplished by using plastic mulch (often clear) on the soil during sunny, hot seasons. The temperature of the soil is closely monitored to ensure that it is high enough to control pests by placing a soil thermometer 10 to 15 cm deep into the soil. Planting media can also be solarized either in bags or flats covered with transparent plastic or in layers 7.5 to 22.5 cm wide sandwiched between two sheets of plastic. In warmer areas of California, soil inside black plastic sleeves can reach 70°C during solarization, equivalent to or above target temperatures for soil disinfestation by aerated steam. At these temperatures, the soil is solarized within 1 week. A double layer of plastic can increase soil temperature by up to 10°C.

Composting   Incorporating certain composted materials into potting mixes can suppress plant disease. This is especially effective if the compost has been inoculated with appropriate beneficial organisms during the curing phase of composting.

Heat   Heating the soil (often with steam) is very effective for controlling soil diseases and has the advantage over chemical treatment in that the soil can be planted immediately after cooling. Many plant pathogens are killed by short exposures to high temperatures (for 30 minutes at 60°C). These treatments kill most plant pathogens; however, some weed seeds and viruses may survive and available ammonium and magnesium may be at toxic levels to plant growth. Where weed seeds are a problem, a higher treatment temperature is often required.

Although pure steam at sea level is at 100°C, the temperature at which steam is used to treat soil is usually about 82°C because of air that is present in the steam or in the soil being treated. If air is mixed with steam, the temperature of the steam-air mixture can be closely controlled, depending on the ratio of air to steam. Some diseases, such as Rhizoctonia damping-off, are much less severe in soil that has been treated at 60°C rather than at 82°C.

Soil fumigants   The most useful soil fumigants are methyl bromide and chloropicrin. Methyl bromide is a gas at temperatures over 4°C and escapes rapidly from the soil if not applied under a gas-proof cover. Polyethylene sheeting is commonly used to confine methyl bromide, although the gas does slowly escape through polyethylene. Methyl bromide is probably the most versatile of the soil fumigants because of its ability to diffuse rapidly through the soil and kill many kinds of organisms, weeds, and seeds. Soil generally can be planted in just a few days after removal of plastic covers, although there are exceptions. A few plants such as *Allium* spp., carnations, and snapdragons are sensitive to and may be damaged by inorganic bromide that remains in the soil following fumigation. Leaching the bromide with water before planting is helpful in reducing the amount of bromide in the rooting area. Methyl bromide may be injected by chisels if the soil is covered immediately by plastic.

Chloropicrin (trichloronitromethane) is another gas that is also injected into soil and is the best fumigant for controlling *Verticillium dahliae*. It is generally combined with methyl bromide in various mixtures depending on the organisms in the soil. If used alone, a water seal may be used to confine the gas. The odor of chloropicrin is very objectionable and irritating (it is more commonly known as tear gas). Both methyl bromide and chloropicrin are restricted-use materials.

Treatment of containers and equipment   Because debris, soil, and plant material cling to containers and equipment, the used containers and equipment is thoroughly washed to remove all soil or planting mix particles. Heat treatment with steam is effective in killing the plant pathogens that adhere to containers or that are in the debris. Where steam is not available, hot water is often effective. Plastic containers are treated with hot water at temperatures that cause the minimal softening of the plastic. The minimum effective water temperature is about 60°C. Treatment time can be as short as 1 minute. Longer treatment times are more reliable because the container or equipment must reach at least 60°C for the treatment to be effective.

Sodium hypochlorite (bleach) is effective in killing some types of fungal spores and bacteria. It is effective only as a surface disinfectant because it penetrates clinging soil and plant material very poorly. As a consequence, containers and tools must be free of soil and plant material before treatment. Sodium hypochlorite is generally used as a surface disinfectant at 0.5%. Quaternary ammonia compounds are excellent bactericides and viricides and are effective in killing some kinds of fungal spores.

Summary

A soil that supports a large number of organisms is often considered a "healthy" soil. Microorganisms, generally found in greater numbers in the soil than macroorganisms, include nematodes, protozoa, algae, fungi, actinomycetes, and bacteria. An important symbiotic fungal association with roots of higher plants is the mycorrhiza. Infection of roots of certain host plants with mycorrhiza fungi has been reported to increase seedling growth, afford the plant protection from pathogenic fungi, provide the host plant with certain growth hormones, and increase plant phosphorus uptake. The two major groups of mycorrhizae based on their morphology are ectomycorrhizae (which form a network of fungal hyphae around the root cortical cells) and endomycorrhizae (in which the hyphae invade the cells of the cortex of the plant root).

Of the macroorganisms in the soil, the earthworm is probably the most important for influencing plant growth, and high earthworm activity is often associated with productive soils. The shallow-dwelling worms (known as redworms, grayworms, fishworms, among others) live primarily in the top 0.3 m of soil, and the deep-dwelling worms (also known as nightcrawlers) build large, vertical, permanent burrows that can extend 1.5 to 1.8 m deep. Earthworms prefer a well-aerated, moist habitat, and they are seasonal in their activity (often most active in the spring and summer in the more temperate climates of the United States). Earthworm casts, the organic materials and mineral soils that have been partially decomposed during passage through the earthworm gut, often contain organic nutrients that have been converted to more usable forms for the plant. The earthworms also create channels in the soil, which aid water and air flow as well as root development.

In general, agricultural systems tend to reduce species diversity in the soil. Monocultures typically provide a narrow range of plant material to serve as host plants for the soil organisms, resulting in fewer soil organism species, but larger numbers of individuals of the species that remain. Because soil organisms also function as soil-borne pathogen, their control is important to the production of many crops. Management practices to reduce the negative impact of soil pathogens include soil solarization, composting, heat, soil fumigants, and treatment of containers and equipment.

Review Questions

1. How can the roots of higher plants positively influence the growth of plants that are grown in the soil in the future?
2. What are mycorrhizae and what are the differences between ectotrophic and endotrophic mycorrhizal associations?

3. What are the three types or classes of earthworms?
4. In a methyl bromide/chloropicrin fumigant, what role does the chloropicrin serve?
5. What are the three components of organic matter?
6. T or F: Earthworms are indicative of a well-aerated, higher organic matter soil.
7. Name three management practices used to combat diseases in soil.
8. T or F: Nematodes are microscopic worms that can invade plant roots and affect water uptake into the plants.
9. What soil microorganisms give good soils an "earthy smell"?
10. What is the benefit of earthworms in a no-till system?
11. What is a herbivore?
12. T or F: A symbiotic relationship is a mutually beneficial relationship.

Selected References

Brady, N. C. 1974. *The nature and properties of soils.* 8th ed. New York: Macmillan.

Cavigelli, M. A., S. R. Deming, L. K. Probyn, and R. R. Harwood, eds. 1998. *Michigan field crop ecology: Managing biological processes for productivity and environmental quality.* Michigan State University Extension Bulletin E-2646.

Salisbury, F. B. and C. W. Ross. 1978. *Plant physiology.* 2nd ed. Belmont, CA: Wadsworth.

Taiz, L., and E. Zeigler. 1998. *Plant physiology.* 2nd ed. Sunderland, MA: Sinauer.

Selected Internet Sites

http://hortipm.tamu.edu/   Hort IPM Web site, Texas A&M University.

http://ippc.orst.edu/cicp/Vegetable/veg.htm   Internet resources on IPM on vegetables.

http://nrcs.usda.gov/technical/pest.html   Pest management technical resources, including ecological sciences pest management and national agricultural pesticide risk analysis. Natural Resources Conservation Service, USDA.

http://nysipm.cornell.edu   Integrated pest management information, Cornell University.

http://soils.usda.gov/sqi/soil_quality/soil_biology/soil_biology_primer.html   Soil biology primer with good information on soil organisms. Soil Quality Institute, USDA, Natural Resources Conservation Service.

20

Allelochemicals

Allelopathy is the production of substrates or compounds by one organism that is injurious to another organism or their progeny. The term *allelopathy* is derived from the Greek meaning "to suffer from each other." Allelochemicals are the organic compounds involved in allelopathy. The most well-known and widespread of allelopathic materials used by humans are antibiotics.

Allelochemicals become stressful only when they are toxic or when they affect the growth and development of plants in such a way as to render them more susceptible to other environmental stresses. They may also benefit the allelochemical-producing plant by reducing competition from other plants. Frequently the difference between beneficial and detrimental effects on a plant is the concentration of the allelochemical in the environment and the length of exposure of the plant to the allelochemical.

Many materials within a plant or tree may be beneficial or detrimental to other organisms if they were released. Many of these materials are not released by the plant and, thus, do not have a significant effect on the environment. A few chemicals are released by plants and do have an ecological effect. These released allelopathic chemicals can produce major changes in the survival, growth, reproduction, and behavior of other organisms. Plant allelochemicals can affect seed germination, root growth, shoot growth, symbiotic effectiveness, microorganism-based soil transformations, pathological infections, insect injury, and impacts from other environmental stresses.

This chapter illustrates allelopathy in plant systems by describing allelopathic effects that have been attributed to the black walnut tree and plants growing in the California chaparral. Discussions of the classification and chemical nature of allelochemicals, some methods of allelochemical action, and common sources of allelochemicals follow this. The chapter concludes with a brief discussion on how some agricultural plants may be using allelopathy for weed control and how allelochemicals may have a possible role in observed situations where the agricultural plant may be killing or injuring itself (autotoxicity).

Background Information

Examples

Allelochemicals and allelopathy might be best explained by describing some examples. The allelopathic nature of the black walnut tree and plants of the California chaparral are probably two of the best examples.

Black Walnut Tree (Juglone) Black walnut (*Juglans nigra* L.) is a valuable hardwood lumber tree that is also grown in the home landscape as a shade tree (Fig. 20.1) and, occasionally, for its edible nuts. The roots, leaves, and nuts of black walnut produce a substance known as juglone (5-hydroxy-alphanapthaquinone). Small amounts of juglone are released by live roots, and juglone-sensitive plants may exhibit toxicity symptoms anywhere within the area of these roots. However, greater quantities of juglone are generally present in the area immediately under the canopy, due to greater root density and the accumulation of juglone from decaying leaves and nut hulls. Because decaying roots still release juglone, toxicity can persist for some years after a tree is removed. The amount of juglone present in a soil depends on soil type, drainage, and soil microorganisms.

Other trees closely related to black walnut also produce juglone, including butternut, English walnut, pecan, shagbark hickory, and butternut hickory. However, these trees produce such limited quantities of juglone compared to black walnut that toxicity to other plants is rarely observed. In addition, not all plants are sensitive to juglone (Table 20.1). Many trees, vines, shrubs, groundcovers, annuals, and perennials will grow in close proximity to a walnut tree.

FIGURE 20.1 The black walnut tree.
Courtesy of Dr. Dennis Wolnick, Penn State University.

TABLE 20.1 *Plant sensitivity to juglone as suggested by Dana and Lerner*

Juglone Sensitivity in Plants

The following lists were compiled from published sources. They are based largely on observations of native woodlands, gardens, orchards, ornamental plantings, and forest plantations. Few plants have been experimentally tested for tolerance or sensitivity to juglone. Thus, the lists should be used for guidance, but not regarded as definitive.

Plants Observed to Be Sensitive to Juglone

Vegetables: asparagus, cabbage, eggplant, pepper, potato, rhubarb, tomato.

Fruits: apple, blackberry, blueberry, pear.

Landscape plants: black alder, azalea, basswood, white birch, ornamental cherries, red chokeberry, crabapple, hackberry, Amur honeysuckle, hydrangea, Japanese larch, lespedeza, lilac, saucer magnolia, silver maple, mountain laurel, pear, loblolly pine, mugo pine, red pine, scotch pine, white pine, potentilla, privet, rhododendron, Norway spruce, viburnum (few), yew.

Flowers and herbaceous plants: autumn crocus (Colchichum), blue wild indigo (Baptisia), chrysanthemum (some), columbine, hydrangea, lily, narcissus (some), peony (some), petunia, tobacco.

Field crops: alfalfa, crimson clover, tobacco.

Plants Observed to Be Tolerant to Juglone

Vegetables: lima bean, snap bean, beet, carrot, corn, melon, onion, parsnip, squash.

Fruits: black raspberry, cherry.

Landscape plants: arborvitae, autumn olive, red cedar, catalpa, clematis, crabapple, daphne, elm, euonymus, forsythia, hawthorn, hemlock, hickory, honeysuckle, junipers, black locust, Japanese maple, maple (most), oak, pachysandra, pawpaw, persimmon, redbud, rose of sharon, wild rose, sycamore, viburnum (most), Virginia creeper.

Flowers and herbaceous plants: astilbe, bee balm, begonia, bellflower, bergamot, bloodroot, Kentucky bluegrass, Spanish bluebell, Virginia bluebell, bugleweed, chrysanthemum (some), coral bells, cranesbill, crocus, Shasta daisy, daylily, Dutchman's breeches, ferns, wild ginger, glory-of-the-snow, grape hyacinth, grasses (most), orange hawkweed, herb robert, hollyhock, hosta (many), hyacinth, Siberian iris, Jack-in-the-pulpit, Jacob's ladder, Jerusalem artichoke, lamb's ear, leopard's bane, lungwort, mayapple, merrybells, morning glory, narcissus (some), pansy, peony (some), phlox, poison ivy, pot marigold, polyanthus primrose, snowdrop, Solomon's seal, spiderwort, spring beauty, Siberian squill, stonecrop, sundrop, sweet Cicely, sweet woodruff, trillium, tulip, violet, Virginia waterleaf, winter aconite, zinnia.

Source: Dana, M. N., and B. R. Lerner. 1994. *Black walnut toxicity.* Purdue University Cooperative Extension Service Leaflet HO-193.

Juglone has been shown to be a respiration inhibitor, which deprives sensitive plants of the energy required for metabolic activity. The juglone toxin occurs in the leaves, bark, and wood of walnut, but these contain lower concentrations than in roots. Juglone is poorly soluble in water and does not move very far in the soil. Mulches of sawdust and bark from walnut are also capable of causing injury.

California Chaparral   The California chaparral vegetation (Fig. 20.2) is comprised of somewhat fire- and drought-resistant evergreen shrubs, annual grasses, and extensive clumps of aromatic plants, particularly purple sage (*Salvia leucophylla*) and

FIGURE 20.2   The California chaparral.
Used with permission from Dr. Sharon Johnson, University of California, Berkeley.

California sagebrush (*Artemisia californica*). Around each patch of shrubs there occurs a characteristic bare zone of one to 2 m wide where plants do not grow, and beyond that a zone 2 to 8 m wide of relatively stunted plant growth zone.

A wide range of possibilities for these regions of poor plant growth, including competition for nutrients, grazing of animals, and washing out of water-soluble toxins, were eliminated. What was not eliminated were volatile toxins, and it was subsequently determined that terpenes were given off by the shrubs. When isolated, these terpenes were capable of inhibiting seedling growth in native grasses. It was further determined that the terpenes were adsorbed by soil, remaining toxic after at least 2 months in the soil. These terpenes could also dissolve in cuticular waxes.

Effects on Plants

Classification and Chemical Nature of Allelochemicals

There are four main allelopathic classes of chemical reactions based on the type of organism providing the allelochemical and the organism that is affected by the allelochemical. Antibiotics are microorganism to microorganism interactions, kolines are plant to plant interactions, marasmins are microorganism to plant interactions, and phytoncides are plant to microorganism interactions.

The classification of plant allelochemicals is generally based on biological activity. These include phytotoxicants, growth promoters, substrates for microorganisms, substances that predispose plants to disease, enhancers of root exudation, and agents for altering soil structure. In each of these categories allelochemicals affect physiological processes in stressful ways that depend on the species and the environmental conditions.

Many of the plant allelochemicals can be classified as terpenes, phenolics, alkaloids, or nitriles. Terpenes often play a defensive role in plants. They are the major components of essential oils in many trees, act as tree scents, and are a component of resins. They are released by vaporization into the air or are leached in small amounts by water. These compounds can cause poor growth.

Phenolics are a large group of aromatic compounds, some of which have been implicated in allelopathy. Phenolics occur in the essential oils and are responsible for plant scents. Phenolic compounds protect trees against infection from microorganisms and injury from higher animals. Tannins are phenolic polymers and are water soluble. Tannins protect trees from animal grazing and injury and act as allelochemicals. Many phenolics occur in combination with sugars.

The alkaloids are a large mixture of nitrogen-containing materials found in many toxins. Plants may develop these compounds as repellents and poisons. Many of these compounds are attached to sugars for storage. An associated group is nitriles, or organic cyanides. Alkaloids and nitriles are ecologically important because prolonged leakage of small amounts can greatly change plant, animal, and microbe distribution. Because of their nitrogen content, alkaloids and nitriles do not last long in the environment.

Physiological Action of Allelochemicals

The physiological action of allelochemicals is generally related to the amount absorbed and translocated to the site where physiological action of a toxic form has a detrimental effect. These actions include inhibition of respiration and oxidative photophosphorylation, photosynthetic carbon fixation, enzyme activity of various kinds, protein synthesis, and cell division and elongation. Allelochemicals cover such a broad range of compounds that the effects on the physiology of the affected plant are also varied and complex.

The points of influence for most allelopathic compounds are at the cell membranes, at the double-wall membranes of cell organelles, and during the energy production steps and energy use processes. A few allelochemicals are specific to one enzymatic step, and usually a number of allelopathic compounds influence a number of plant growth processes.

Many organisms respond quickly to allelopathic attack by breaking up the allelochemicals or transforming them into nondamaging forms. As a general rule, the longer the two organisms have lived together, the less likely allelopathic effects will be observed. New specie combinations, rapid successional changes, and introduced exotic species can generate a large allelopathic effect. Under good growing conditions, allelopathy can represent 5% to 10% of the total interaction between species. As stresses become greater, allelopathy increases in importance.

Species with large allelopathic components usually modify their own rhizosphere and surrounding soils sufficiently to act as a shield from other allelopathic species. A number of allelopathic species can be found growing together because they each successfully control their own interface with the environment while protecting themselves from allelopathic materials of others.

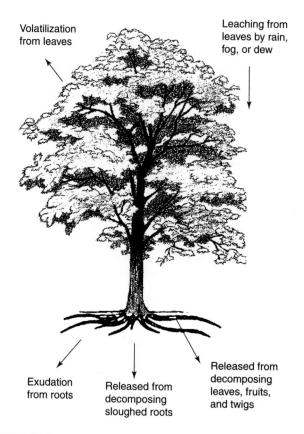

Volatilization from leaves

Leaching from leaves by rain, fog, or dew

Exudation from roots

Released from decomposing sloughed roots

Released from decomposing leaves, fruits, and twigs

FIGURE 20.3   Methods of release of allelochemicals from plants.
Source: Virginia Cooperative Extension Service Publication Number 430–021, *Trees for Problem Landscape Sites—The Walnut Tree: Allelopathic Effects and Tolerant Plants* by Bonnie Appleton, Roger Berrier, Roger Harris, Dawn Alleman, and Lynnette Swanson, 2000.

Sources and Nature of Allelochemicals

Release of allelochemicals from plants occurs by a variety of means (Fig. 20.3). These include leaching by fog, drip, or rain; exudation from roots; degradation of dead plant parts; and volatilization. All plant parts have been shown to contain inhibitors but leaves and roots appear to be the most important sources. Many chemicals are localized in plant organs so that it is not unusual to find different allelopathic compounds in different parts of the same plant.

As tissues approach senescence they are more prone to leaching of metabolites than young tissues. Factors of the environment that affect plant development may also affect leaching by changing the wettability of the surfaces and removal of organic molecules to the surface. Any moisture that wets leaf surfaces can cause leaching. Rain washes materials out of trichomes (plant hairs), cracks, and oil glands on living leaves. Leaves that are relatively large, flat, and pubescent (covered with leaf hairs) are easily wetted and prone to leaching. Leaves that are smooth and have waxy surfaces are not easily wetted and are less likely to be leached. Acid precipitation can significantly increase leaching through its damage to leaf surfaces.

Materials leached from leaves can be reabsorbed by roots, stems, and foliage of the same plant or tree or by any other organism in the area. Specific associations of different organisms are the natural result of rain-wash chemical impacts. The drip line of a tree crown can be a clear line between high and low allelopathic effects.

A wide variety of organic compounds are known to be released from roots into the rhizosphere. The amounts released vary with usual loss of cells and tissues from root surface and root cap as the root grows in length and diameter, injury from a wide variety of sources, stage of development of the root, its type and growth rate, and all environmental factors that affect root processes and activity of microorganisms in the rhizosphere. Release follows the pattern of diffusion, secretion, or sloughage from the root surface into the surrounding soil.

The major source of exudation from roots is in the young, nonwoody roots, which tend to have high concentrations of allelopathic chemicals and which are thinly protected from the environment. The root development zone immediately behind the primary root tip is a primary site of exudation. Exudates are also concentrated in the root hair zone and the area where lateral roots are generated.

Decomposition of roots is another source of allelopathic chemicals. Death and physical damage to parts of the root system allow leakage of cell materials into the environment. The turnover (growth, decline, and death) of absorbing root systems allows significant concentrations of allelochemicals to be shed into the soil.

Death of any plant part and resulting contact with the soil results in the release and degradation of organic matter. In the process of dying, membrane permeability increases. This permits water-soluble inhibitors to be released. Microorganisms affect the release of allelopathic agents from inert materials by decay and respiration. Glycosides (glucose-bonded compounds) are one important class of compounds because microorganisms break apart these materials to use the glucose, leaving the active allelochemical behind. As decay continues, many breakdown products are released, some from the organic material source and some from the decay organism. Years after the death of some trees the allelopathic patterns of the tree in the soil can still be observed. Some of the lignin-based products and humic acid residuals are still present after decades, thus preserving some type of allelopathic heritage.

Stems and branches of deciduous trees are also weathered and leached during the dormant season. This can provide year-round chemical inputs. The stem-flow water is deposited on the soil in a small circular band around the stem base. A number of bark-surface organisms (especially lichens) can also add their own allelopathic agents to the stem flow.

The release of allelopathic volatile compounds is mostly found in arid or semiarid climates. A number of plants have been found to do this, and most of the chemicals that are volatilized are identified as terpenoids. Other sources of allelopathic chemicals are seeds, fruits, and pollen. Pollen allelopathy might be a reason for fertility problems in plants and allergy problems in humans.

Agricultural Technologies That Affect Allelochemicals

Weed Control

Allelopathy has been investigated as a possible method of weed control. For example, wild accessions of cucumber had allelopathic-suppression effects on the growth of millet. Interestingly, modern breeding has virtually eliminated allelopathic properties from modern cucumber cultivars. Also some varieties of oats produce roots that contain high amounts of an inhibitor of root growth, scopoletin. When placed under drought or high-temperature stress, they produced 25 times more scopoletin. Although scopoletin inhibits root growth, the effects are reversed when scopoletin is removed. In the future, crops may be bred that are allelopathic to weeds or natural allelopathic chemicals might be synthesized and applied as herbicides.

Cover Crops

Growers often use the allelopathic nature of barley, rye, sorghum, and wheat when they use these crops as cover crops. Although these cover crops may reduce weed population by limiting light, it has now been shown that these crops produce chemicals toxic to many weeds. This may also result in reduced production of subsequently planted crops sensitive to the allelopathic chemicals.

Plant Autotoxicity

Some plants may exhibit autotoxicity or self-poisoning from allelochemicals that they release. The decline in asparagus beds observed with age may be due to allelochemical increases in the soil from the asparagus plants. Other crops can be inhibited by asparagus plant residue as well. Replanted peach or apple trees in a location can be a problem that may be partially due to autotoxic chemicals remaining in the soil from the original tree as well as pathogens and nematodes that may exist in large numbers.

Summary

Allelochemicals are organic compounds that are produced by one organism that is stressful (or allelopathic) to another organism or their progeny. Examples of plants or communities of plants that exhibit allelopathy are the black walnut tree and plants of the California chaparral. Black walnut tree roots, leaves, and nuts produce juglone, which in small amounts adversely affects the growth of plants sensitive to this chemical. Plants of the California chaparral (particularly *Salvia leucophylla* and *Artemisia californica*) appear to emit volatile terpenes that are toxic to sensitive plants by inhibiting seedling growth. In each of these cases, the allelopathic effects reduced plant competition.

The biological activity of plant allelochemicals include phytotoxicants, growth promoters, substrates for organisms, substrates that predispose plants to disease, enhancers or root exudation, and agents for altering soil structure. Many of the plant allelochemicals are terpenes, phenolics, alkaloids, or nitriles. Each of these chemicals can affect plant physiological processes depending on the plant species and the environmental conditions. The chemical activity observed is generally related to the amount absorbed and translocated to the site where the detrimental physiological affect occurs. Allelochemicals can inhibit respiration and oxidative photophosphorylation, photosynthetic carbon fixation, enzyme activity of various kinds, protein synthesis, and cell division and elongation.

Allelochemicals can be released by plants due to leaching by precipitation, exudation from roots, degradation of dead plant parts, and volatilization. Leaves and roots appear to be the plant organs of greatest importance for containing allelochemicals. Senescent (or dying) tissues are more prone to leaching and release of allelochemicals.

Allelochemicals may function in weed control to reduce competition. Commonly used cover crops (such as barley, rye, sorghum, and wheat) in agricultural systems may reduce weed competition at least partially due to release of allelochemicals. Plant autotoxicity (such as observed in older asparagus beds or in replanted peach and apple orchards) may be also at least in part due to allelochemicals accumulating in the soil.

Review Questions

1. Define allelopathy.
2. How might allelochemicals be beneficial for plant growth and development?
3. Generally the longer two species have lived together, the _____ allelopathy affects their existence. Choose between **more** or **less**.
4. T or F: Physiological actions of allelochemicals are related to the amount absorbed and translocated to the site where physiological action of a toxic form has a detrimental effect.
5. What are the four main allelopathic classes?
6. What types of leaves are easily wetted and prone to leaching?
7. Define plant autotoxicity and name two crops discussed that may exhibit autotoxicity.
8. Black walnut trees can adversely affect other plants growing in the area. What chemical compound is suggested as being the allelochemical from the tree?
9. What ecological advantage is there to the fact that arid and semiarid plants that exhibit allelopathy do so by releasing volatile compounds instead of leaf exudate?
10. How might one incorporate a plant's allelopathic effects in developing a commercial agricultural production system for a different crop?

References

Coder, K. D. 2001. Allelopathy in trees. *Sylvan Communities,* Winter 2001:13–15.

Dana, M. N., and B. R. Lerner. 1994. *Black walnut toxicity.* Purdue University Cooperative Extension Service Publication HO-193.

Fitter, A. H., and R. K. M. Hay. 1987. *Environmental physiology of plants.* 2nd ed. New York: Academic Press.

Funt, R. C. 1993. *Black walnut toxicity to plants, humans and horses.* Ohio State University Extension Fact Sheet HYG-1148–93.

Hale, M. G., and D. M. Orcutt. 1987. *The physiology of plants under stress.* New York: John Wiley.

Silva, E. M. 1997. Plant against plant. *The Virginia gardener newsletter.*

Selected Internet Sites

www.clunet.edu/wf/ Wildflowers of Southern California, including pictures of plants from the California chaparral. Website of Barbara J. Collins, California Lutheran University.

a p p e n d i x

Conversion Factors for International System of Units (SI Units)[1] to and from Non-SI Units

To convert column 1 into column 2 multiply by	Column 1 SI unit	Column 2 non-SI unit	To convert column 2 into column 1 multiply by
Length			
0.621	kilometer, km	mile, mi	1.609
1.094	meter, m	yard, yd	0.914
3.28	meter, m	foot, ft	0.304
0.394	centimeter, cm	inch, in	2.54
3.94×10^{-2}	millimeter, mm	inch, in	25.4
Area			
2.47	hectare, ha	acre	0.405
247	square kilometer, km^2	acre	4.05×10^{-3}
10.76	square meter, m^2	square foot, ft^2	9.29×10^{-2}
Volume			
0.240	cubic kilometer, km^3	cubic mile, mi^3	4.165
9.73×10^{-3}	cubic meter, m^3	acre-inch	102.8
2.84×10^{-2}	liter, L	bushel, bu	35.24
0.265	liter, L	gallon	3.78

[1]The International System of Units (SI units) is the metric system of measurement traditionally used in science and international commerce.

To convert column 1 into column 2 multiply by	Column 1 SI unit	Column 2 non-SI unit	To convert column 2 into column 1 multiply by
Mass			
2.20×10^{-3}	gram, g	pound, lb	454
2.205	kilogram, kg	pound, lb	0.454
1.102	tonne, t	ton (US), ton	0.907
Yield and Rate			
0.893	kilogram per hectare, kg ha^{-1}	pound per acre, lb acre^{-1}	1.12
0.107	liter per hectare, L ha^{-1}	gallon per acre	9.35
893	tonnes per hectare, t ha^{-1}	pound per acre, lb acre^{-1}	1.12×10^{-3}
0.446	tonnes per hectare, t ha^{-1}	tons per acre, tons acre^{-1}	2.24
2.24	meter per second, m s^{-1}	mile per hour, mph	0.447
Pressure			
9.90	megapascal, MPa	atmosphere, atm	0.101
10	megapascal, MPa	bar	0.1
1.45×10^{-4}	Pascal, Pa	pound per square inch, lb in^{-2} (PSI)	6.90×10^{3}
14.22	kilogram per square cm, kg cm^{-2}	pound per square inch, lb in^{-2} (PSI)	0.0703
Temperature			
(9/5C) + 32	Celsius, C	Fahrenheit, F	5/9(F − 32)
Energy, Work, Quantity of Heat			
9.52×10^{-4}	joule, J	British thermal unit, BTU	1.05×10^{3}
0.239	joule, J	calorie, cal	4.19
0.735	joule, J	foot-pound	1.36
1.43×10^{-3}	watts per meter square, w m^{-2}	calorie per square centimeter minute, cal cm^{-2} min^{-1}	698
Plane Angle			
57.3	radian, rad	degrees (angle)	1.75×10^{-2}
Water Measurement			
9.73×10^{-3}	cubic meter, m^{3}	acre-inches, acre-in	102.8
4.40	cubic meter per hour, m^{3} hour^{-1}	gallon per minute, gal min^{-1}	0.227
97.28	hectare-meters, ha-meters	acre-inches	1.03×10^{-2}

Concentrations

1	milligram per kilogram, mg kg^{-1}	parts per million, ppm	1

Selected Internet Site

http://physics.nist.gov/cuu/Units/introduction.html The U.S. National Institute of Standards and Technology (NIST) reference site on constants, units, and uncertainty. The Physics Laboratory is one of the major operating units of the NIST, and its mission is to support U.S. industry by providing measurement services and research for electronic, optical, and radiation technologies.

a p p e n d i x
B

Conversion Factors for Some Commonly Used Non-International System of Units (non-SI Units)[1]

To convert column 1 into column 2 multiply by	Column 1	Column 2	To convert column 2 into column 1 multiply by
Temperature			
1.00 (K−273)	Kelvin, K	Celsius, C	1.00(C + 273)
Pressure			
14.50	bar	pound per square inch, lb in^{-2} (PSI)	6.90×10^{-2}
Concentrations			
0.1	gallons per kilogram	percent, %	10
Plant Nutrient Conversions			
	Elemental	Oxide	
2.29	P	P_2O_5	0.437
1.20	K	K_2O	0.830
1.39	Ca	CaO	0.715
1.66	Mg	MgO	0.602

[1]The International System of Units (SI units) is the metric system of measurement traditionally used in science and international commerce.

Selected Internet Site

http://physics.nist.gov/cuu/Units/introduction.html   The U.S. National Institute of Standards and Technology (NIST) reference site on constants, units, and uncertainty. The Physics Laboratory is one of the major operating units of the NIST, and its mission is to support U.S. industry by providing measurement services and research for electronic, optical, and radiation technologies.

appendix

C

Some Physical Constants or Values

Quantity	Symbol	Numerical value[1]
Speed of light (in vacuum)	c	3.00×10^8 m s^{-1}
Gravitational constant	G	6.67×10^{-11} N m^2 kg^{-2}
Universal gas constant	R	8.31 J K^{-1} mole^{-1}
Planck's constant	h	6.63×10^{-34} J s
Acceleration of gravity	g	9.81 m s^{-2}
Standard atmospheric pressure		1 atm
		1.10×10^5 Pa
		760 mm Hg
		760 torr
Absolute zero		0°K
		-273.15°C
Speed of sound		331 m s^{-2}
Mean orbital speed of earth around sun		2.98×10^4 m s^{-1}
Mass of earth		5.98×10^{24} kg
Volume of earth		1.09×10^{21} m^3
Earth to sun, mean distance		1.50×10^{11} m

[1]Numerical values in most cases have been rounded off to three significant figures.

Selected Internet Site

http://physics.nist.gov/cuu/Units/introduction.html   The U.S. National Institute of Standards and Technology (NIST) reference site on constants, units, and uncertainty. The Physics Laboratory is one of the major operating units of the NIST, and its mission is to support U.S. industry by providing measurement services and research for electronic, optical, and radiation technologies.

appendix

D

Some Prefixes Used to Define Multiples of International System of Units (SI Units)[1] Conversions

Fraction	Prefix	Symbol
10^{-12}	pico	p
10^{-9}	nano	n
10^{-6}	micro	μ
10^{-3}	milli	m
10^{-8}	centi	c
10^{-1}	deci	d
10	deka	da
10^8	hecto	h
10^3	kilo	k
10^6	mega	M
10^9	giga	G
10^{12}	tera	T

[1]The International System of Units (SI units) is the metric system of measurement traditionally used in science and international commerce.

Selected Internet Site

http://physics.nist.gov/cuu/Units/introduction.html The U.S. National Institute of Standards and Technology (NIST) reference site on constants, units, and uncertainty. The Physics Laboratory is one of the major operating units of the NIST, and its mission is to support U.S. industry by providing measurement services and research for electronic, optical, and radiation technologies.

appendix

E

Compilation of Internet Resources

http://www.nrcs.usda.gov/technical/ Technical resources and information including tools and data on soil science education, soil surveys and maps, soil geography, and soil use and management. Natural Resources Conservation Service, USDA.

http://www.wcc.nrcs.usda.gov/wcc.html Government weather site containing observations, forecasts, maps and models, weather safety, and education. National Water and Climate Center, USDA.

http://www.nws.noaa.gov/ Government weather site containing observations, forecasts, maps and models, weather safety, and education. National Weather Service, National Oceanic and Atmospheric Administration.

http://hcs.osu.edu/plants.html Plant dictionary created and maintained by Tim Rhodus, Ohio State University.

http://www.nrcs.usda.gov/technical/water.html Water resources information on water and climate data, water management documents, and water supply forecasts. Natural Resources Conservation Service, USDA.

http://www.usgs.gov/ Water resources of the United States including water data, publications and products, technical resources, and programs. U.S. Geological Survey.

http://water.usgs.gov/pubs/acidrain/ Trends in precipitation chemistry in the United States, 1983 to 1994. U.S. Geological Survey.

http://minerals.usgs.gov/ Current information on the occurrence, quality, quantity, and availability of mineral resources. U.S. Geological Survey.

http://www.nrcs.usda.gov/technical/nutrient.html Nutrient management information including ecological sciences nutrient management, animal waste management, manure management planner, nutrient and pest management, and phosphorus index. Natural Resources Conservation Service, USDA.

http://www.wcc.nrcs.usda.gov/nutrient   Nutrient management resources on nitrogen, phosphorus, and documents and fact sheets. National Water and Climate Center, USDA.

http://hortipm.tamu.edu/   Hort IPM Web site, Texas A&M University.

http://www.ippc.orst.edu/cicp/Vegetable/veg.htm   Internet resources on IPM on vegetables.

http://www.nrcs.usda.gov/technical/pest.html   Pest management technical resources including ecological sciences, pest management, and national agricultural pesticide risk analysis. Natural Resources Conservation Service, USDA.

http://nysipm.cornell.edu   Integrated pest management information. Cornell University.

http://soils.usda.gov/sqi/soil_quality/soil_biology/soil_biology_primer.html   Soil biology primer with good information on soil organisms. Soil Quality Institute, USDA, Natural Resources Conservation Service.

http://www.clunet.edu/wf/   Wildflowers of Southern California including pictures of plants from the California chaparral. Site of Barbara J. Collins, California Lutheran University.

http://www.chm.bris.ac.uk/motm/brassinolide/brassinolideh.htm   Brassinolide page created by Martin A. Iglesias-Arteaga at the University of Havana.

http://www.plant-hormones.bbsrc.ac.uk/education/Kenhp.htm   Plant hormone information page created by Ken Maas at Northern Illinois University.

http://www.nrcs.usda.gov/technical/ECS/   Technical information on ecological sciences including agricultural ecology, aquatic ecology, ecological climatology, forestry & agroforestry, range & grazing land ecology, understanding ecosystem, and wildlife management. Natural Resources Conservation Service, USDA.

http://www.peak.org/~mageet/tkm/ecolenv.htm   Internet resources related to the science of ecology and the state of the environment. Peak Organization.

http://www.westminster.edu/staff/athrock/ECOLOGY/Botlinks.htm   Internet resources for botany and plant ecology. Westminster College, New Wilmington, PA.

http://www.epa.gov/   Home site of the Environmental Protection Agency (EPA).

http://www.epa.gov/oar/   Government site providing information on air pollution, clean air, and air quality information. Good links for acid rain, ozone depletion, and climate change. Office of Air and Radiation, EPA.

http://www.eia.doe.gov/   Official energy statistics of the U.S. government. Energy Information Administration, DOE.

http://www.nrcs.usda.gov/technical/airquality.html Air quality information and report of Agricultural Air Quality Task Force. Natural Resources Conservation Service, USDA.

http://www.epa.gov/air/criteria.html Information on the National Ambient Air Quality Standards. EPA.

http://physics.nist.gov/cuu/Units/introduction.html The U.S. National Institute of Standards and Technology (NIST) reference site on constants, units, and uncertainty. The Physics Laboratory is one of the major operating units of the NIST, and its mission is to support U.S. industry by providing measurement services and research for electronic, optical, and radiation technologies.

Glossary

Abscisic acid (ABA)   a sesquiterpenoid (15-carbon) compound that functions as a growth inhibitor in plants, often counteracting the growth-promoting effects of auxins and gibberellins

Abscission   the shedding of plant parts from a living plant

Absorption spectrum   graphical presentation of the relative absorption of the different wavelengths by a substance

Acid deposition   process by which acidic components of atmospheric air pollutants are deposited to the earth by means of rain, sleet, snow, fog, or as dry particles

Actinomycetes   microorganisms that resemble molds and are similar to bacteria and are important in the liberation of nutrients of soil organic matter as a result of decomposition

Action spectrum   graphical presentation of the relative effectiveness of the different wavelengths of light at generating a biological response

Adequate (or sufficient) nutritional concentration range   nutrient concentration range where essential elements are provided in suitable amounts for normal plant growth

Aerenchyma   gas-filled chambers in roots that create an internal gas exchange channel from an aerobic shoot to the hypoxic or anoxic root

Aerial environment   that portion of the ecosphere that exists predominantly above the soil surface in relative proximity to the growing plant and that has the potential to affect the physiology and growth of the plant

Aerobic respiration   respiration that occurs in the presence of oxygen

Agroecology   the study of the interactions among the many biological, environmental, and management factors that make up and influence agriculture

Agronomy   branch of agriculture that studies the principles and practice of crop production and field management

Air pollutant   gaseous or particulate substance released into the atmosphere in sufficient quantity or concentration to cause injury to plants, animals, or humans

Algae   most are chlorophyll-containing organisms that are capable of photosynthesis, though a few forms obtain their energy largely from organic matter

Allelochemical   organic compound involved in allelopathy

Allelopathy   production of substrates or compounds by one organism that is injurious to another organism or their progeny

Anaerobic respiration   respiration that occurs in oxygen-deficient environments

Annual   a plant that lives one year or less

Anoxia   absence of oxygen

Antagonism or amensalism   interaction that depresses one organism while the other remains stable

Anthesis   the opening of flowers with appropriate plant parts available for pollination

Antitranspirants   materials that are applied to plants to reduce water loss from transpiration by covering the surfaces of the foliage with a barrier impermeable to water

Apical dominance   the inhibitory effect of one bud on the growth of other buds of a plant

Apical meristem   a mass of undifferentiated cells at the tips of shoots or roots that are capable of division

Asymmetrical pressure   pressure that has a directional characteristic (i.e., the intensity of the pressure is not uniform in all directions

Autotrophs (autotrophic organisms)   organisms that capture light energy and use simple inorganic substances (mineral nutrients) through the process of photosynthesis to develop carbohydrates and other complex substances

Auxins   compounds that are characterized by their capacity to induce elongation in shoot cells and chemically resemble indole-3-acetic acid, which is the only known naturally occurring auxin, in physiological action

Bacteria   single-celled organisms, one of the simplest and smallest forms of life, involved in all of the organic transformations of the soil, as well as the three basic enzyme transformations of nitrification, sulfur oxidation, and nitrogen fixation

Biennial   a plant that completes its life cycle within two seasons that are often separated by a cold period sufficient to initiate flower and fruit formation

Biogeochemistry   the study of how, when, and in what form soil nutrients become available to plants, microbes, and other organisms

Biological mechanical disturbances   adverse influences from humans and other organisms (insects, diseases, and animals)

Bioindicators (biomonitors)   plants used for monitoring environmental conditions

Biome   a group of ecosystems with similar biological features, though the same species are often not involved

Biosphere   the zone of the earth that contains living organisms, extending from the crust into the surrounding atmosphere

Bolting   premature elongation of the stem usually associated with flowering of a rosette plant

Bonsai   art form that originated in China and was further developed by the Japanese that creates the illusion of dwarf trees

Botany   scientific study of plants

Boundary layer   a thin film of still air at the leaf surface

Brassinosteroids   consist of approximately 60 steroidal compounds that appear to be widely distributed in the plant kingdom, with some of their effects on plant growth and development including enhanced resistance to chilling, disease, herbicides, and salt stress; increased crop yields, elongation, and seed germination; decreased fruit abortion and drop; inhibition of root growth and development; and promotion of ethylene biosynthesis and epinasty

C3 pathway CO_2 fixation pathway in which the first stable product is the 3-carbon acid 3-phosphoglycerate

C4 pathway CO_2 fixation pathway in which the first stable product is the 4-carbon compound, oxaloacetic acid

Calorie   the amount of energy required to raise the temperature of 1 gram of water 1°C

Cereals   grasses for edible seeds

Chemical potential   quantitative expression of the free energy associated with a chemical that is available to do work

Chilling injury   crop injury at temperatures at or slightly above freezing

Climate (macroclimate)   the weather pattern for a particular location over several years

CO_2 fixation reactions of photosynthesis   non-light-requiring reactions that occur in the stroma in which electrons and hydrogen atoms are added to CO_2 to form sugar

Cold frame   a protected ground bed, usually sunken, with a removable glass or plastic roof that has no artificial heat source

Color   the distinctive distribution of wavelengths from a radiation or reflective source

Commensalism   interaction that stimulates one organism but has no effect on the other

Communities   collective organisms within a location

Competition   mutually adverse effects to organisms that utilize a common resource in short supply

Conduction   transfer of kinetic energy (heat) in a fixed media

Convection   transfer of energy in fluid motion

Cover crop   a crop that is not harvested for sale but is grown to benefit the soil and/or other crops in a number of ways

Crassulacean acid metabolism (CAM) CO_2 fixation pathway during which stomata are open during the night to allow diffusion of CO_2 into the plant (stomata are generally only open during the daylight for C3 plants) and 4-carbon compounds (OAA and malate) are produced

Crop   an agricultural plant

Crop ecology   part of agroecology that specifically addresses crop production

Crop growth rate (CGR)   the rate of increase in plant weight/unit area of ground or the NAR x LAI

Crop removal values   estimated nutrient content values of plants and fruits expressed on a hectare basis

Crop rotation   the growing of two or more crops in a sequence, on the same land, during a period of time

Cryptochrome   pigment involved in the plant responses that have been attributed to radiation in the blue portion (400 to 500 nm) of the electromagnetic spectrum

Cultivar term designating certain cultivated plants that are clearly distinguishable from others and they retain their distinguishing characteristics when they reproduce

Cytokinins adenine-resembling molecules that stimulate cell division

Dew point temperature temperature at which air reaches water vapor saturation, condensation begins, and dew forms

Differentiation the process by which cells become specialized

Disease the process by which living or nonliving entities (pathogens) interfere, over a period of time, with an organism's physiology and metabolism

Diurnal recurring or repeating each day

Dormancy living condition in which growth is not occurring

Drought meteorological term defined as a period of time without sufficient rainfall

Dryland farming growing crops in the field without irrigation

Dry weight weight of desiccated tissue

Earthworm an important soil macroanimal that can be grouped into three major behavioral classes: the litter dwellers, shallow soil dwellers, and deep burrowers

Ecology scientific study of the interactions between organisms and their environment

Ecophysiology scientific study of the controls over the growth, reproduction, survival, abundance, and geographical distribution of plants as they are affected by the interactions between plants and their physical, chemical, and biotic environment

Ecosphere (biosphere) thin shell of air, water, and soil on or near the earth's surface in which life exists

Ecosystem sum of the plant community, animal community, and environment in a particular region or habitat

Electromagnetic spectrum graphical representation of the known distributions of electromagnetic energies arranged according to wavelengths, frequencies, or photon energies

Element a substance that cannot be divided or reduced by any known chemical means

Environment sum of all the external abiotic (physical) and biotic (biological) forces that affect the life of an organism

Equinox the time of the year when the sun's noon rays are perpendicular to the equator and the circle of illumination passes through both poles and cuts all latitudes exactly in half (i.e., day length = night length)

Espalier French term used to describe a plant that is trained to grow along a flat plane, such as a wall or fence

Essential element an element that is required by plants for normal growth and production

Ethylene a hydrocarbon gas that is a product of plant metabolism with effects on plant growth and development including stimulation of ripening of fleshy fruits, stimulation of leaf abscission, inhibition of root growth, stimulation of adventitious root formation, inhibition of lateral bud development, epinasty of leaves, flower fading, flower initiation in bromeliads, and root geotropic responses

Evapotranspiration the total loss of water by evaporation from the soil and by transpiration from plants for a given area

Evolution the development of a species, genus, or larger group of plants or animals over a long period of time

Farm a human-managed ecosystem with relatively few species generally designed to produce as much harvestable and/or marketable biomass (yield) as environmental conditions will allow

Fiber crops plants grown for their fiber, which is used in fabrics and ropes

Field capacity (field moisture capacity) the percentage of water remaining in a soil 2 or 3 days after having been saturated and after drainage due to gravity has ceased

Flooding water in excess of soil's field capacity

Flux density the rate of transport of a substance across a unit area per unit of time

Forages crops cultivated and used for hay, pasture, fodder, silage, or soilage

Forestry study of forest management with the goals of conservation and the production of timber, and associated with nonfood tree crops and their products

Freezing injury injury that occurs when plants are cooled slightly below 0°C and ice forms within the cells and disrupts the structure of the protoplasm

Fresh weight weight of living tissue

Frost pocket low-lying areas in a valley that can be substantially cooler than surrounding high ground

Fungi microorganisms that contain no chlorophyll and depend on organic matter in the soil for their energy and carbon that are important in the chemical transformation of soil constituents and as pathogens

Gas form of matter that expands without limit to fill the available space of a container

Germination resumption of active plant embryo growth after a dormant period

Gibberellins isoprenoid compounds capable of stimulating cell division or cell elongation, or both

Glazing formation of ice on tree limbs

Gravitropism (geotropism) growth movements of plants toward (positive) or away from (negative) the earth's gravitational pull

Greenhouse effect the gradual trend in global warming since about 1880 in both the Northern and Southern hemispheres that has been attributed to an increase of selected atmospheric gases commonly referred to as greenhouses gases

Greenhouse gases gases released into the atmosphere as a consequence of increasing consumption of fossil fuels that absorb infrared radiation and influence global warming

Green roofs contained green space on top of human-made structures

Growing season number of consecutive frost- or freeze-free days in a calendar year

Growth   an irreversible increase in mass (weight) or size (volume) of the plant due to the division and enlargement of cells

Guard cells   a pair of specialized cells that surround the stomatal pore and regulate its opening and closing

Hardening   processes that increase the ability of a plant to survive the impact of an unfavorable stress

Hardiness zones   rating system developed by the U.S. Department of Agriculture in which zones based on the average annual minimum temperatures are used to help in determining appropriate plant material to grow in an area

Heat units   the calculated cumulative amount of time above a certain minimum temperature a crop receives to determine harvest maturity and harvest dates

Heliotropism   the movement of leaves in response to the moving direction of sunlight

Herbals   writings that listed common plants and often supplied their medicinal usage

Herbivory   soil organisms that feed on plants and are involved in degradation of higher plant tissue into organic matter

Heterotrophs (heterotrophic organisms)   organisms that acquire their nourishment from others though digestion and/or decomposition, and the resulting re-arrangement of complex materials

High tunnel   season-extending structure similar to what has been used by ornamental perennial growers as Quonset polyhouses and polyhuts (strong polyethylene-covered structures) that has no and very low sophisticated heating systems

Horticulture   science and art of cultivating fruits, vegetables, and ornamental plants

Hot bed   heated cold frame that is used for season extending and is most often used to give an early start to warm-season vegetables

Hydrophilic gels   moisture-holding polymer compounds that absorb many times their weight in water

Hydroponics   the science of growing plants without using soil

Hydrostatic pressure (turgor pressure)   the intercellular pressure in plant cells that occurs as a consequence of normal water balance in the plant

Hypoxia   low oxygen levels

Ideal gas law   the volume of the gas is proportional to the number of moles of the gas at a constant pressure and temperature

Imperfect flower   a flower that lacks either stamens or petals

Incipient wilting   recoverable plant wilting that occurs under weather conditions favoring rapid transpiration, even when soil moisture is at field capacity

Integrated pest management (IPM)   a sustainable approach used by crop producers to manage pests by combining biological, cultural, physical, and chemical tools in a way that minimizes economic, health, and environmental risks

Irradiance (light, visible radiation)   the radiant flux density on a given surface from the visible and neighboring wavelengths portion of the electromagnetic spectrum

Jasmonates   class of cyclopentanone compounds with activity similar to (-)-jasmonic acid and/or its methyl ester with reported growth effects on plants, including inhibition of growth and germination; and promotion of senescence, abscission, tuber formation, fruit ripening, pigment formation, and tendril coiling; they also appear to have important roles in plant defense by inducing proteinase synthesis

Kinetic energy   the motion of a molecule that contributes to the energy of the molecule

Lake effect   the influence of large bodies of water that serve as heat reservoirs in fall and heat sinks in spring

Latent heat   the energy required to cause a change in physical state or phase (ice <-> liquid <-> solid) of a molecule or substance without a change in temperature

Leaf area index (LAI)   the amount of land surface area covered by green foliage, measured on a small plot of land as the amount of leaf area per ground area at any point in time or estimated over a larger area using satellite data and analysis

Light compensation   the amount of light for a plant at which photosynthetic uptake of CO_2 is equal to the amount of CO_2 released by respiration

Light reactions   light requiring reactions of photosynthesis that trap light energy and cleave water molecules into hydrogen and oxygen and serve as electron and proton transfer reactions

Light saturation point   the amount of light above which further increases in light intensity does not result in increased photosynthetic rates

Macroconsumer   primarily animals that ingest other organisms or particulate matter

Macronutrient   an essential element needed by plants in relatively large quantities

Matrix potential   the force with which water is adsorbed onto surfaces such as cell walls, soil particles, or colloids

Mesosphere   atmospheric zone located above the stratosphere from 50 to 80 km where temperature falls greatly with increasing height above the earth

Microclimate   little weather variations that exist in a location or field that are often easily changed or modified

Microconsumer   primarily bacteria and fungi that break down the complex compounds of dead cells, absorbing some of the decomposition products and releasing inorganic nutrients that are then usable by the producers

Micronutrient   an essential nutrient needed by plants in smaller amounts than the macronutrients

Molds   distinctly filamentous, microscopic, or semimacroscopic organisms that respond to soil aeration with increasing numbers and are important in decomposing organic residues

Molecule   usually consists of two or more atoms and is the smallest portion of an element or a compound that retains its chemical identity

Morphogenesis (development)   the organized and systematic formation of plant organs from differentiated cells

Mulch   any material used at the surface of a soil primarily to prevent the loss of water by evaporation or to reduce weed establishment and growth

Mutualism   an obligate interaction where the absence of the interaction depresses both partners

Mycorrhizae   mutualistic relationship of certain fungi and plant roots

Necrosis   type of cell death

Nematodes   mostly microscopic and slender organisms that are round in cross section and spindle-shaped and are generally characterized on the basis of their food demands, including those that live on decaying organic matter; those that are predatory on other nematodes, bacteria, algae, and protozoa; and those that are parasitic, attacking the roots of higher plants to pass at least a part of their life cycle attached to such tissue

Net assimilation rate (NAR)   the rate of increase in plant weight per amount of leaf area

Nitrogen-fixing bacteria   soil bacteria that form symbiotic relationships with plant roots of certain species and reduce N_2 to NH_4^+ during the process of nitrogen fixation

Nutrient cycles   chemical element transformations that circulate in characteristic paths from environment to organisms and back to the environment

Nyctoperiod   length of the night period

Organ   a part of a plant that has a particular function

Osmotic potential   the chemical potential of water in a solution due to the pressure of dissolved materials

Oxidation   removal of electrons from a molecule

Ozone hole   ozone-depleted zone that occurs over Antarctica

Ozone layer   stratospheric layer that absorbs a proportion of the ultraviolet radiation from the sun

Parthenocarpic fruit   fruit that develop without sexual reproduction and do not contain mature functioning seed

Pathogen   an organism that can cause a disease

Perfect flowers   flowers that contain female and male parts

Permanent wilting point   when water uptake ceases and plant tissue and cells are damaged

Pesticide   substance for preventing, destroying, repelling, or mitigating any pest

Phenotype   the external physical appearance of an organism

Phloem   tissues that transport the products of photosynthesis

Photometry   measurement of flux typically utilized by the human eye

Photomorphogenesis   light-regulated plant growth and development that is independent of photosynthesis

Photoperiod (day length)   length of the light period

Photoperiodism   the growth responses of a plant to photoperiod

Photosynthesis   physio-chemical process often carried out in the chloroplasts of leaves by which CO_2 and H_2O are transformed in the presence of light into carbon-containing, energy-rich organic compounds such as sugars and starches

Photosynthetically active radiation (PAR)   the measured flux between 400 to 700 nm that is utilized by plants during the carbohydrate-producing reactions of photosynthesis

Phototropism (phototaxis)   response of plants to directional light rays

Phytochrome   pigment in plants in small concentrations that exists in two photoreversible forms, the red form and the far-red form

Phytoremediation   the use of plants to clean up contaminated soils

Pigment   any substance that has its own characteristic absorption spectra and is often colored in appearance

Plant anatomy   study of the internal structure of a plant

Plant autotoxicity   self-poisoning from allelochemicals released

Plant breeding   the systematic improvement of plants

Plant cytology   study of the plant cell structure, function, and life history

Plant ecology   study of the underlying order of plant species and vegetation

Plant ecophysiology   science that seeks to describe the physiological mechanisms that underlie ecological observations

Plant genetics   study of the mechanisms of heredity of plant traits

Plant geography   study of the geographical distribution of plants and factors that determine this distribution

Plant growth regulator   organic compound other than a nutrient that in small amounts (< 1 mM) promotes, inhibits, or qualitatively modifies growth and development

Plant hormone   organic substance other than a nutrient that is active in very small amounts (< 1 mM and often < 1 µM) and formed in certain parts of plants and usually (though not always) translocated to another part of the plant where it evokes specific biochemical, physiological, and/or morphological responses

Plant pathology   study of the cause, nature, prevalence, severity, and control of plant diseases

Plant physiology   study of plant function that is mostly concerned with the individual plant and its interactions with the environment

Plant population density   spacing between neighboring plants

Plant propagation   study or practice of producing numerous plants either through sexual or asexual means

Plant science   specialized area of study of botany that emphasizes the use of plants in agricultural situations

Poikilotherms   organisms that assume the temperature of their environment

Pollination   transfer of pollen from a stamen to a stigma of a flower

Polyamines   widespread in all cells and appear to be essential in growth and cell division

Potential energy   the possibility of a molecule to acquire kinetic energy

Pressure chamber   instrument that measures the water potential in the xylem of the plant

Protozoa   one-celled single organisms that are divided into three groups: amoebae, ciliates, and flagellates

Pruning   the judicious removal of plant parts to reduce competition between the parts of a single plant or among neighboring plants

Psychrometer   instrument used to determine atmospheric humidity by comparing the readings of a dry-bulb thermometer and a wet-bulb thermometer

Quantum theory   the quantum energy of a wavelength is directly proportional to its frequency and inversely proportional to the length of the wavelength

Radiation   energy in the form of either electromagnetic waves or discrete packets that travels at the speed of light (3×10^8 m s^{-1}) in a vacuum and slightly slower in a medium such as the atmosphere

Radiometry   measurement of radiant flux

Redox reactions   chemical reactions involving simultaneous oxidation and reduction of molecules

Reduction   addition of electrons or hydrogen atoms to a molecule

Relative humidity   the amount of water present in the air as a percentage of what could be held at saturation (100% relative humidity) at the same temperature and pressure

Respiration   process used by organisms (plants and animals) that utilizes the energy-rich fuel products of photosynthesis to perform energy-requiring activities such as producing new tissue or organs

Rhizosphere (plant root zone)   the portion of the soil in the immediate vicinity of plant roots in which microbial populations and biochemical reactions are influenced by the presence of the roots

Root girdling   a plant root that entwines around another large root or the base of the plant and prevents or hinders water and nutrient uptake

Root pruning   cutting the root ball with vertical superficial cuts prior to transplanting

Root suckering   initiation of new shoots from the root system in some shrubs and trees

Row covers   flexible, transparent coverings that are installed over single or multiple rows of horticultural crops to enhance plant growth by warming the air around the plants in the field

Salicylates   class of compounds having activity similar to salicylic acid (ortho-hydroxybenzoic acid) including thermogenesis (plant development in response to temperature) in Arum flowers, enhanced plant pathogen resistance, enhanced longevity of flowers, and inhibition of ethylene biosynthesis and seed germination

Seasons divisions of the calendar year that are based in most areas on changes in temperature and light

Seismomorphogenesis plant growth responses to vibrational or shaking stress

Senescence physiological aging process in which tissues deteriorate and organisms finally die

Sensible heat the energy gain or loss from molecules by transfer of kinetic energy to adjacent molecules

Sexual reproduction development of new plants by the process of meiosis and fertilization in the flower to produce a viable embryo in a seed

Soil the solid portion of the earth's crust made up of a complex of physical, chemical, and biological substrates where energy and matter are captured and transformed by plants, animals, and microbes

Soil analysis determination of the nutrient content in a soil sample from the root zone

Soil compaction the packing effect of a mechanical force or gravity and time on a soil

Soil horizons horizontally stratified zones that are enriched or depleted in clay, organic matter, and nutrients relative to the material from which the soil formed

Soil pH a measurement of the concentration of hydrogen ions, H^+, in the soil

Soil solarization use of the sun's energy to generate high soil temperatures to destroy living organisms

Soil textural triangle chart used to classify a soil into one of 12 different categories, each of which has different physical and chemical properties

Soil texture proportion of sand, silt, and clay in a soil

Solar noon (local noon) the maximum solar angle for that location, time, and day

Solar zenith angle the angle of sun away from vertical

Specie group of morphologically and ecologically similar populations that may or may not be inbreeding

Specific heat the number of calories needed to change the temperature of 1 g of a substance by 1°C

Spectroradiometry measurement of the radiant flux dependent on the wavelength of the radiation

Stress any environmental condition that is potentially unfavorable to living organisms for normal growth and development

Stress ethylene increased ethylene levels in and around a plant resulting from a stress or stresses

Stomate consists of two guard cells that can rapidly change their aperture and regulate the amount of CO_2 that may enter the leaf

Stratosphere atmospheric layer located above the troposphere that extends to about 50 km above the earth's surface

Surface tension the force per unit length that can pull perpendicular to a line in a plane of the surface

Symmetrical pressure   a pressure that is the same in intensity regardless of direction

Symptomatology   description of the stress-induced injury to plants

Temperature   measure of the average kinetic energy of molecules

Temperature acclimation   development of tolerance to injury from either low or high temperatures

Temperature coefficient (Q_{10})   a measure of the effect of temperature on a chemical reaction process

Temperature inversion   a layer of cold air close to the earth's surface is topped by a layer of warm air and as a result the air temperature is cooler near the earth's surface than in an adjacent upper layer (the inverse of what normally occurs)

Tensile strength   the ability to resist a pulling force

Tensiometer   sealed, water-filled tube with a porous ceramic tip on the lower end and a vacuum gauge on the upper end to measure the amount of water in a soil

Thermal blankets (thermal fabrics)   temperature protective materials used for overwintering containerized perennials

Thermal radiator   light source that has a continuous spectral distribution of energy with respect to the absolute temperature of the source

Thigmomorphogenesis   plant growth in response to contact stress

Tissue analysis   determination of nutrient content of a plant or plant part

Topiary   form of plant sculpture that originated with the ancient Romans and became popular in European formal gardens

Training of plants   method of orientating the plant growth in a distinct way in space

Transgenic plant   a plant expressing a foreign gene introduced by genetic engineering

Transpiration   the evaporative loss of water (as water vapor) from plants, mostly through their stomata

Tree well   area surrounding the tree within the drip line that is left unmodified

Troposphere   the lower layer (~ 9 to 12 km) of the earth's atmosphere, where temperatures decrease with height

Turgor   firmness of a cell resulting from hydrostatic pressure

Ultraviolet (UV) radiation   nonionizing electromagnetic radiation of wavelengths just shorter than those normally perceived by the human eye

Urban heat island   the warming of air temperature of a city as compared to the surrounding countryside

Vascular tissue   plant tissue specialized for the transport of water (xylem) and photosynthetic products (phloem)

Vegetative reproduction   method of producing physiologically independent individuals without sexual reproduction

Vernalization   the induced or accelerated (premature) flowering that occurs in certain plants due to exposure to low temperatures

Visible light   a relatively small region (from about 380 nm to about 770 nm) of the electromagnetic spectrum to which the average light-adapted human eye is sensitive

Water potential   chemical potential of water in a specified part of the system compared with the chemical potential of pure water at the same temperature and atmospheric pressure

Weather   immediate, day-to-day composite of the temperature, rainfall, light intensity and duration, wind direction and velocity, and relative humidity of a specific location for a set amount of time

Weed   any plant growing in the wrong place

Wilting   reduction in the rigidity (turgidity) of the plant

Windbreak   planting of vegetation often perpendicular to principal wind direction to protect growing plants and soil from effects of the wind

Wind chill   the cooling effect of moving air on a warm body that is expressed in terms of the amount of heat lost per unit area per unit time

Xylem   specialized cells through which water and nutrients move upward from the soil through the plant

Index